MEMENTO
PUBLISHING

Copyright © 2012 Memento Publishing

ISBN: 978-0-615-68570-0
Editors: Mike DeVries, Paul Galan
Design & Layout: Tyler Clinton, Mike DeVries, Paul Galan
Writer: Jinxi Caddel
Artist Research & Collections: Jinxi Caddel
Cover Design: Tyler Clinton

www.MementoPublishing.com

All rights reserved. No part of this book may be
copied or reproduced in any manner whatsoever: electronic, mechanical photocopy,
recording, etc. Without written permission from the publisher. Out of respect for the artists
and the collectors, no photos from this book may be copied or reproduced in any shape or form;
rather use the contents and photos for enjoyment and inspiration.

Printed & Bound in China

TATTOO EXTREMITIES

Artistic Focus on the Head, Hands, Neck and Feet

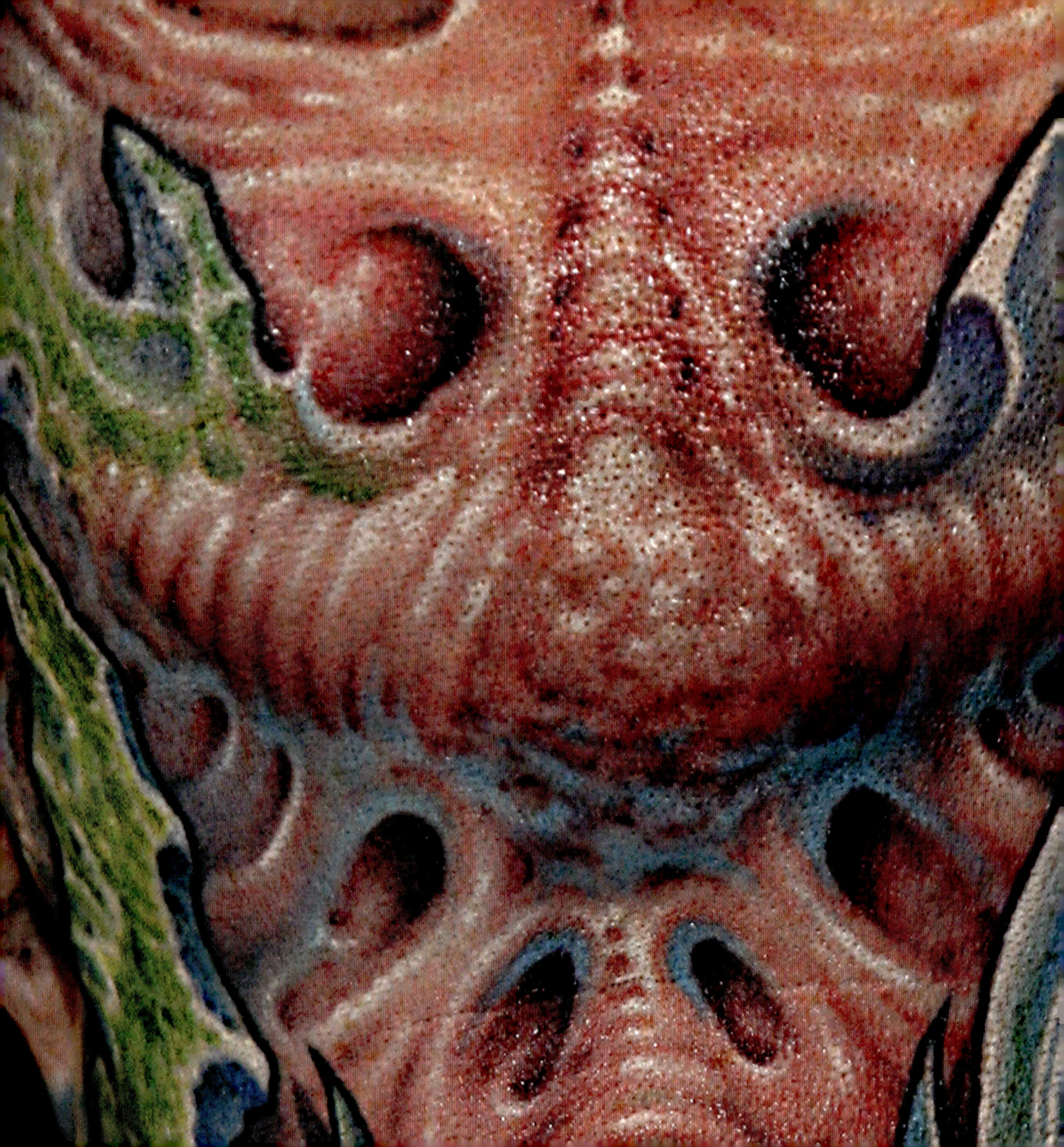

Nick Baxter

DEAR READERS

There is no doubt that tattoo art has taken many exciting twists and turns throughout history. From humble beginnings with rudimentary tools, when tattoos were used to mark the skin for therapeutic reasons, to the mind-blowing designs that are executed on a daily basis in today's industry, each piece of ink adds to that incredible journey.

Not only has the artwork itself progressed and changed throughout the years, but so have the areas of the body to which tattoo art has been applied. While there are many common areas where collectors choose to display their ink, there certainly is no rhyme or reason to where one decides to place such permanent artwork. As it's becoming more evident, the extremities of one's body can be ideal spots to display designs. It was our goal to collect and showcase a wide range of these pieces that are showing up on the far-reaching limbs of tattoo collectors all around the globe.

We were lucky to receive impressive submissions from tattooists around the world and took great care in selecting the images that you will see within this book. One of the exciting things about featuring hand, foot, and head-related tattoo art is the truly diverse range of artistic styles and genres that were presented to us. It is evident that the level of talent within the tattoo world today is awe-inspiring and constantly advancing.

This project would not have been possible without the many people involved in it and we want to extend a special thanks to each of the artists who contributed their artwork and time. We also send our gratitude to the collectors who wear these impressive tattoos on their skin and for being a part of such an exciting era of tattoo culture.

This project has been a labor of love and we have greatly enjoyed bringing Tattoo Extremities to life. It is our sincere hope that you find inspiration within these pages and enjoy reading and perusing the artwork as much as we enjoyed putting it all together for you. Thanks for joining us on the ride!

With Gratitude,
Mike DeVries & Jinxi Caddel

Nick Baxter

TABLE OF CONTENTS

Introduction
8

Chapter One
The Head & Face
11

Chapter Two
The Neck
37

Chapter Three
Hands & Palms
81

Chapter Four
The Feet
179

Conclusion
240

Jeff Ensminger

TATTOO EXTREMITIES

While there is continuing prevalence of tattoos in today's society and more people than ever before are deciding to ink their skin with permanent designs, the concept of tattooing certain areas of the body is still bewildering for some. While in the tattoo world we see the beauty and importance of each tattooed creation, there is still a certain amount of stigma apparent when it comes to tattooing extremities. But as we can see with a jaunt down memory lane, this was not always the case; in fact, tattooing of the hands, feet, and head has been well documented throughout history and has had special significance to many different cultures throughout time.

Tattoos displaying decorative elements have been found on many prehistoric mummies all over the world. In 1991, Otzi the Iceman, a five thousand year-old well-preserved mummy, was found in the Otz Valley in the Alps and bore 57 tattoos on his skin. Lines found on his skin were 15 centimeters long above his kidneys and there were numerous parallel lines on the ankles joints; as well as a cross on the inside of his knee. The position of the tattoos showed that they may have been applied for therapeutic reasons, perhaps for arthritis.

An archaeological team led by Maria Anna Pabst of Austria's Medical University of Graz looked at the tattoos on the hands and neck of a female Chiribaya Alta mummy found in southern Peru. The Chiribaya were farmers who lived from 900 to 1350 AD. Like the Otzi findings, they have speculated that the circular tattoos on her neck and upper part of her back were placed on her skin for therapeutic reasons. The ornamental tattoos which they discovered on her hands and leg displayed birds, reptiles, apes, and symbols. In addition, rings were tattooed on four of her fingers.

Chinchorro mummies include the remains of people from the South American Chinchorro culture, found in what is now northern Chile and southern Peru. These mummified human remains date to thousands of years before the Egyptian mummies and are believed to have first appeared around 5000 BC. Tattoos from the Chinchorro culture have shown a thin, pencil mustache design which was tattooed on the upper lip of an adult male.

Evidence of tattoos in ancient Egypt were found on a well-preserved female mummy from Thebes known as Amunet. She was a priestess of the goddess Hathor of Thebes during the Dynasty XI and was excavated in 1891, by Eugene Grebaut. When Amunet was found, all of the surrounding tombs had been pillaged, but buried within the Pharaoh's mortuary temple precinct, in a triangular court in the north corner, was hers. Amunet wore many collars and necklaces, and her body was tattooed. Because she was probably a concubine, as well as Hathor's priestess, her bandages were marked with the names of the King of Upper and Lower Egypt; Mentuhotep, the Son of Re; as well as other women's names. Her body was very well-preserved and featured tattoos in abstract patterns on her abdomen, arms, legs, dorsum of the feet, and pelvic region. There were dashes and dots tattooed on her body, and an elliptical pattern was found on her lower abdomen, beneath the navel. She had a cicatrix pattern over her lower pubic region. The placement of these tattoos is thought to validate the belief that the ancient Egyptians used these markings for reproductive protection.

The practice of tattooing has had immense historical significance in Polynesian cultures, where it is believed that a person's spiritual life force, or mana, is displayed through their tattoos. The Maori in New Zealand believed that the moko, or tattoo, reflects their artistic erudition; the full-face moko being a sign of prestige and superiority of status. A tattooed face was the sign of a warrior and was believed to make him more attractive to women and more threatening during wartime. Women often had their lips tattooed in an outline fashion, had solid blue facial ink applied, and oftentimes had their chins, cheeks, and foreheads tattooed as well.

The islands known as the Marquesas Islands are located a little over a thousand miles from Peru and are legendary for their influence on symbolic motifs and tattoo artwork. Hand tattooing is an important element of Marquesan tattoo history, where the designs contain special themes and patterns that symbolize certain characteristics and iconography. There are Marquesan designs for both females and males, with some of the motifs used for both genders.

In Samoa, tattooing by hand is associated with rank, title and prestige, and there are tattoo ceremonies at the onset of puberty that celebrate young chiefs perseverance, endurance, and dedication to their cultural heritage and tradition. The Hawaiian culture's traditional tattoo art is known as 'kakau' and is said to protect the wearer's spiritual well-being and health, as well as offering distinction and ornamental beauty. Many men wear detailed patterns that are shaped from natural forms and designs. This artwork can be seen on legs, arms, torsos and faces; while women are often tattooed on fingers, wrists, and hands.

Thailand also carries a rich tattoo culture, where it was believed that tongue tattoos would improve relationships and one's ability to communicate with others. In addition, tattoos on the back of the head were supposed to fill one's head with blessings to protect one's soul, because it was believed that the closer a tattoo was to one's head (where the soul resides), the greater the power of the inked design.

In addition to the historical significance of tattooing one's extremities for therapeutic and traditional reasons, the idea of tattooed artwork as an entertainment value flourished during the late 19th century and early 20th century, when circuses began hiring tattooed men and women to travel with them. While many of those wearing the tattoos adorned their skin for personal reasons, an added benefit was that they were also able to expose the public attending the exhibitions to alternative appearances that they were not used to seeing. Early in the circus era, in 1841, P.T. Barnum opened the Barnum American Museum in New York and included James F. O'Connell as one of the main attractions. O'Connell had been tattooed in the Caroline Islands where he had lived for many years and upon returning to the United States, he began telling stories about his tattoo collection acquisition. Another famous circus performer was George "Prince" Constantine (aka Captain Costentenus), who had most of his body tattooed in Burmese-influenced designs, including his face, head, and genitals. With the invention of the electric tattoo machine, the opportunity to acquire tattoos on all areas of the body became more prevalent and the rapid progression of talent continued to flourish with each new era. Our industry forefathers have paved the way and helped those owning tattoo art today to wear it proudly like a badge of honor, no matter which spot on the body one chooses to display it.

As tattooing has gained more popularity and acceptance throughout the years, it is no wonder that the areas of the body which are tattooed have evolved as well. While placement of tattoo designs is a completely personal decision, the majority of work is usually seen on limbs and areas of the body which are not categorized as "extremities." This can be due to a number of reasons; whether they be work-related, aesthetic preferences, pain tolerance levels, harmony with the shape of the tattoo design, etc.

Many artists enjoy the challenge of designing a tattoo to grace the alignment of a neck, the fun of piecing together a foot project, or the gratification of inking clean, bold knuckle tattoos. And while there are still people not accustomed to seeing bald heads dripping in inked masterpieces or decorated hands peeking out from jacket sleeves, the reality is that the concept is catching on and becoming much more acceptable than in years gone by.

This book aims to celebrate the far-reaching limbs of tattoo collectors all around the globe and to honor the artists responsible for such exquisite artwork. We only need to look to the history of our tattooed friends, from cultures far and wide, to see that the tattooing of extremities is a tradition that should be respected, cultivated, and continued for as long as tattoos exist.

Stefano Alacantara

Mike DeVries

"What a wee little part of a person's life are his acts and his words! His real life is led in his head, and is known to none but himself."
— Mark Twain

The magnificence of the human head can be demonstrated by exploring some of the incredible statistics about it. Containing 22 bones, which consist of eight cranial bones and 14 facial bones, the head is quite a detailed and intricate element. The adult human cranium is formed of fused skull bones: the parietals, temporals, ethmoid, sphenoid, frontal, and occipital; with the skull protecting the brain and most of the chief sensory organs, like the eyes, ears, nose, and tongue.

In addition to serving as the armor for such important sensory equipment, the skull has also served as an artistic icon throughout history, with symbolic meanings, and representations of power, protection, mortality, rebellion, and more. Given such noteworthy significance, it is no wonder that utilizing such an integral body part has become a part of tattoo culture.

Many societies throughout the world have historically used the head as a canvas for tattoo artwork, with different meanings and special significance equated to each civilization. Some have noted that because the head is where the soul resides, that the closer a tattoo was to the head, the greater the power of the inked design; while others have seen the placement of tattoos on one's head as a sign of strength and prestige.

In today's society, collectors who tattoo their skulls have many reasons for doing so as well, and there is no denying that choosing this route is a bold gesture, and one that cannot be easily concealed without hair growing or hat wearing. This chapter provides a peek at some of the astounding tattoo work that has been done upon the domes of tattoo wearers from around the world.

John Anderton

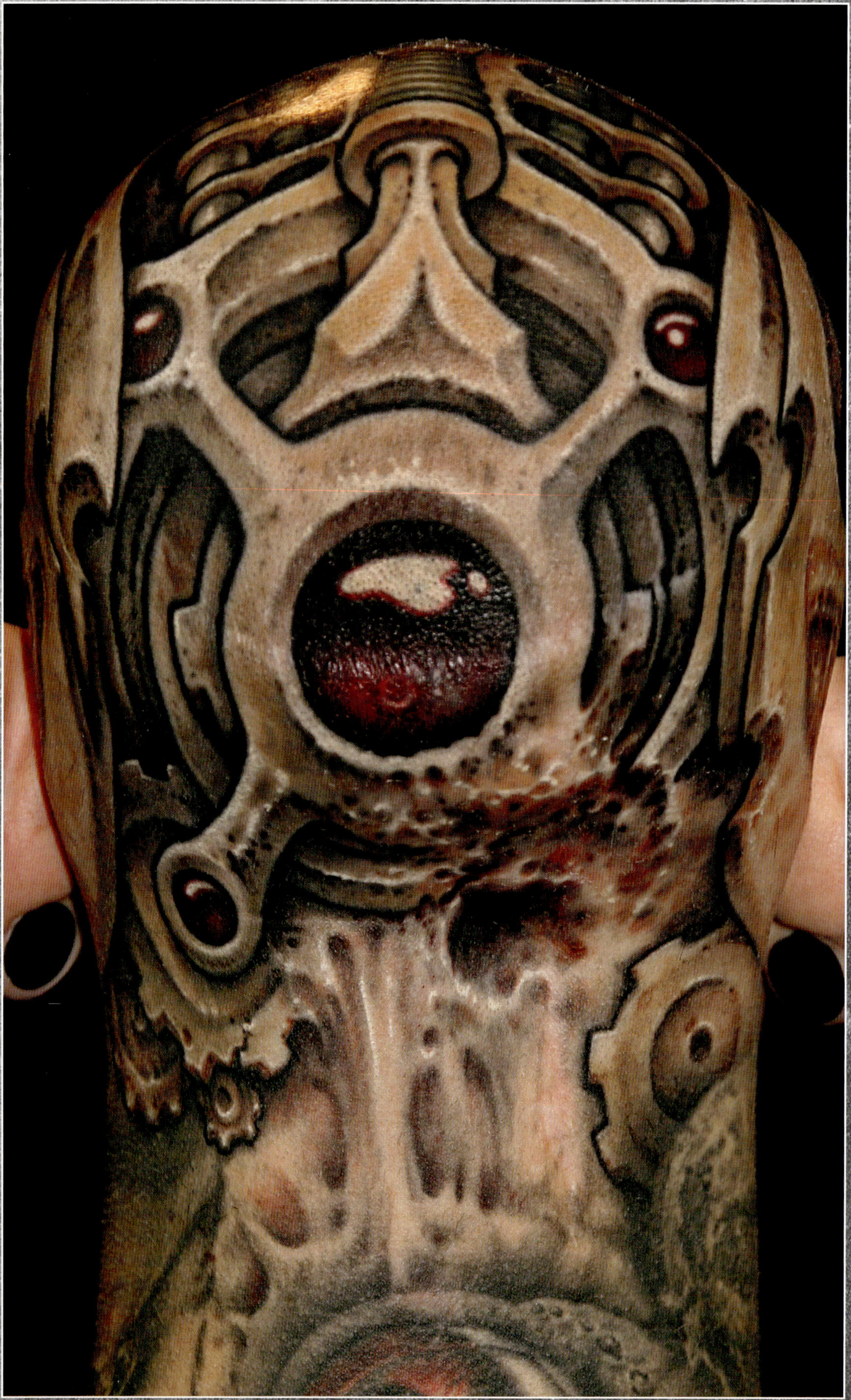

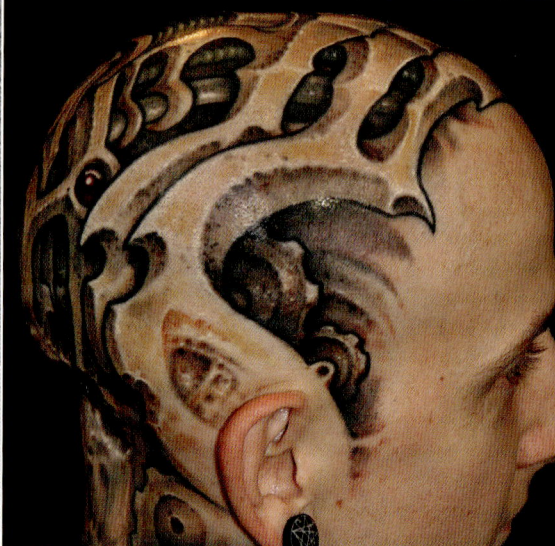
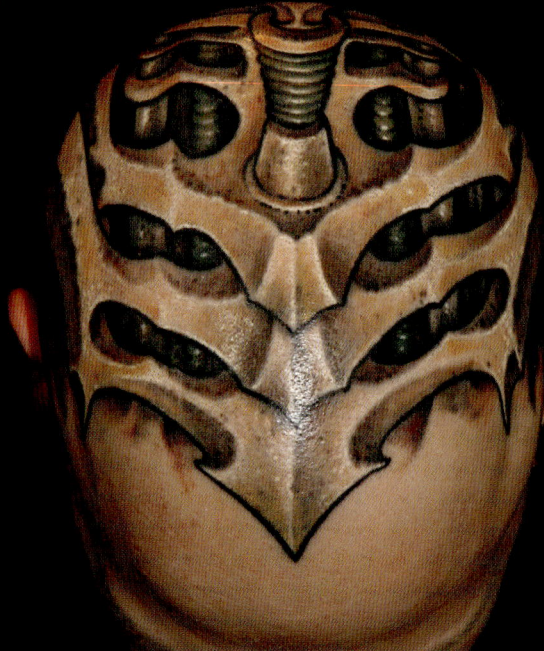
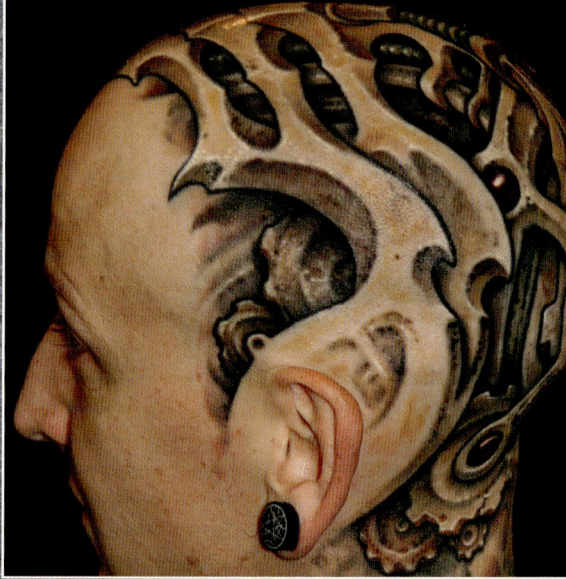

Nick Baxter

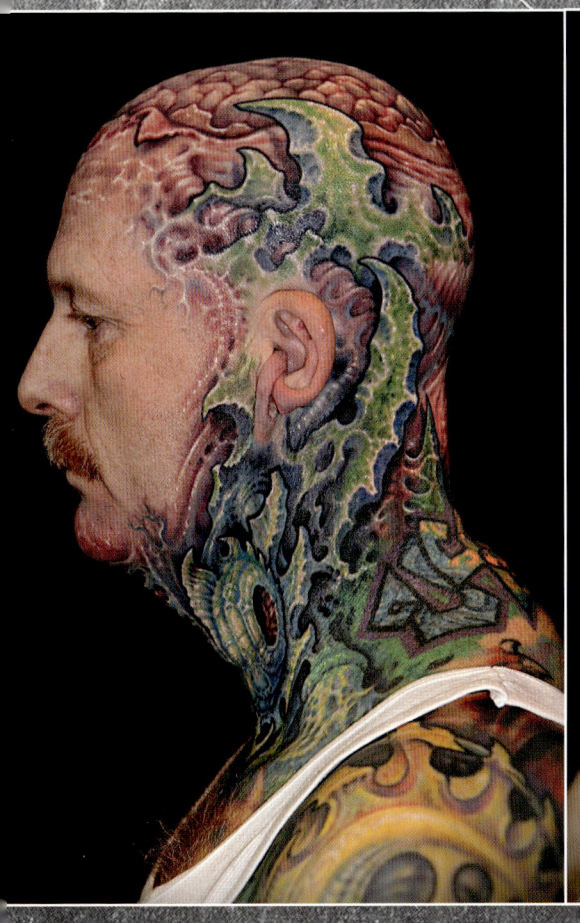
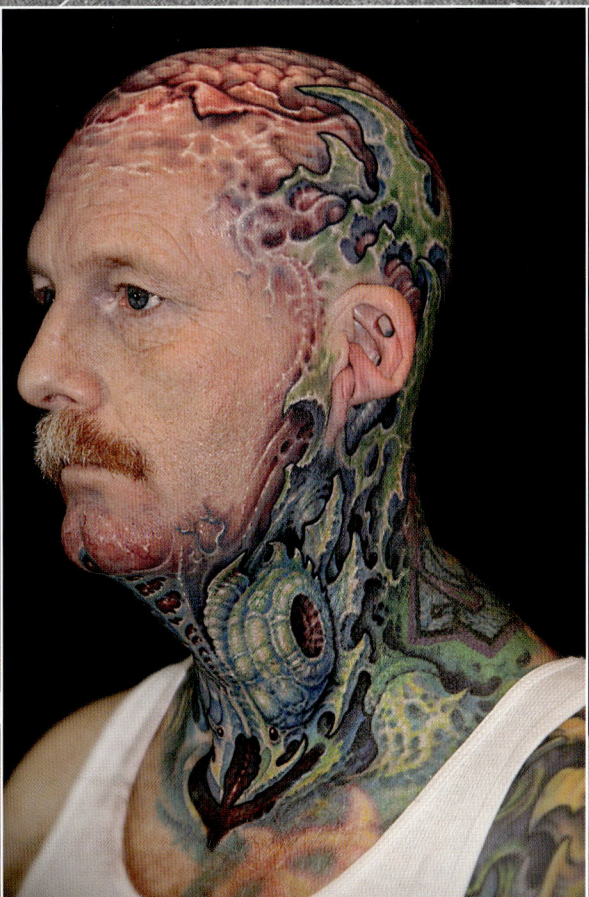
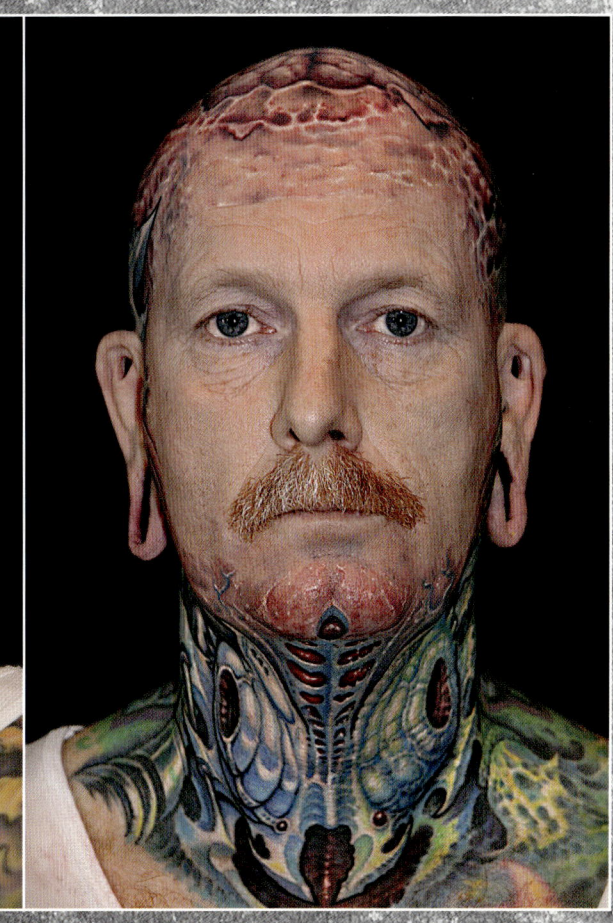
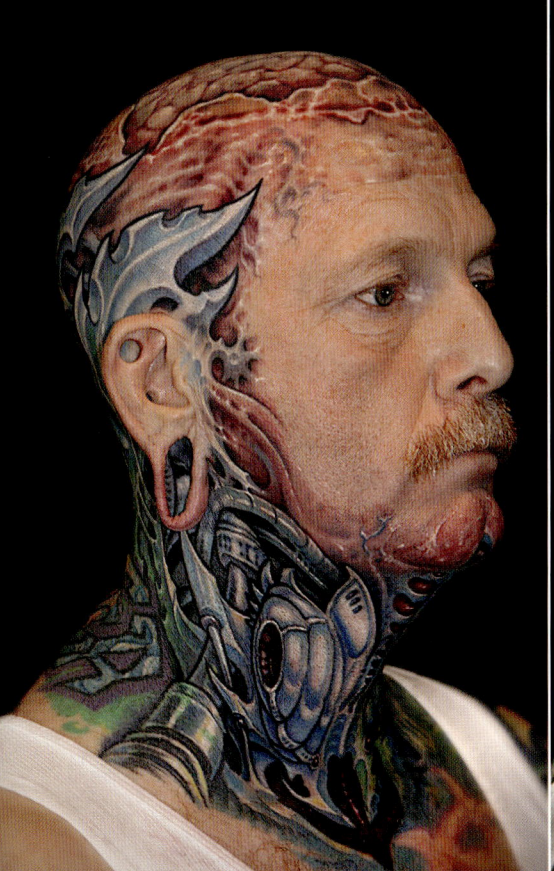
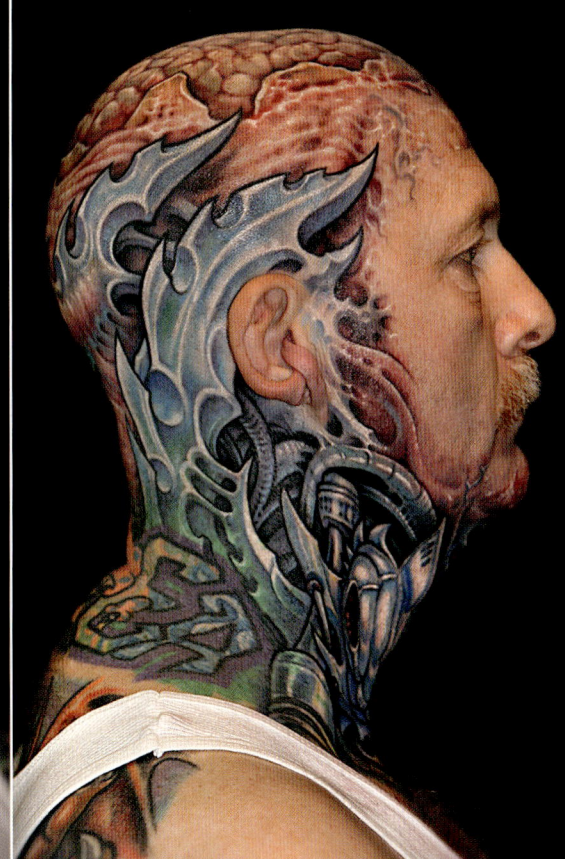
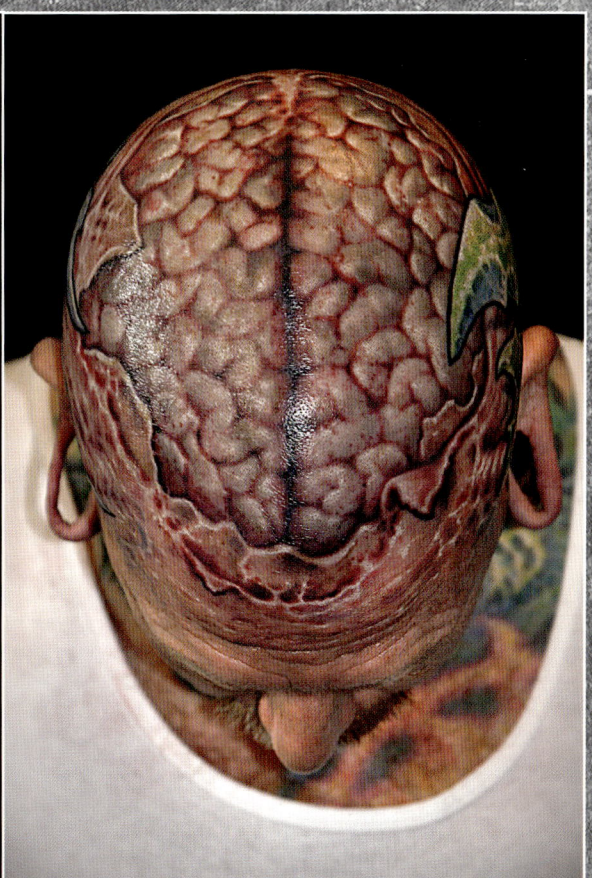

Nick Baxter

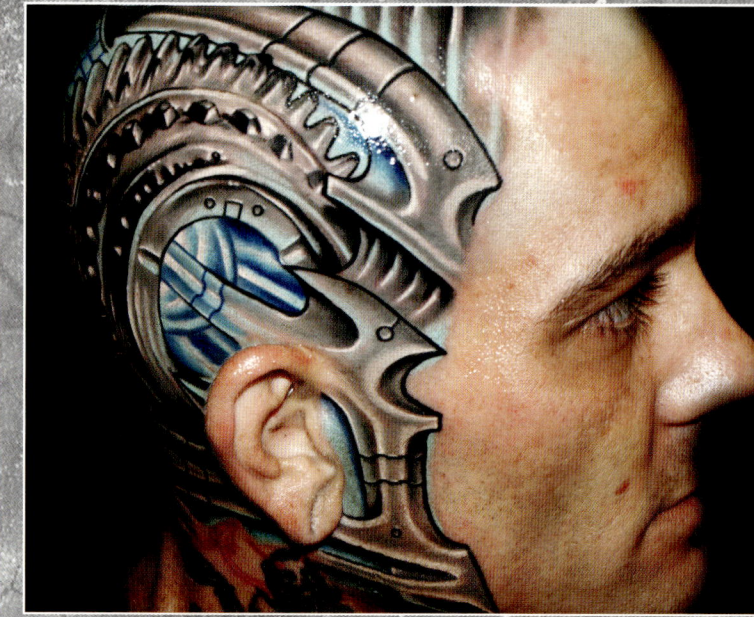
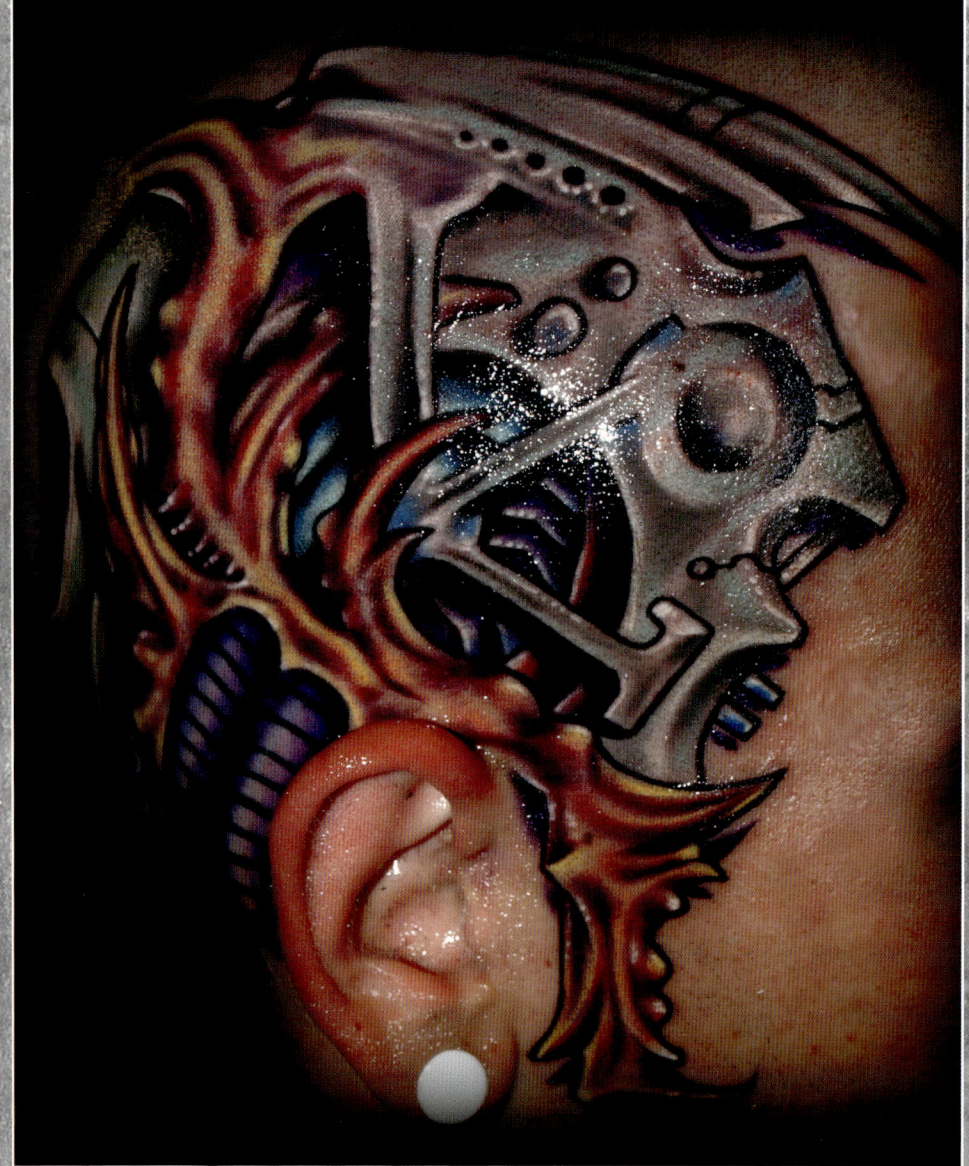

Roman Abrego

Stefano Alcantara

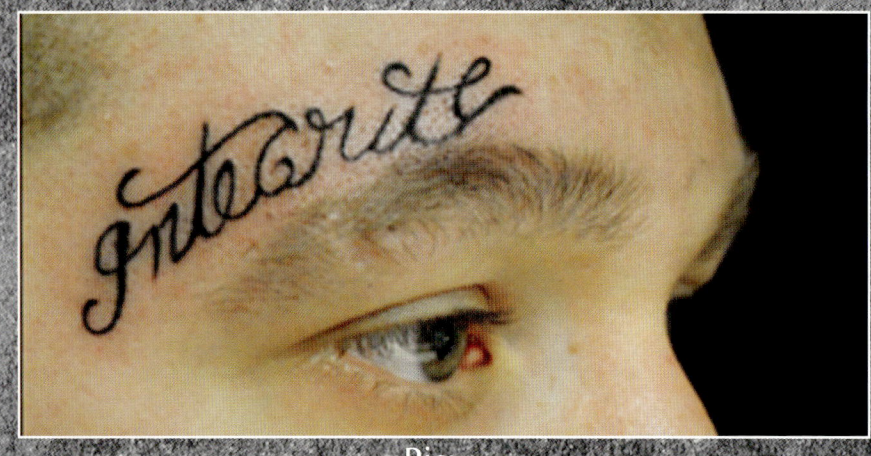
Piew

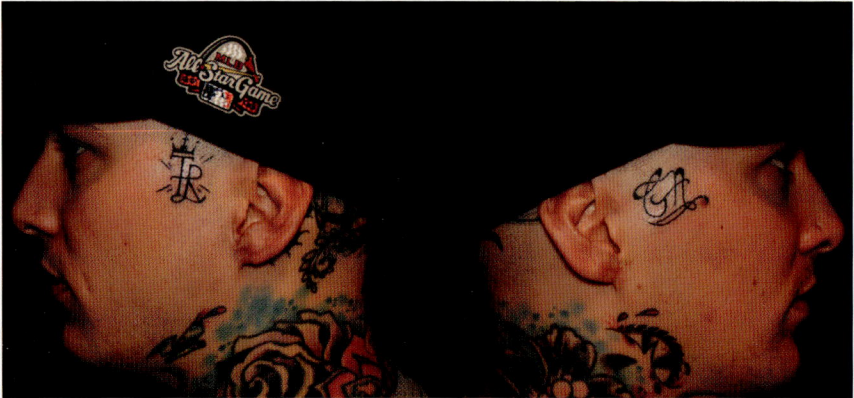
Piew

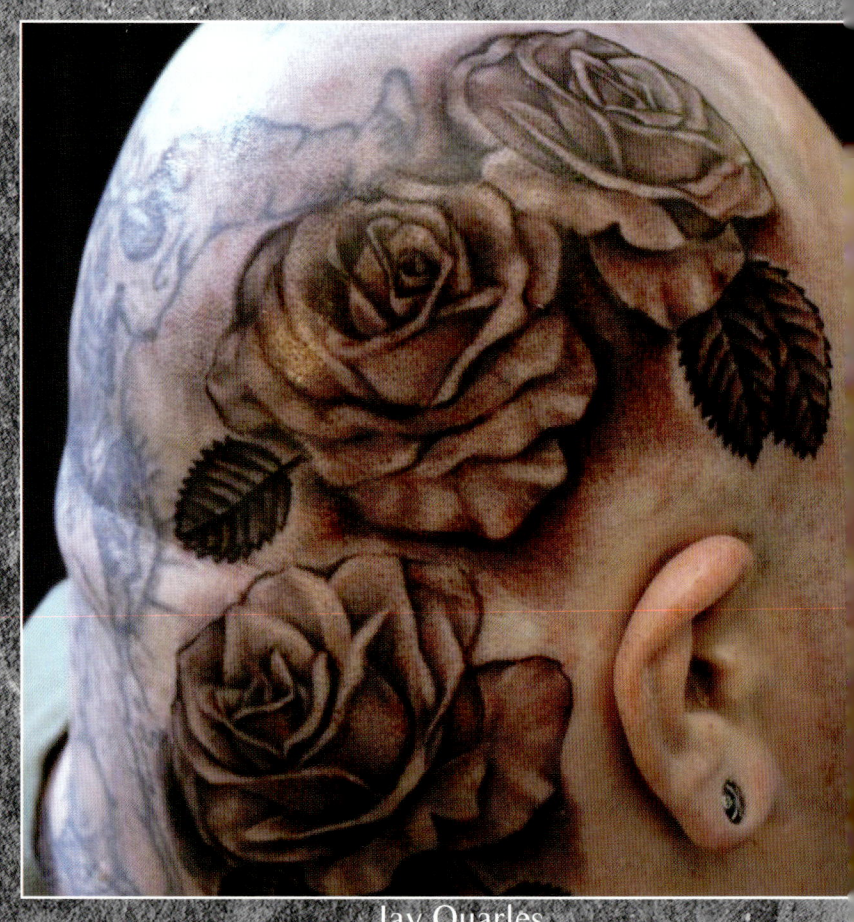
Jay Quarles

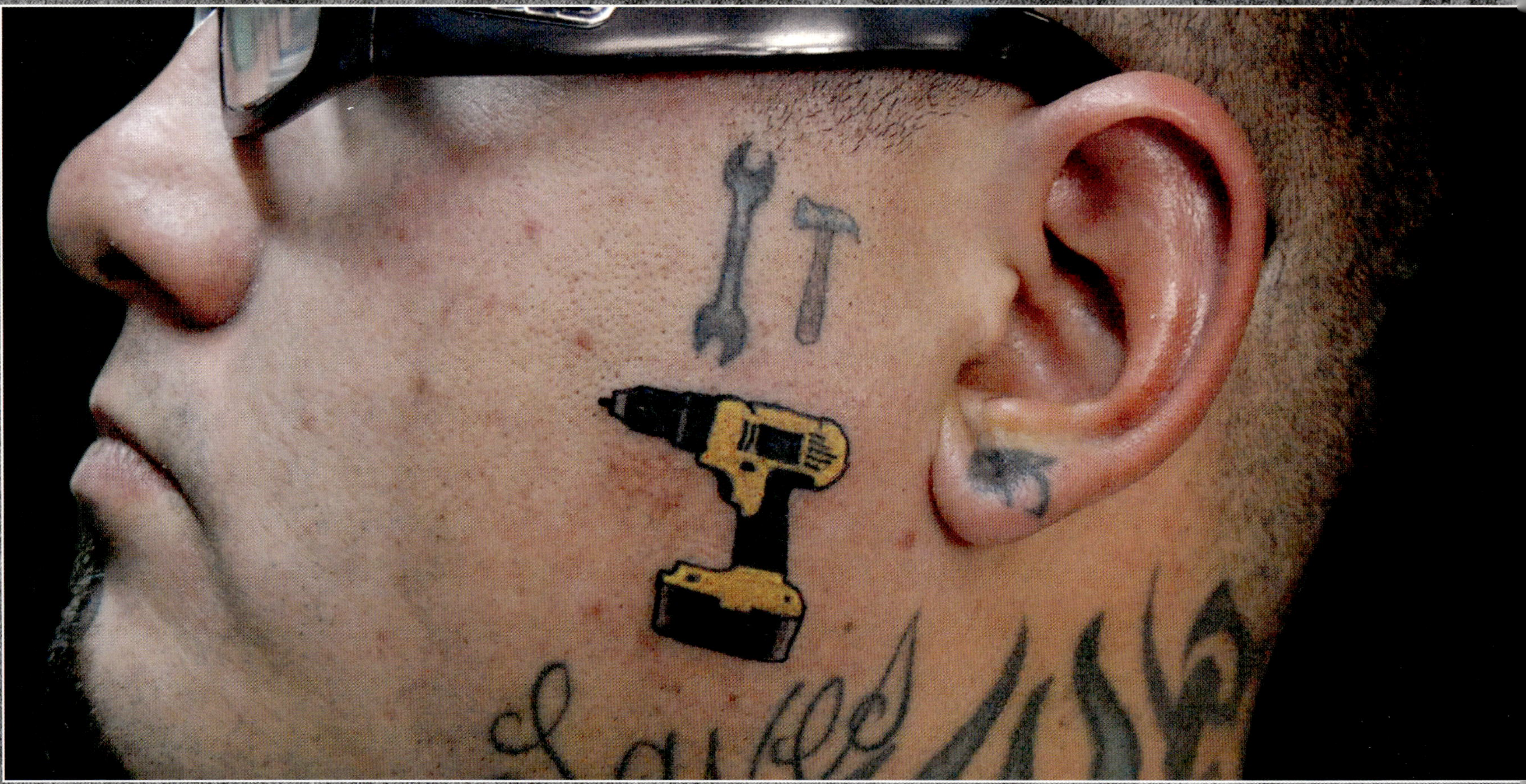

Jim Sylvia

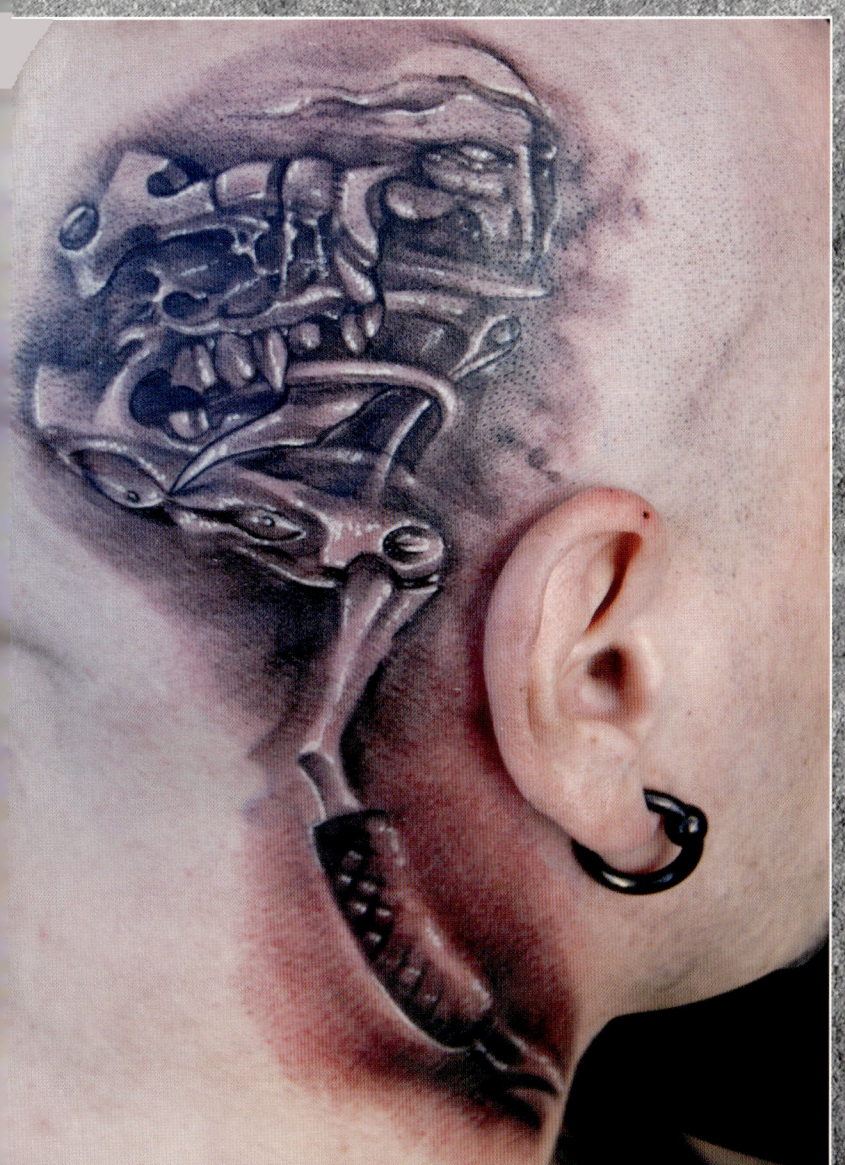

Eric Kueh

DJ Minor

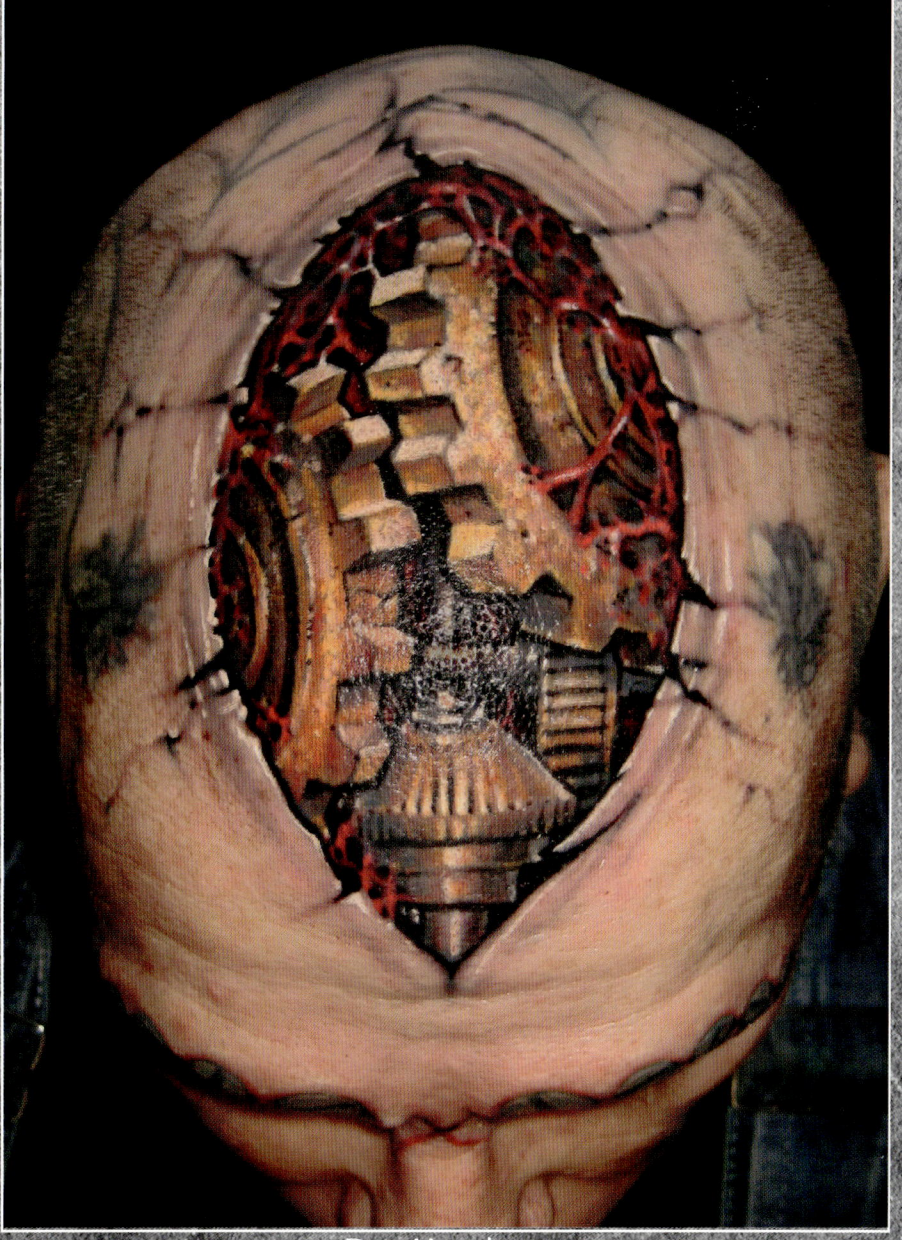

Dan Hazelton

Dan Hazelton

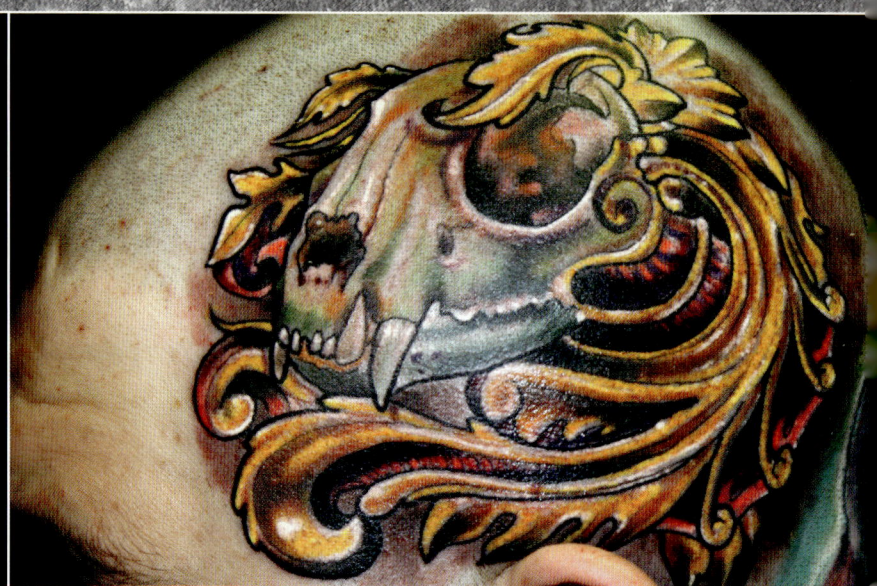

Sean Herman

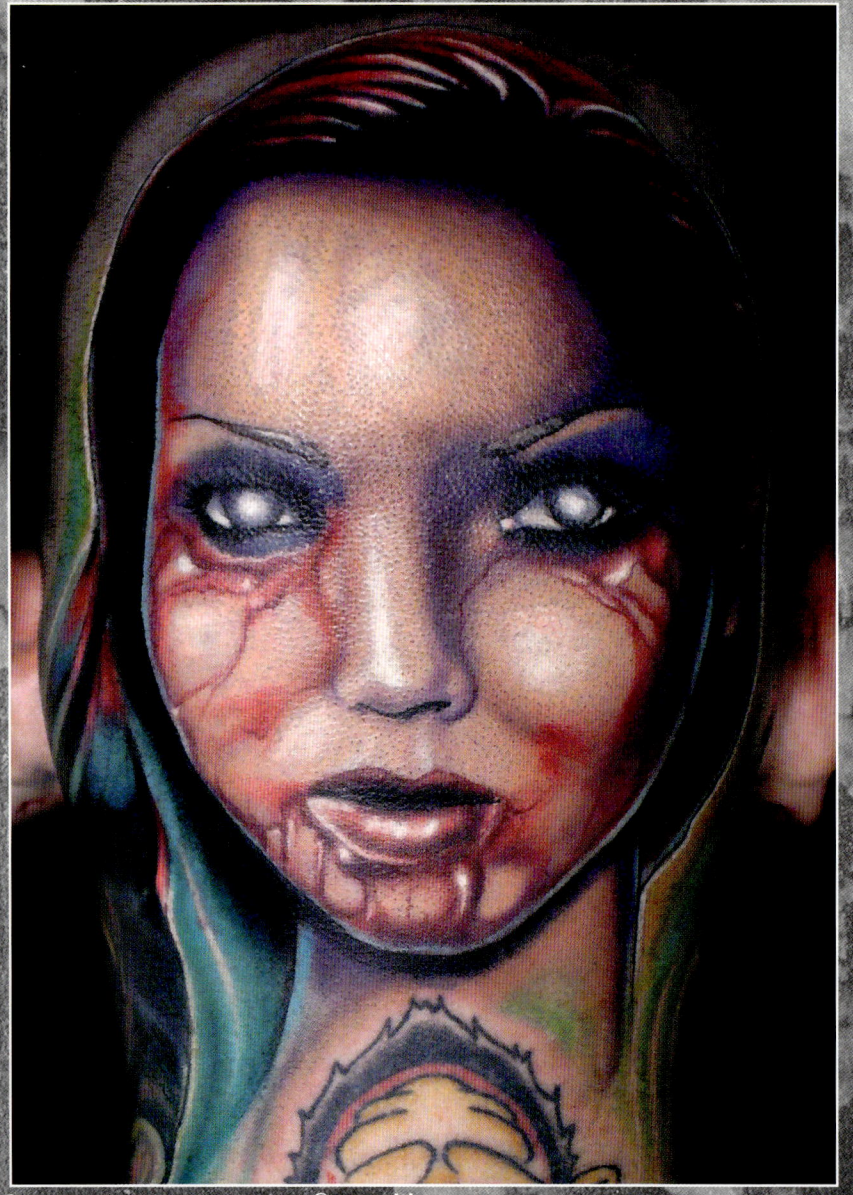

Sean Herman · Jeremy Miller

Paul Booth

Bob Tyrrell

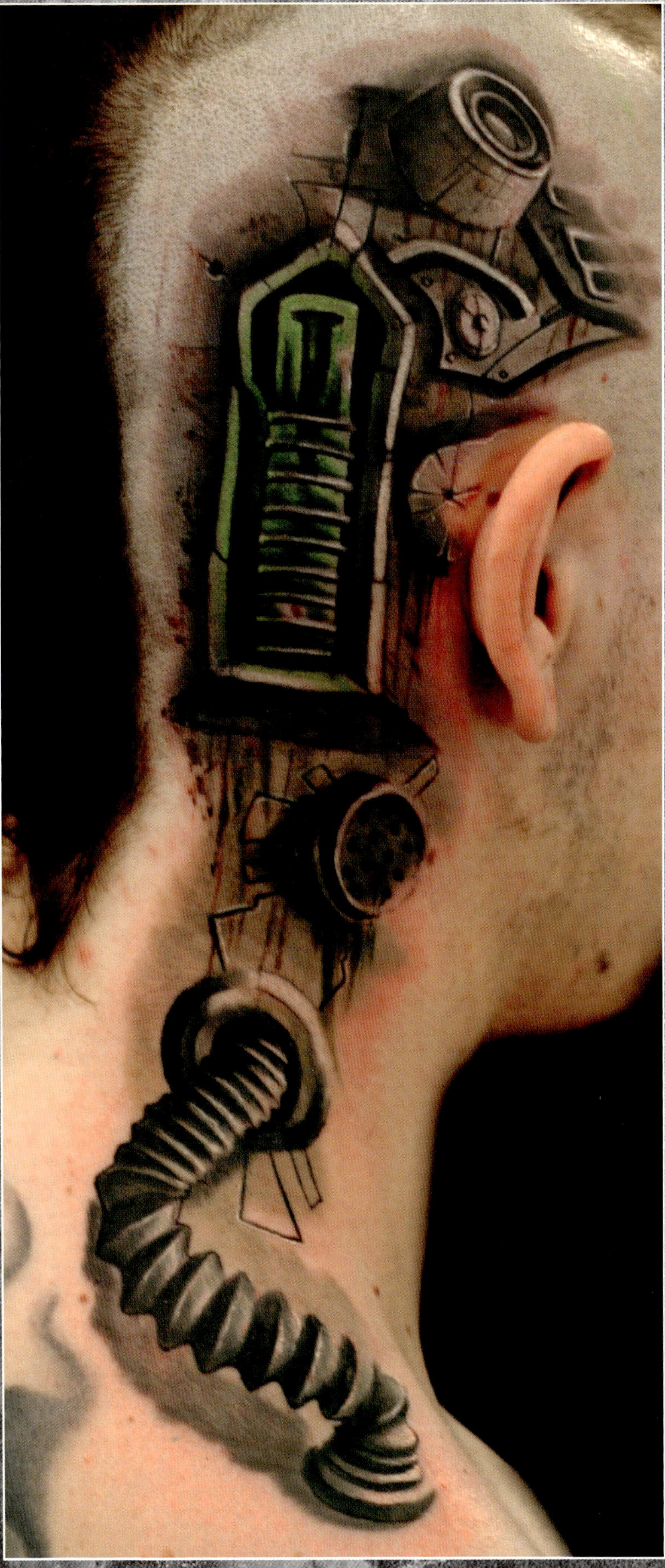

Kazimierz "KOSA" Rychlikowski

Indio Reyes

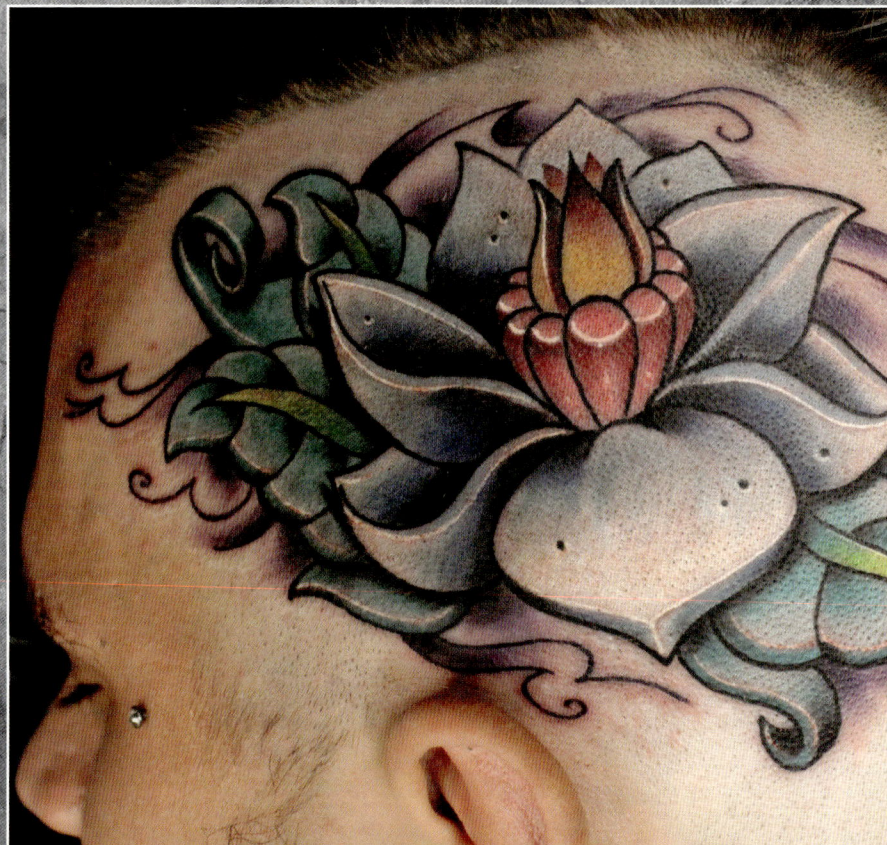
Jeremy Miller

Juan Salgado

Jasmin Austin

Indio Reyes

André Tenório

Don Mcdonald

Steve Byrne

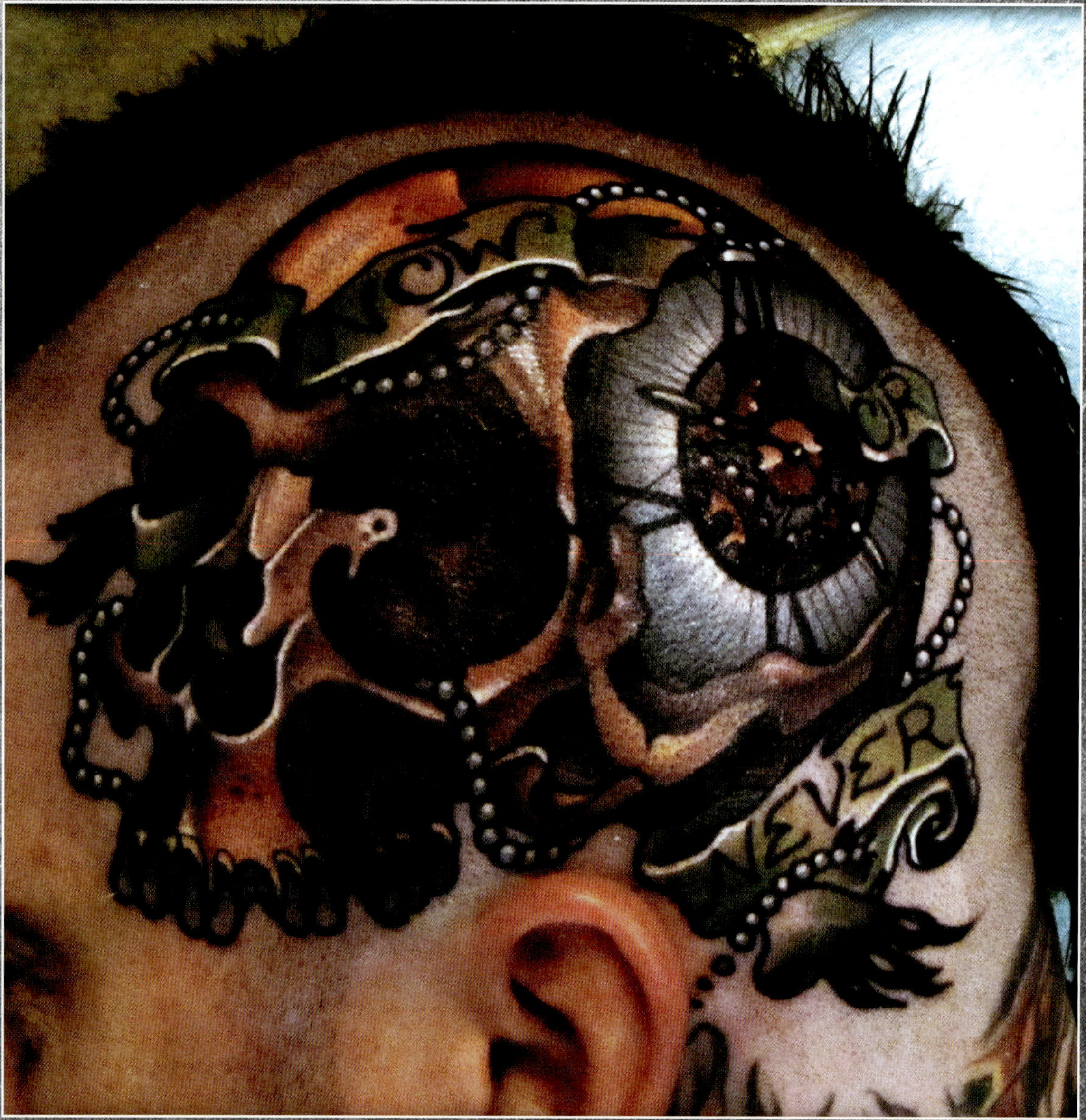

Timmy B

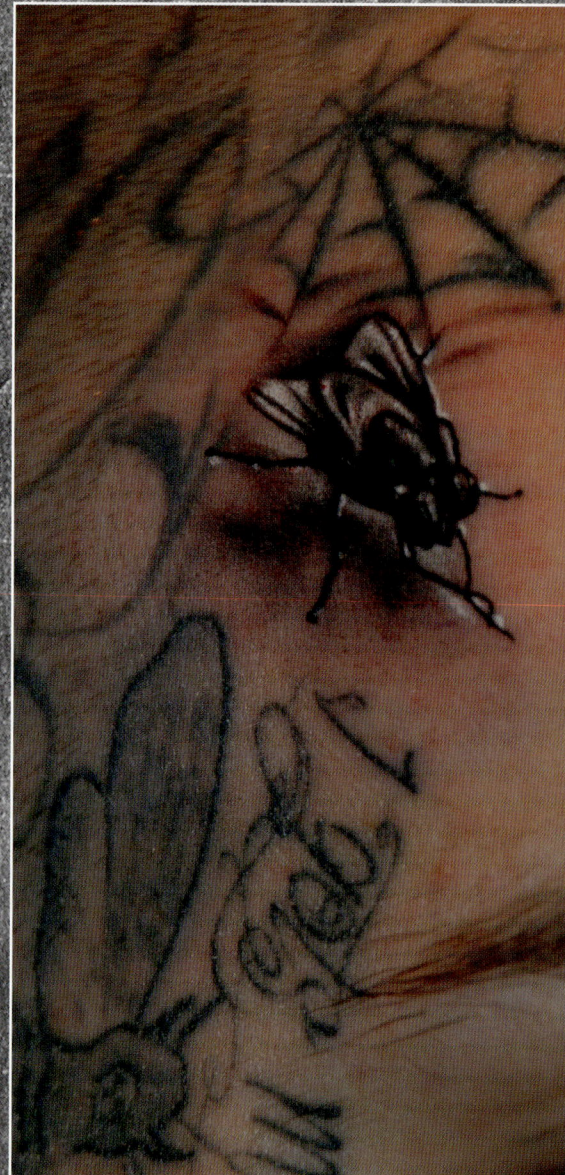
Kyle Cotterman

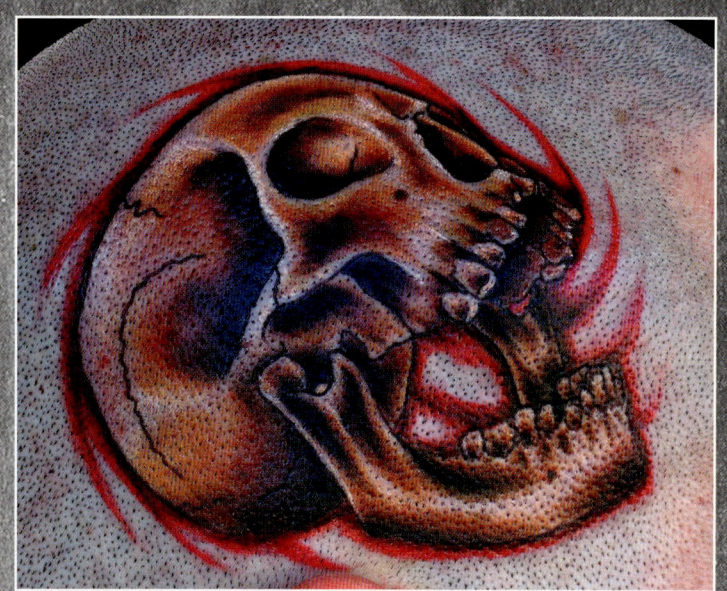
Danny Fugate

Eva Huber

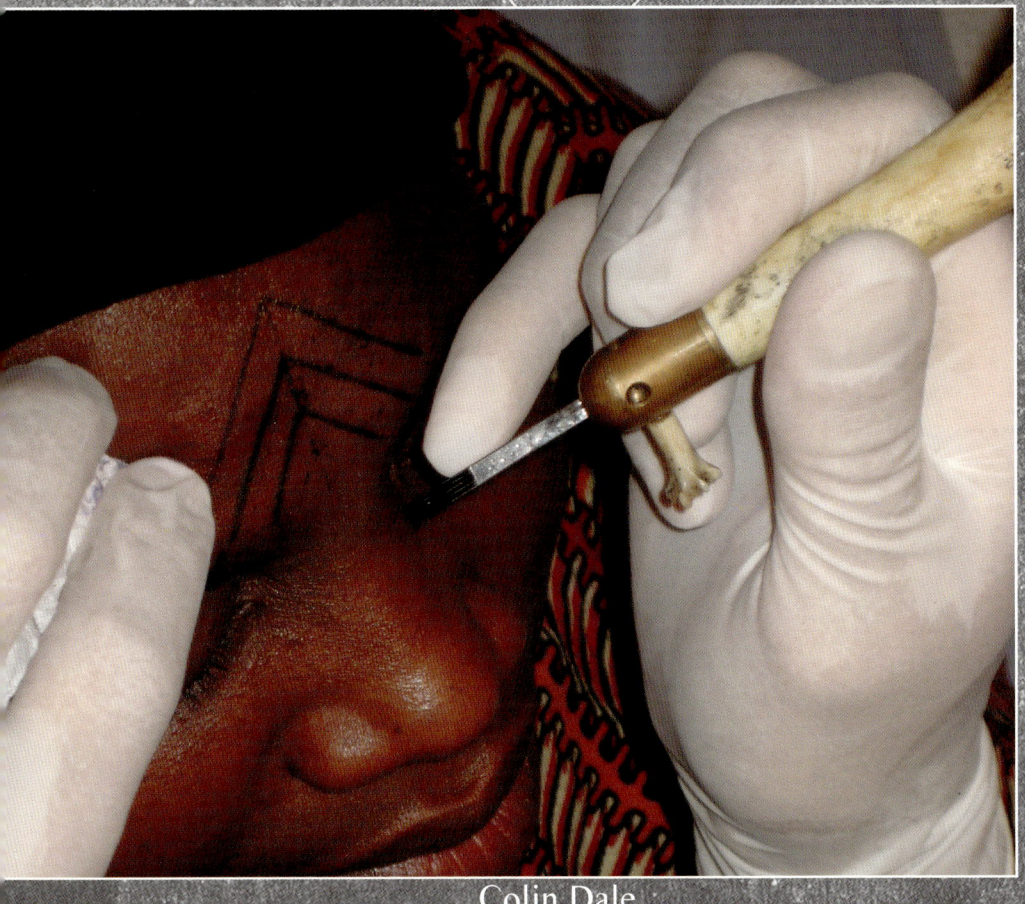
Colin Dale

Colin Dale

Colin Dale

Larry Brogan

Myke Chambers

Myke Chambers

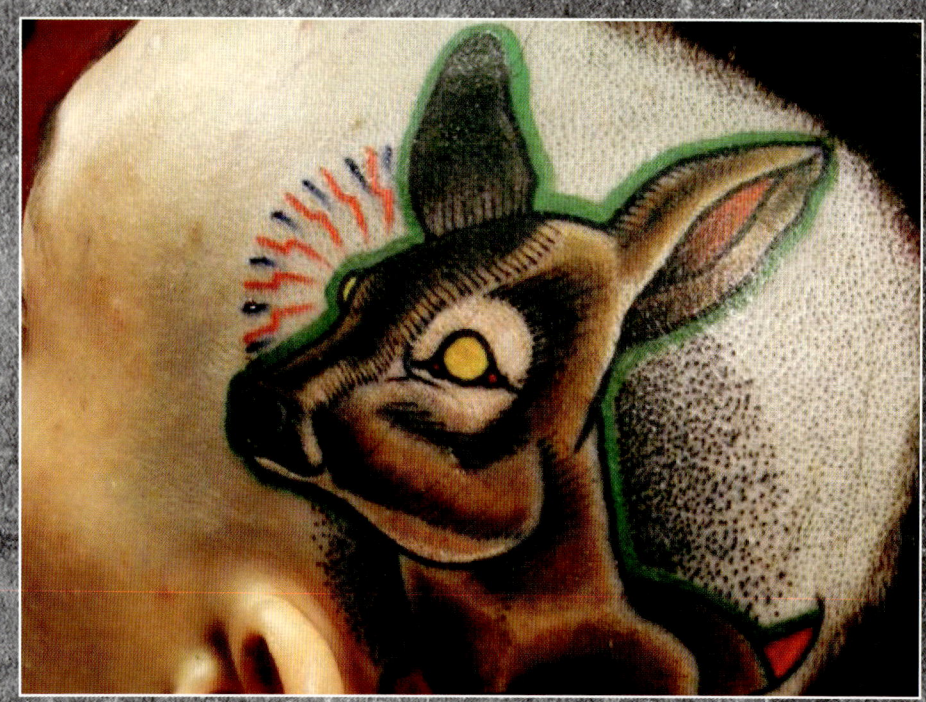

Destroy Troy

Khan

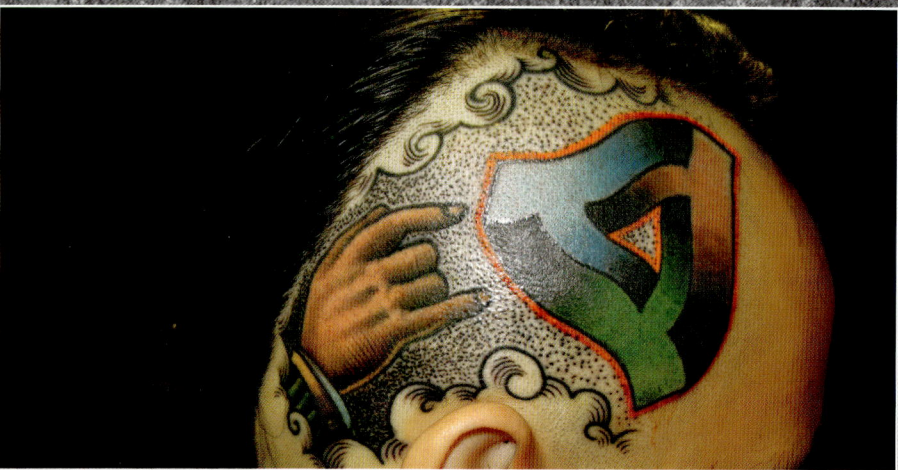

Destroy Troy

Richard Holt

Steve Byrne

Mike DeVries

Jeremiah Barba

Tim Pangburn

Tim Pangburn

Mil Martinez

Mil Martinez

Mathew Clarke

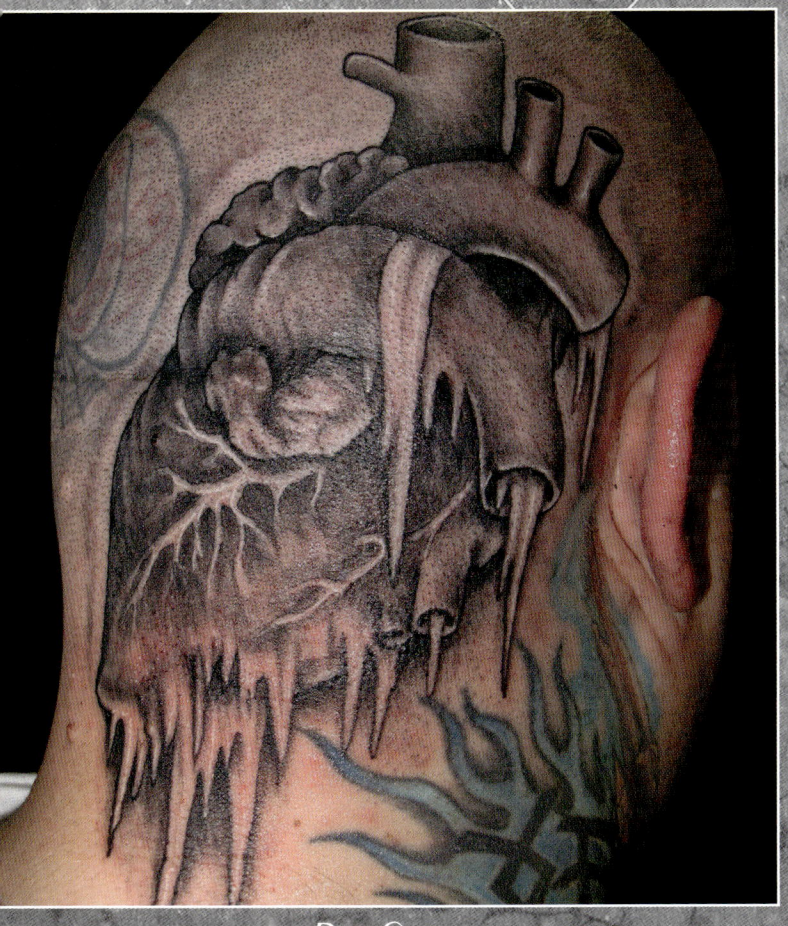
Ben Corn

Stefano Alcantara

David Bennett

Larry Brogan

30

Carson Hill

Matt Griffith

John Anderton

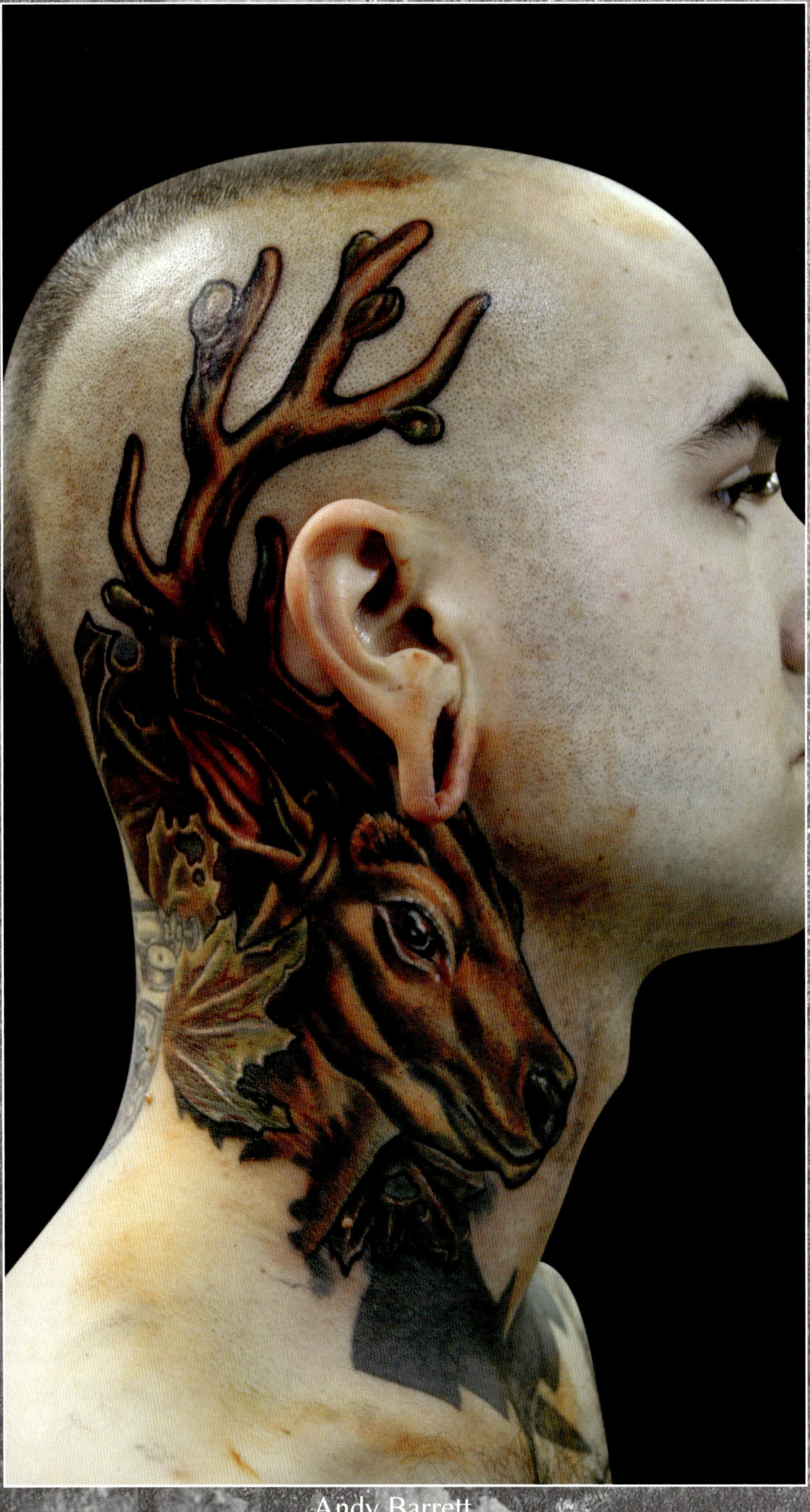

Andy Barrett

Cory Norris

Cory Norris

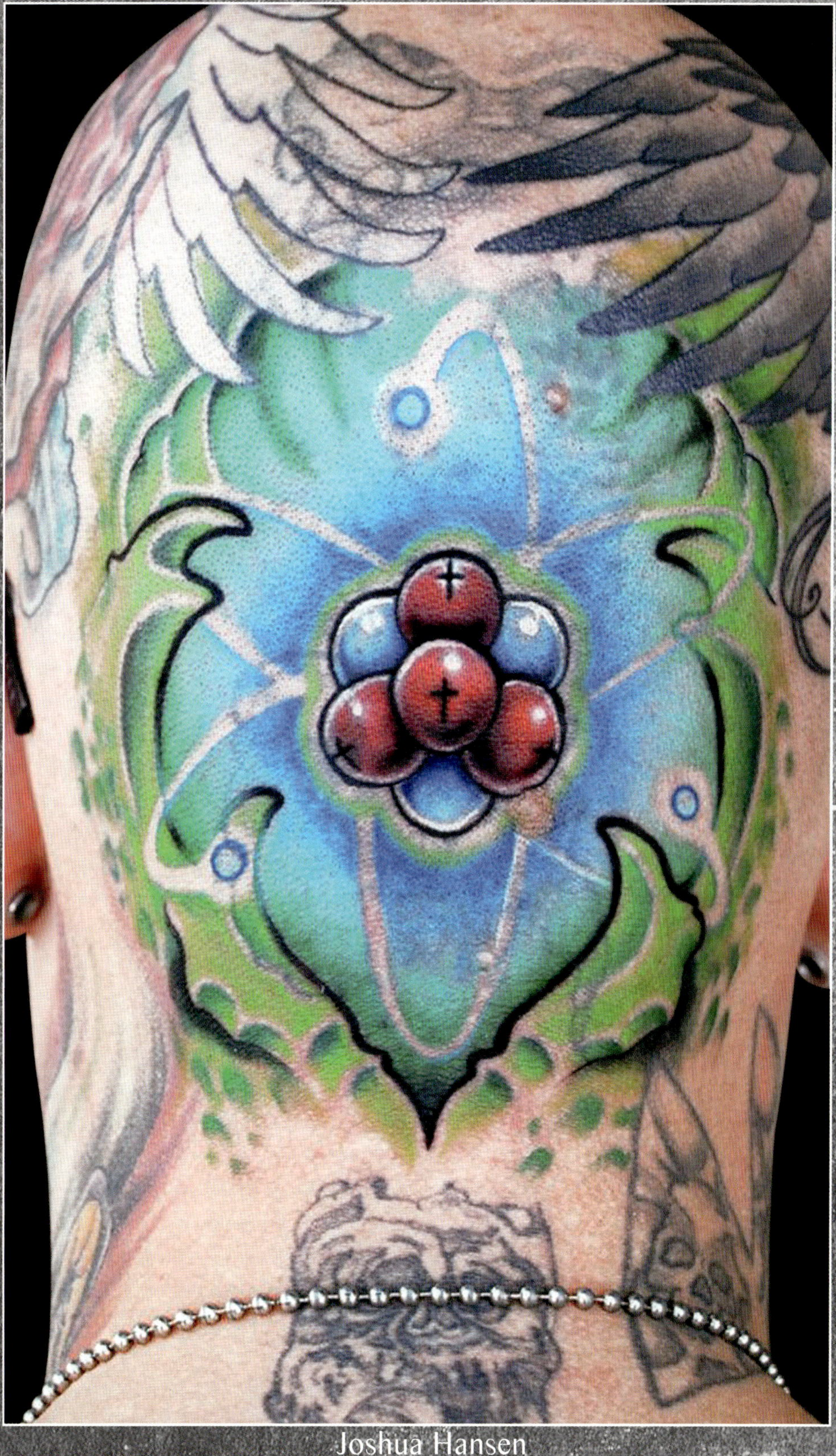

Joshua Hansen

Tim McEvoy

Tim Kern

THE NECK

"Behold the turtle. He makes progress only when he sticks his neck out." - James Bryant Conant

The neck is a fascinating body part and an excellent front-and-center choice for tattoo artwork of all types. The neck supports the weight of the head and protects the nerves that carry sensory and motor information from the brain down to the rest of the body. While neck tattoos are becoming increasingly popular in today's society, they have also been regularly seen throughout history.

Neck tattoos have been found on mummies like the female Chiribaya Alta mummy discovered in southern Peru by an archaeological team from Austria's Medical University of Graz. The mummy featured circular tattoos on her neck that are believed to have been applied for therapeutic reasons. And while neck tattoos in today's culture most likely serve more of an aesthetic purpose, it's fascinating to see the usage of this area of the body from an archaeological perspective as well.

Given the tricky angles and sitting positions necessary for tattooing this spot of the body, adding an inked treasure the neck area is not the easiest of spots for artists to work on; but as this chapter proves, the results are definitely worth the effort. From side angles on both the left and the right to collarbone divots; from front-and-center throat spots to the napes of necks - this chapter focuses on neck tattoos in all of their grandeur.

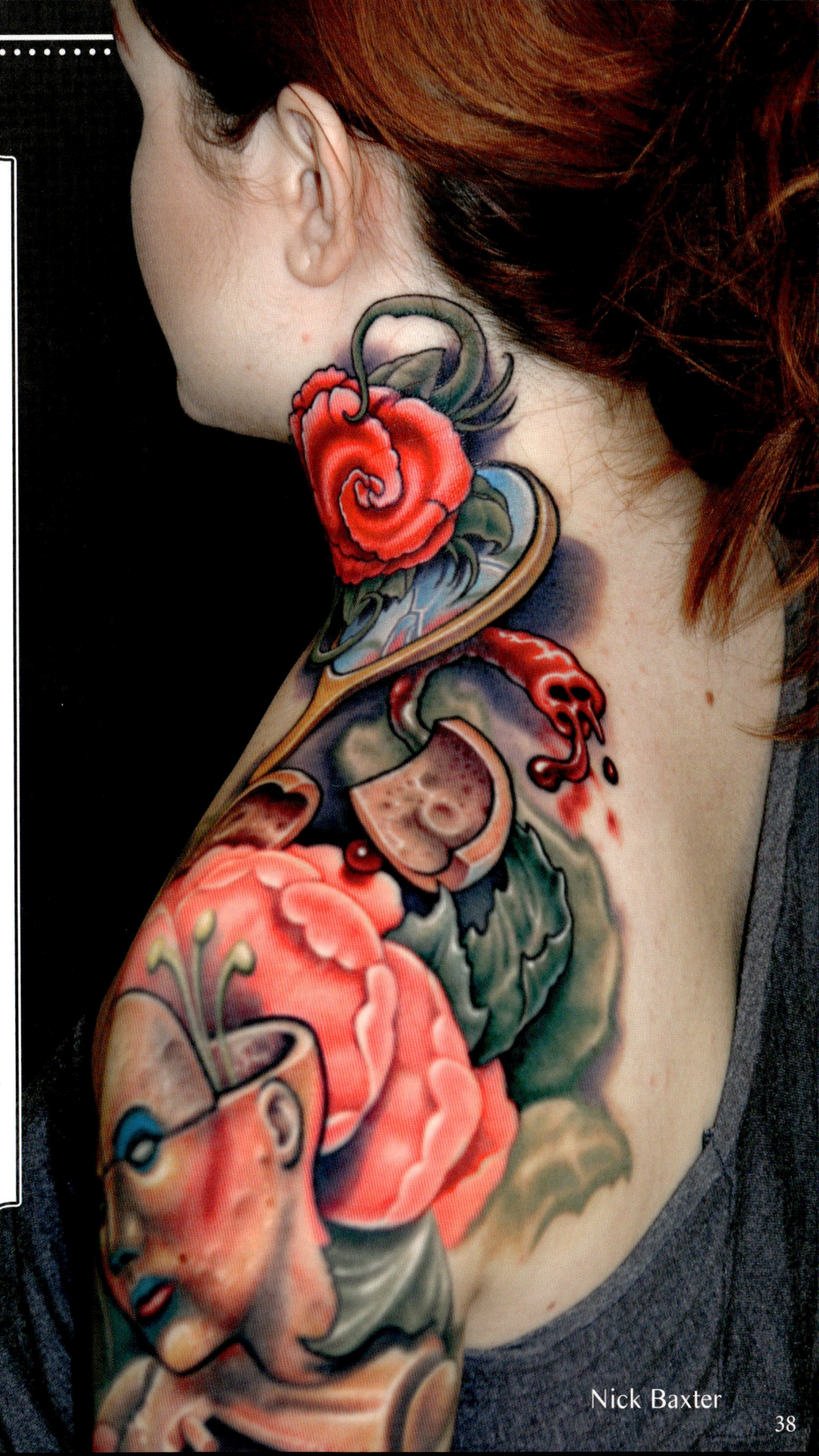

Nick Baxter

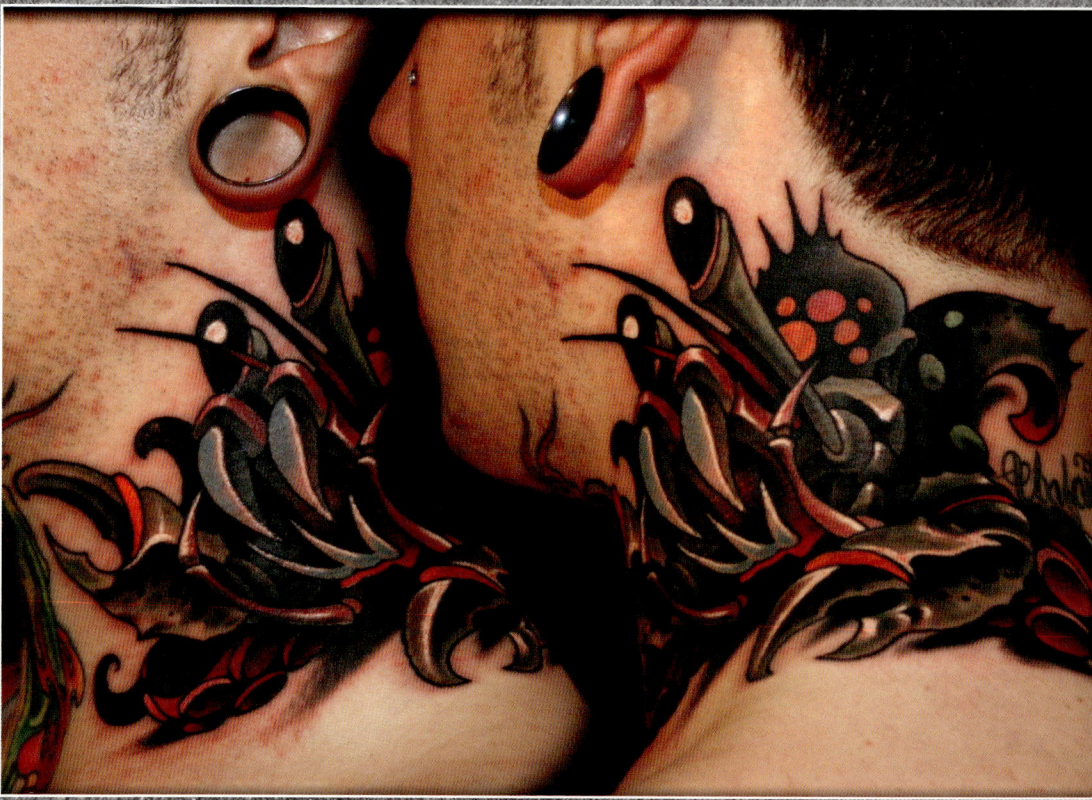

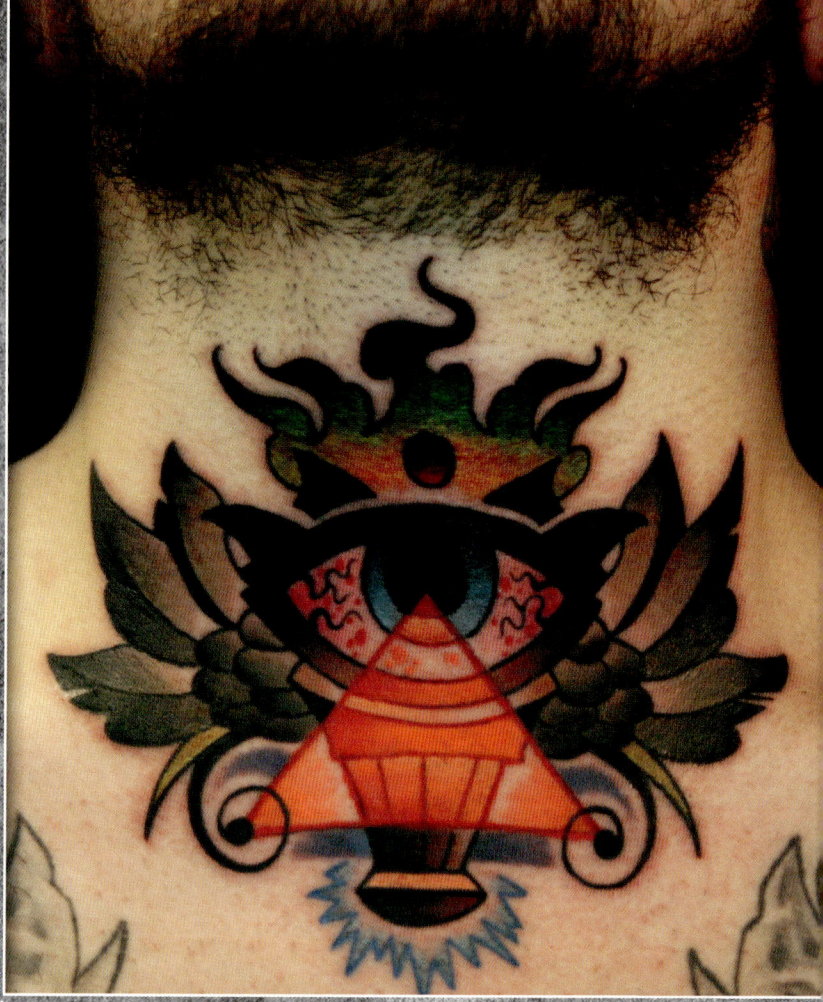

Jeff Ensminger

Jeff Ensminger

Jonathan Montalvo

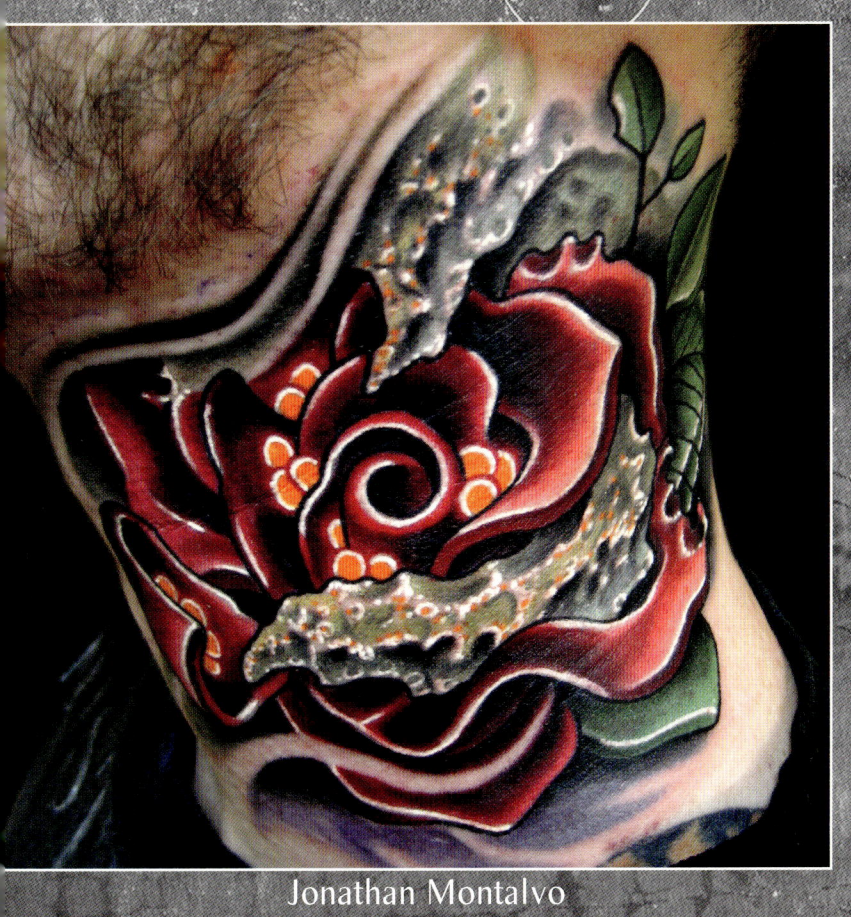

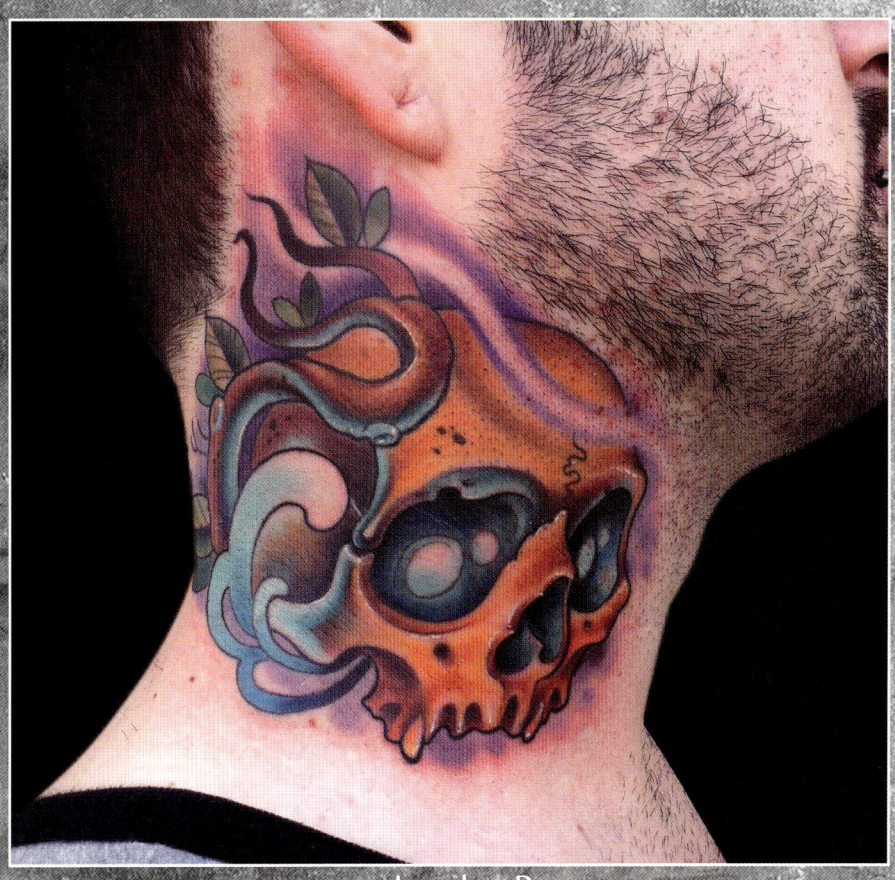

Jonathan Montalvo

London Reese

Jonathan Montalvo

43 Russ Abbott

Stefano Alcantara

Leigh Oldcorn

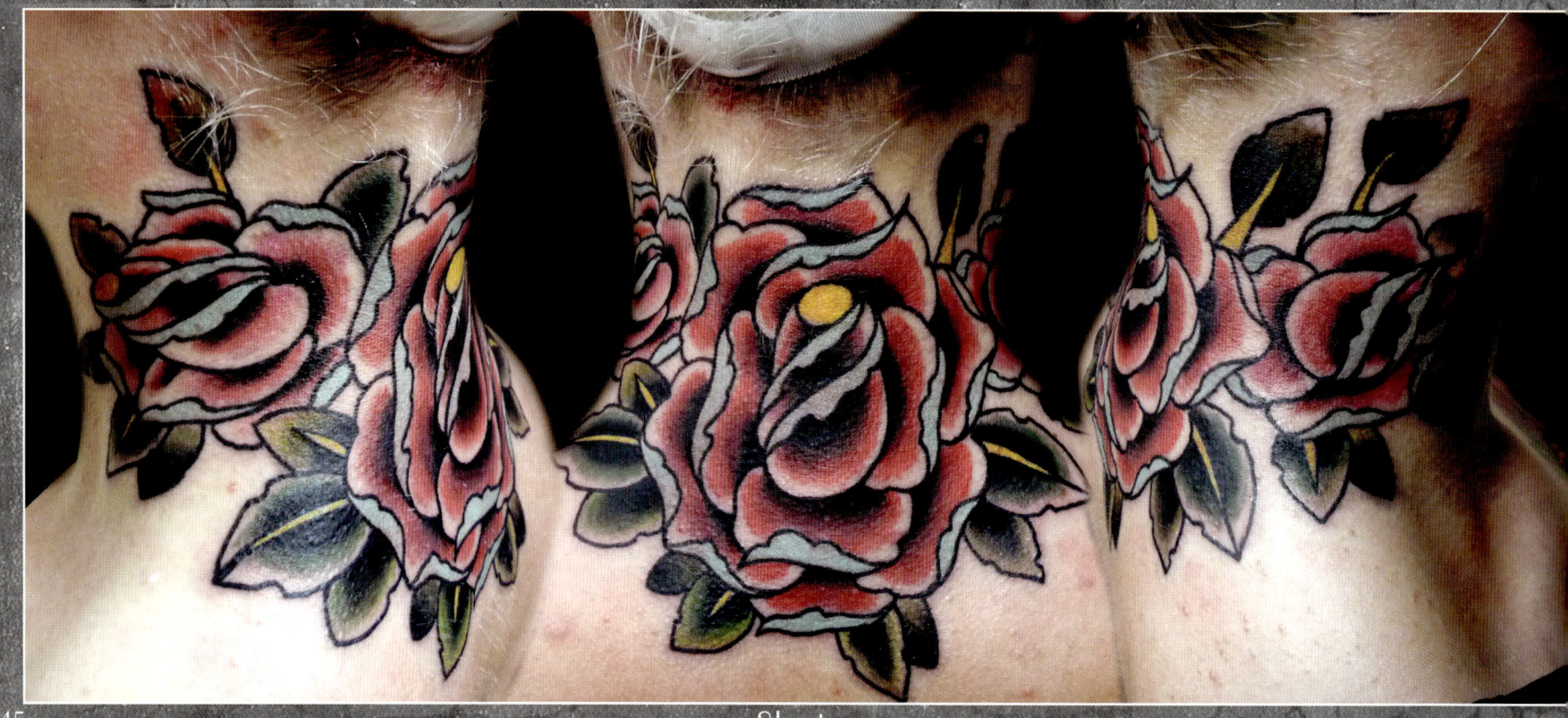

Short

Kazimierz "KOSA" Rychlikowski

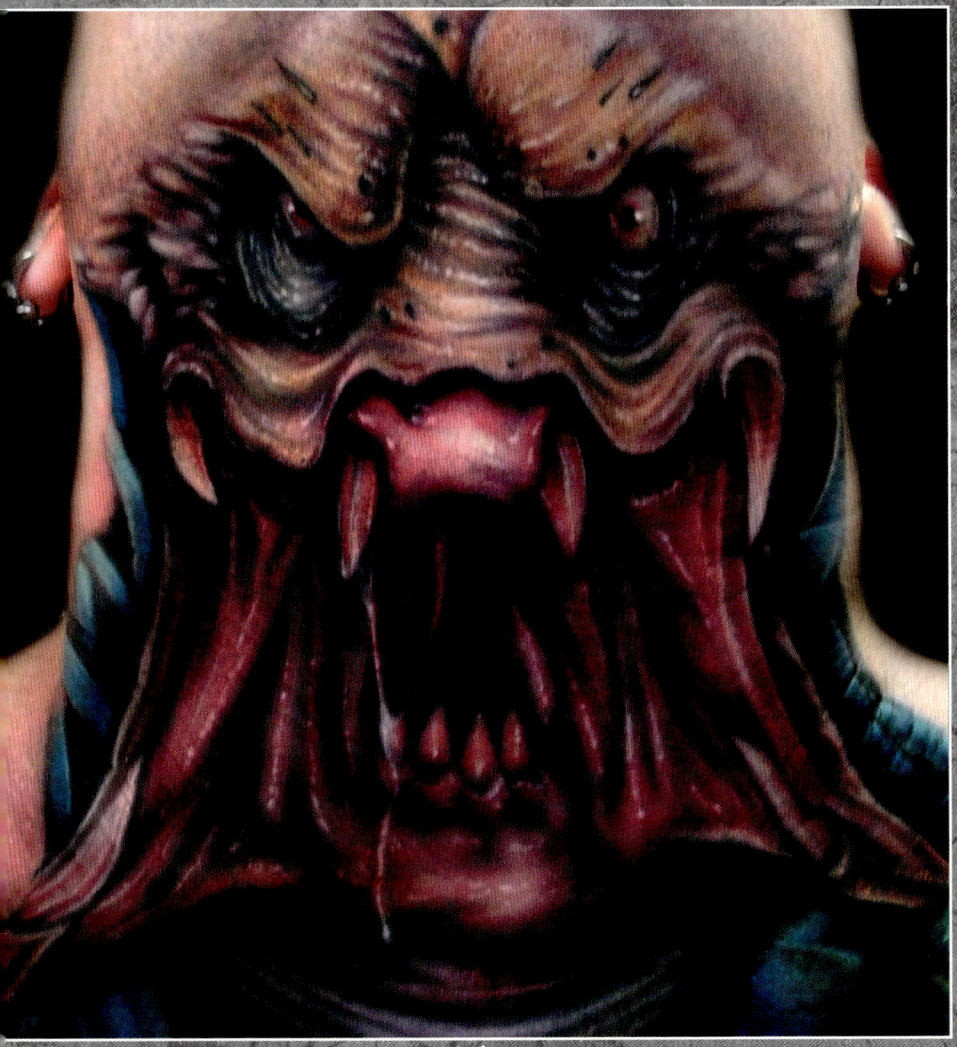

Khan

Mason Williams

46

Christian Perez

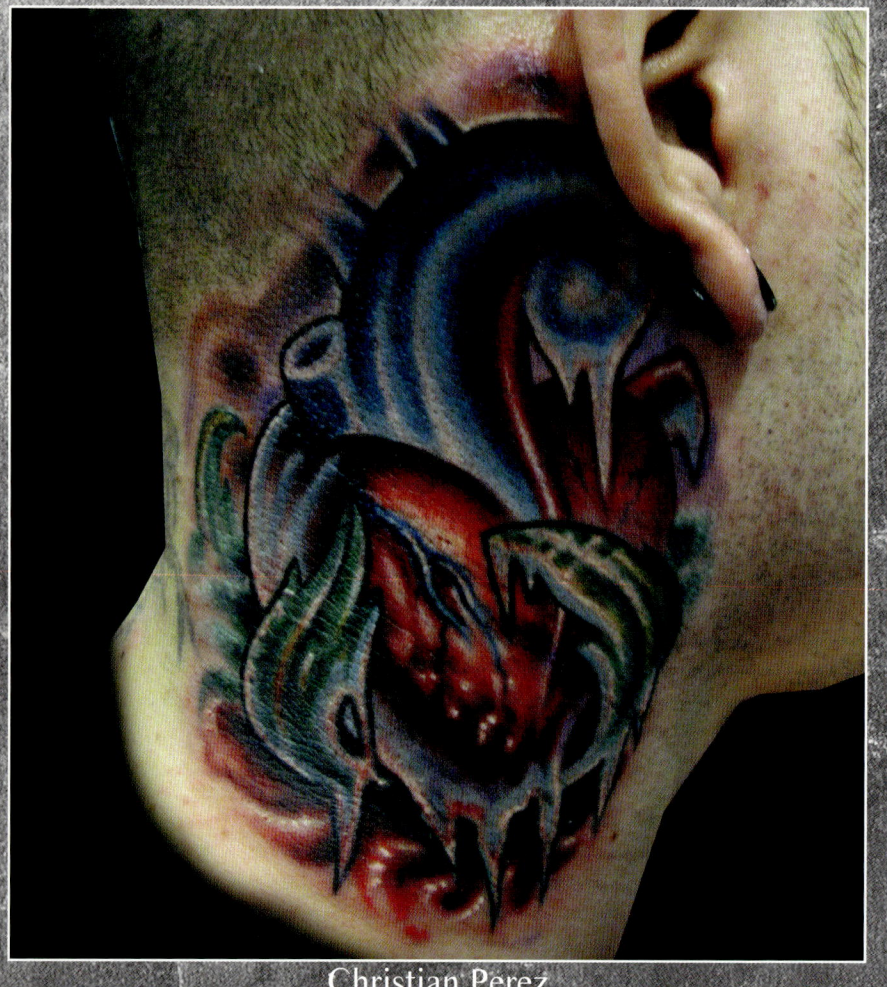

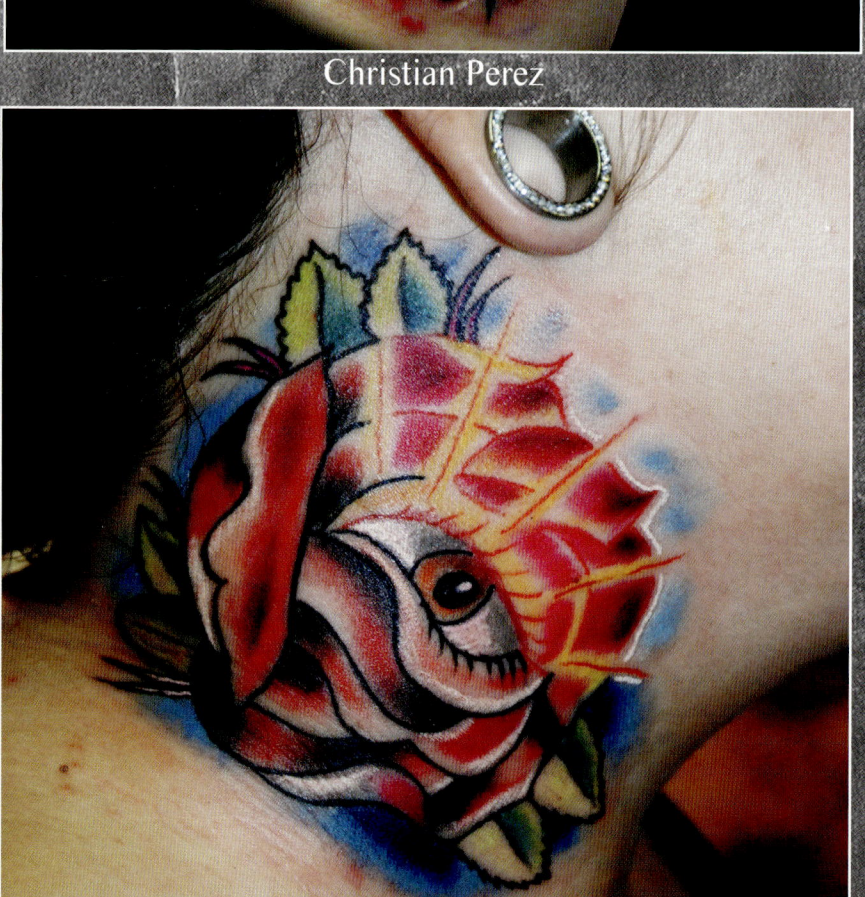

Christian Perez

Johnny Jackson

Johnny Jackson

Christian Perez

Christian Perez

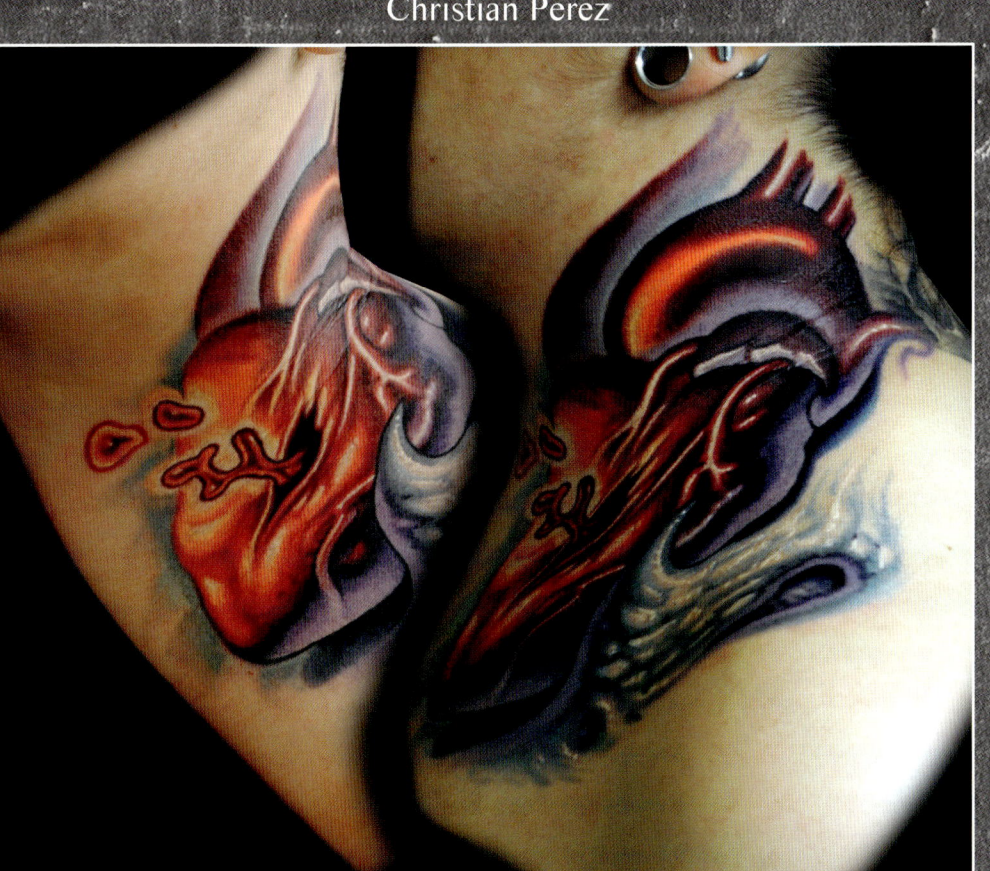

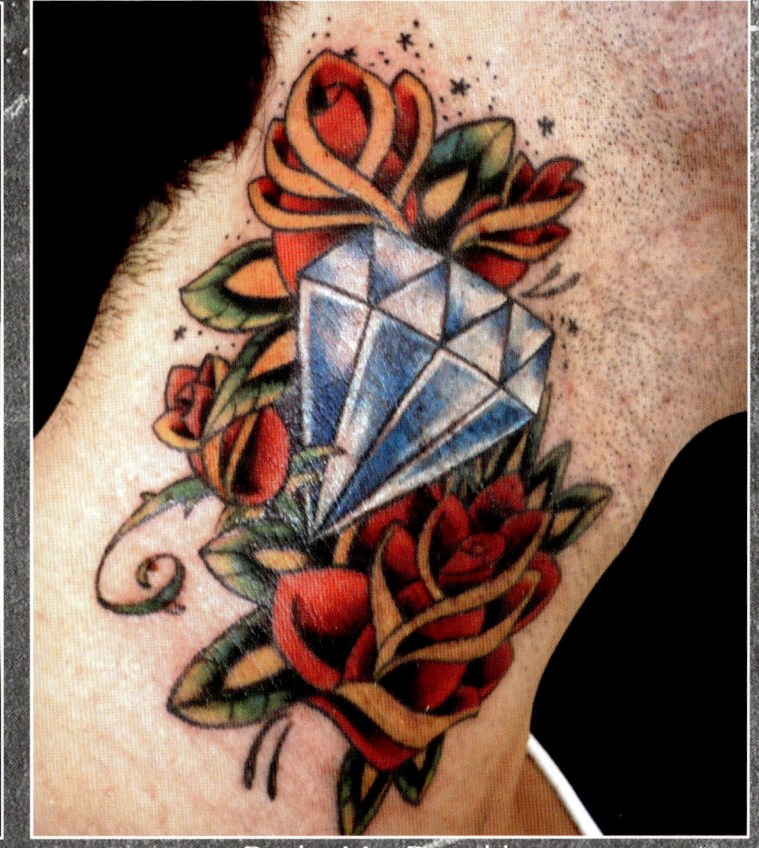

Christian Perez

Becka MacDonald

48

Mike DeVries

Mike DeVries

Mike DeVries

Mike DeVries

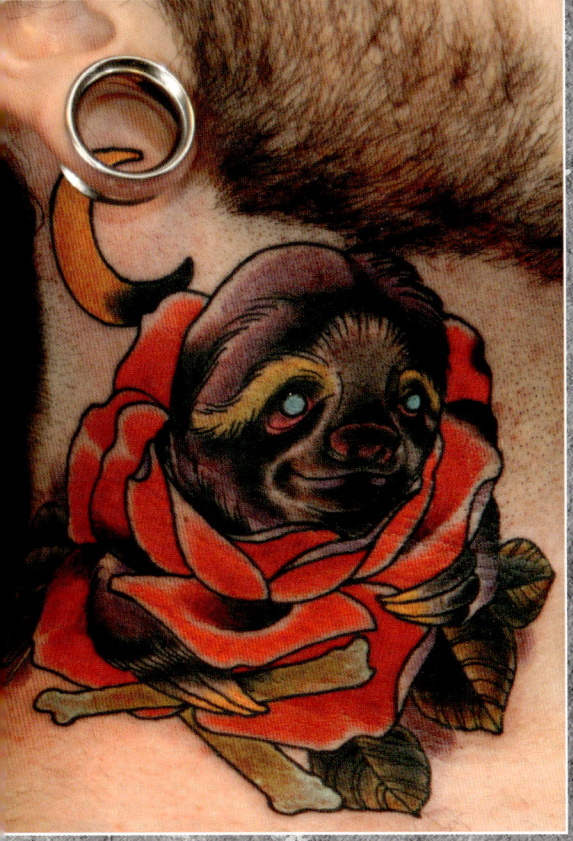

Uncle Allan

Uncle Allan

52

Mike DeVries

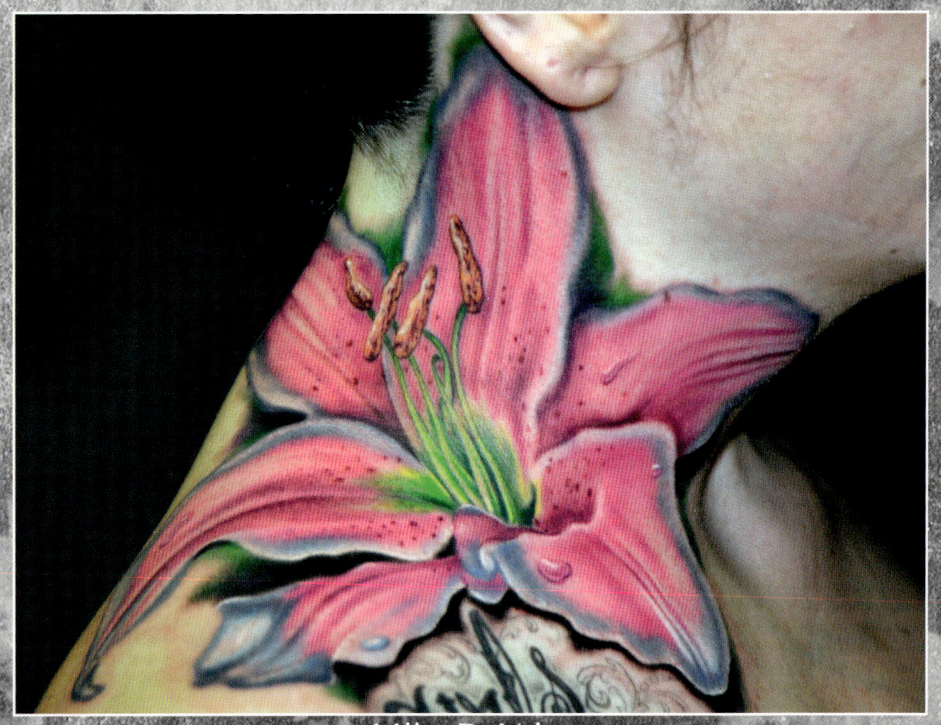

Nick Baxter

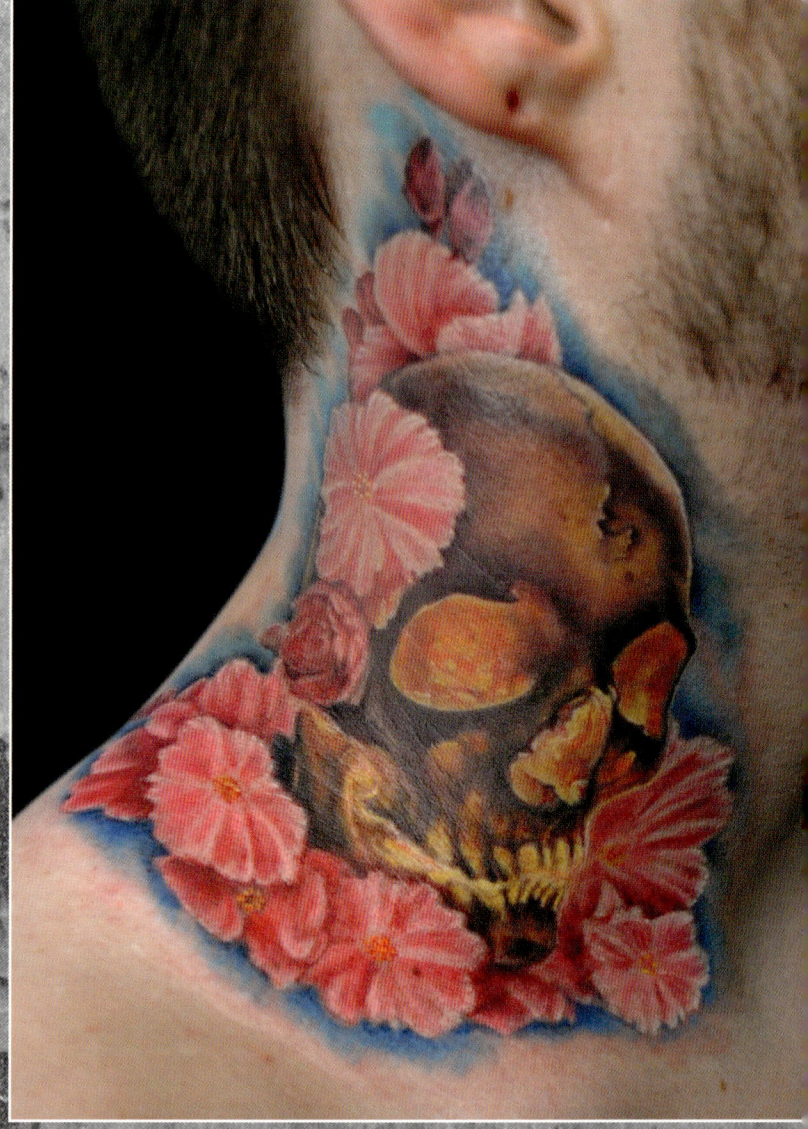

Mike DeVries

Steve Morris

Nick Baxter

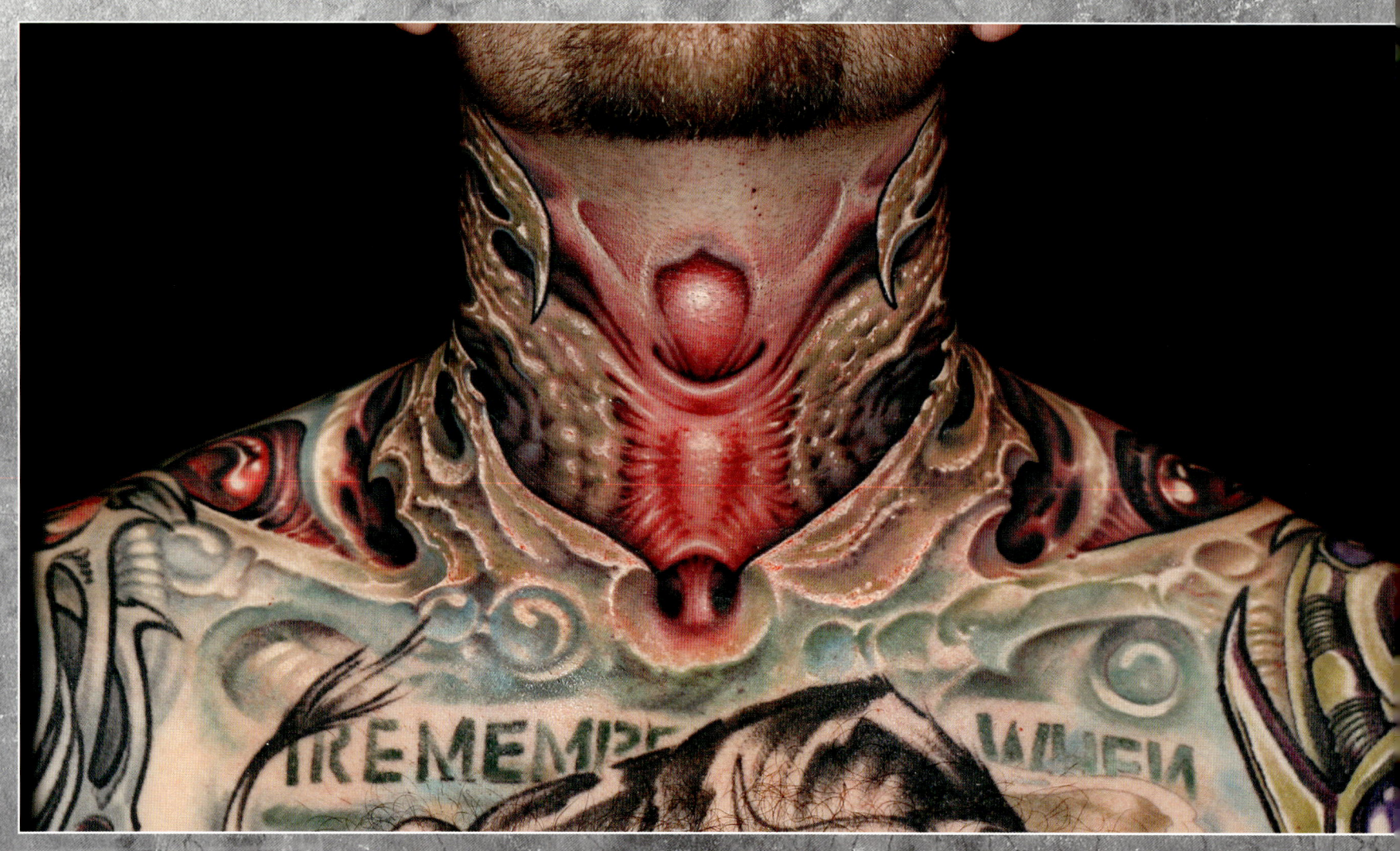

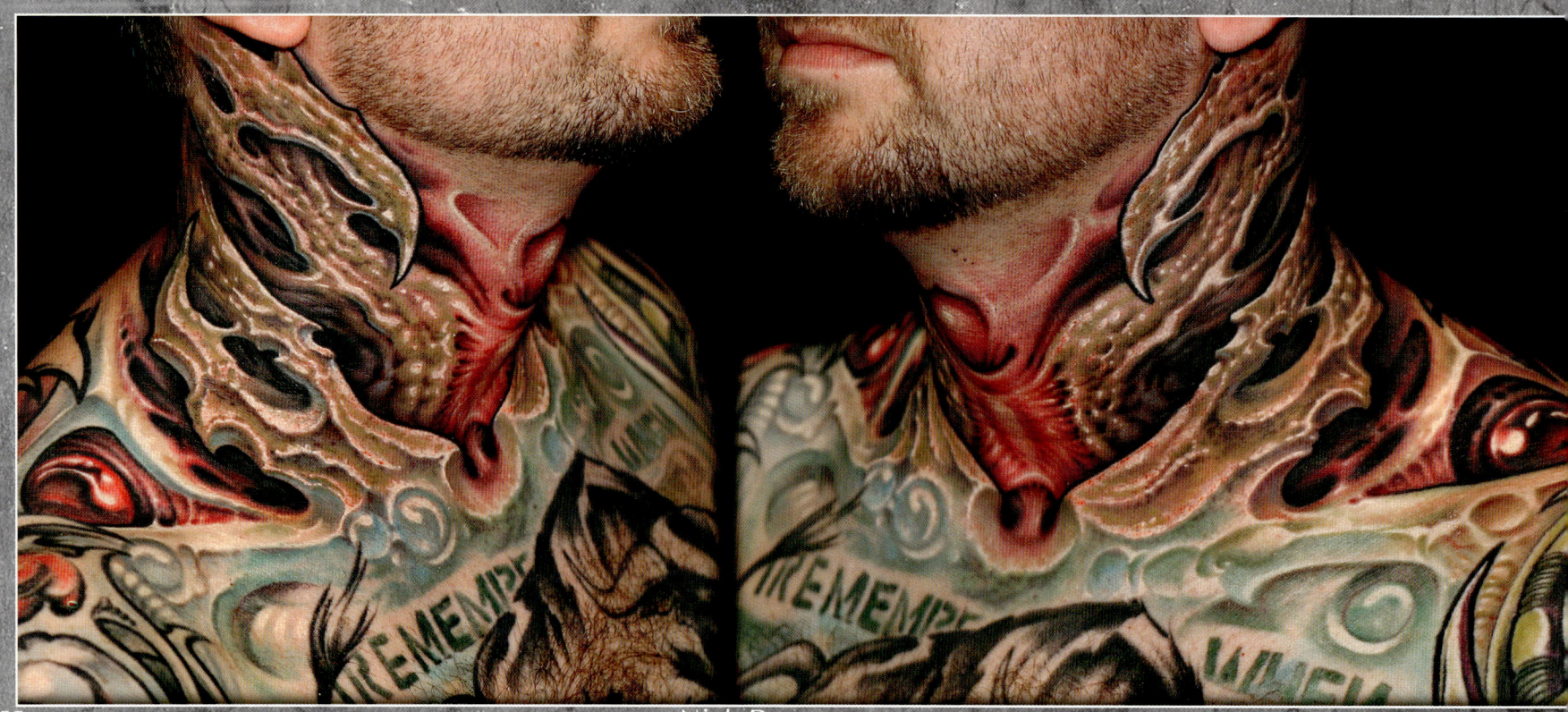

Nick Baxter

Nick Baxter

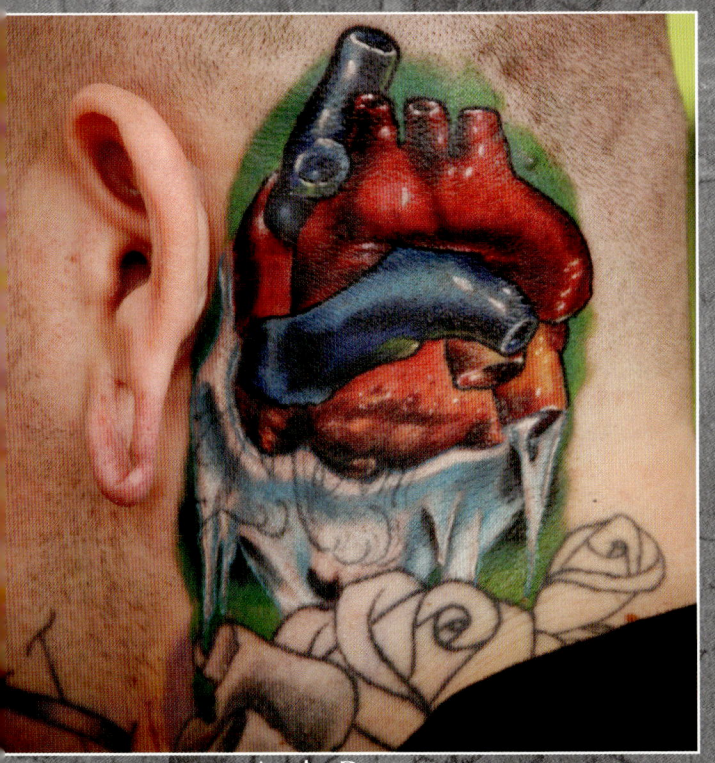

Andy Barrett

Deno

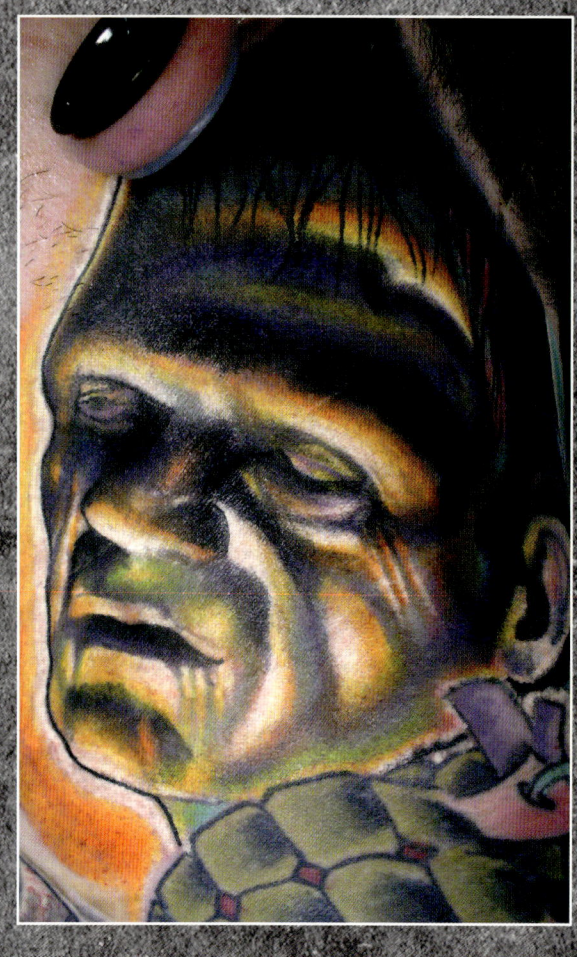

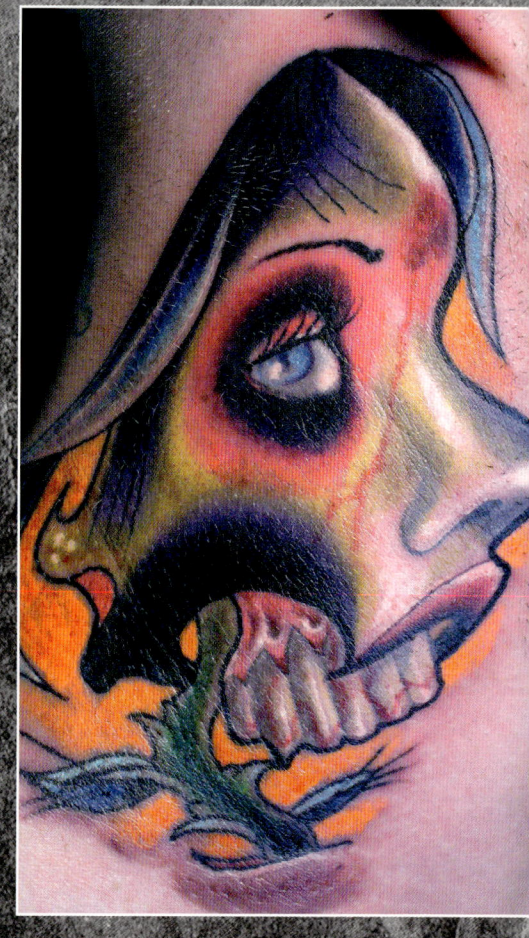

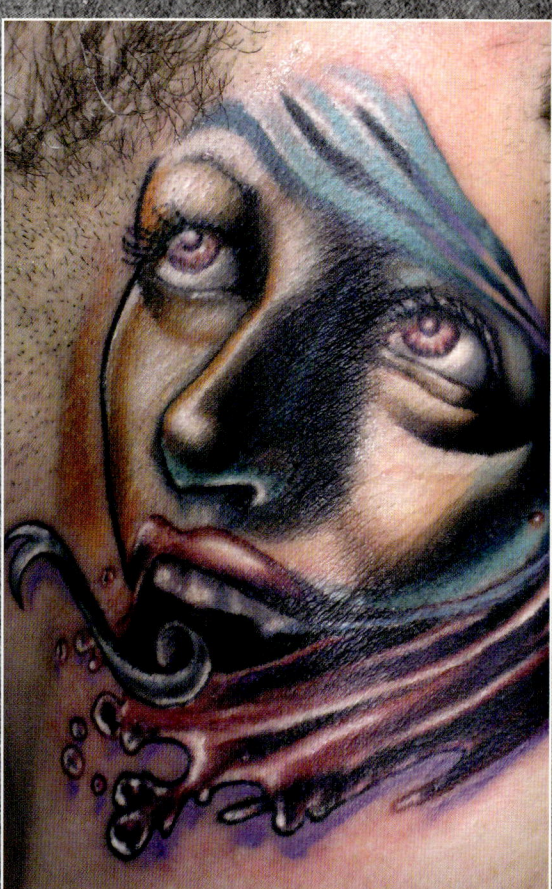

Sean Herman

Sean Herman

Sean Herman    Sweet Laraine   58

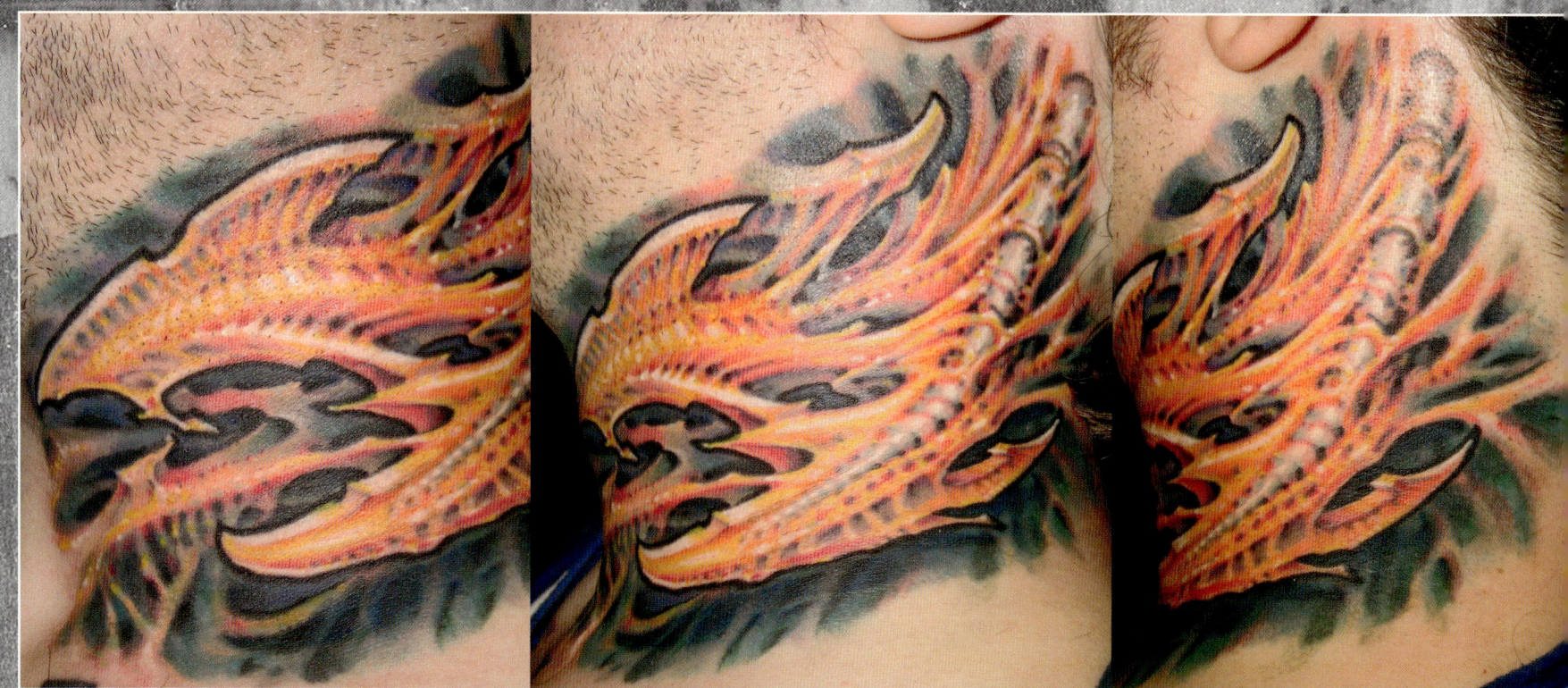

Dan Hazelton

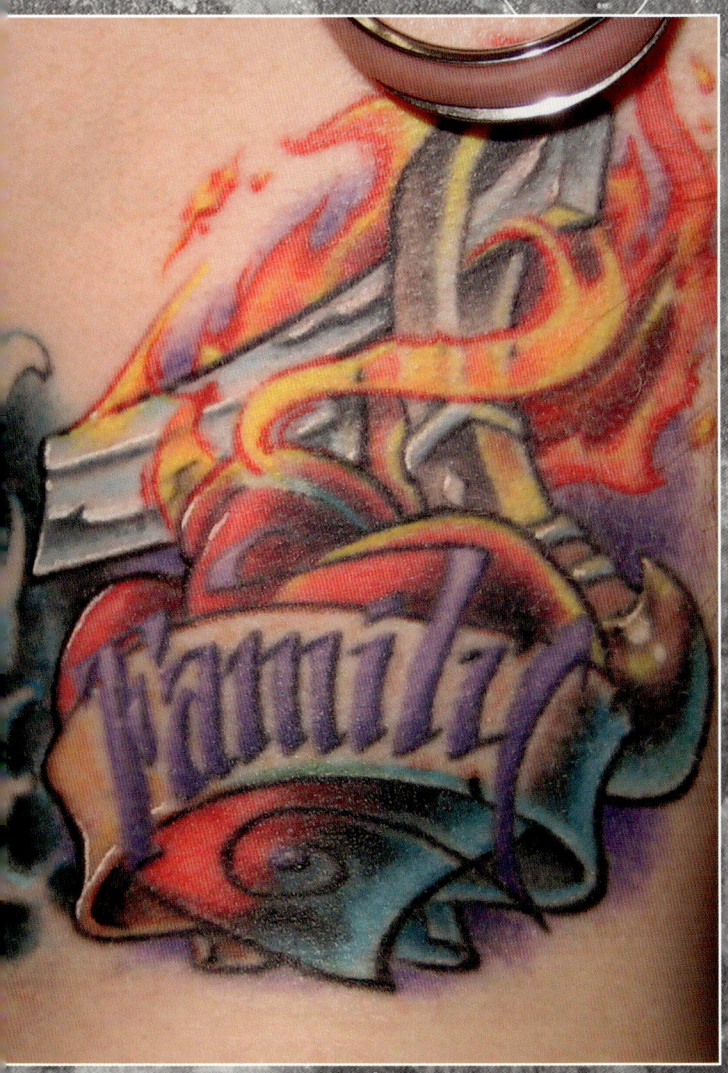

Dan Hazelton

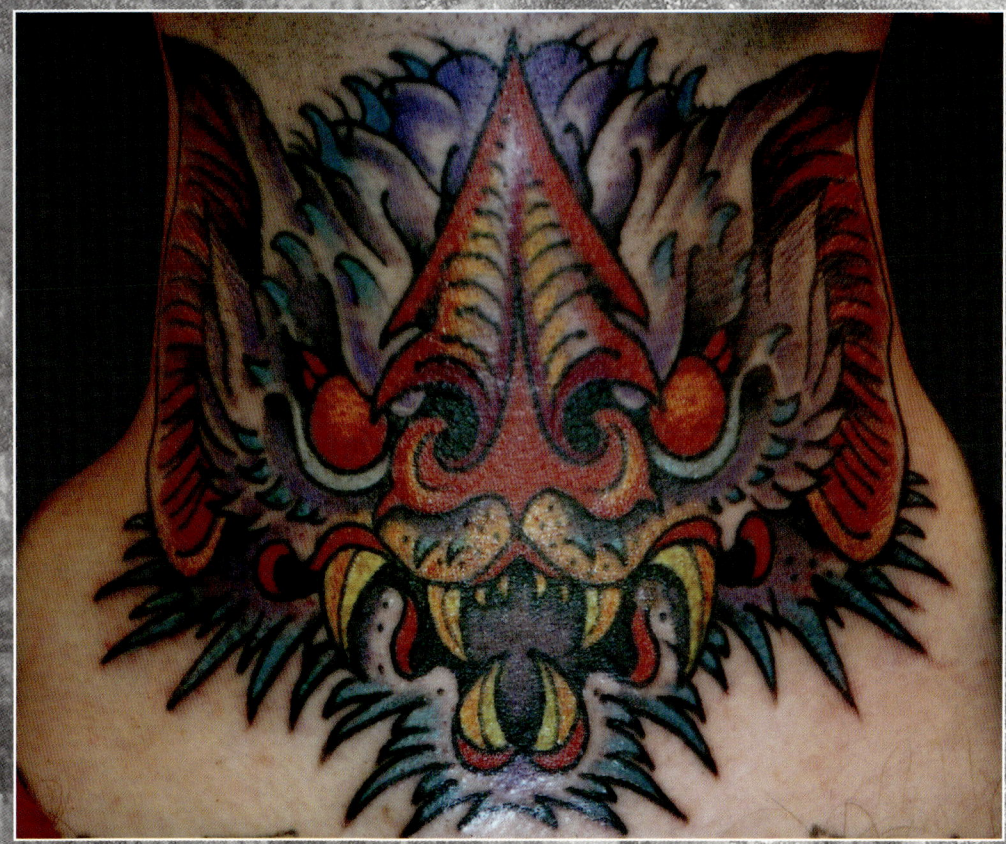

Jeff Zuck

Jeff Zuck

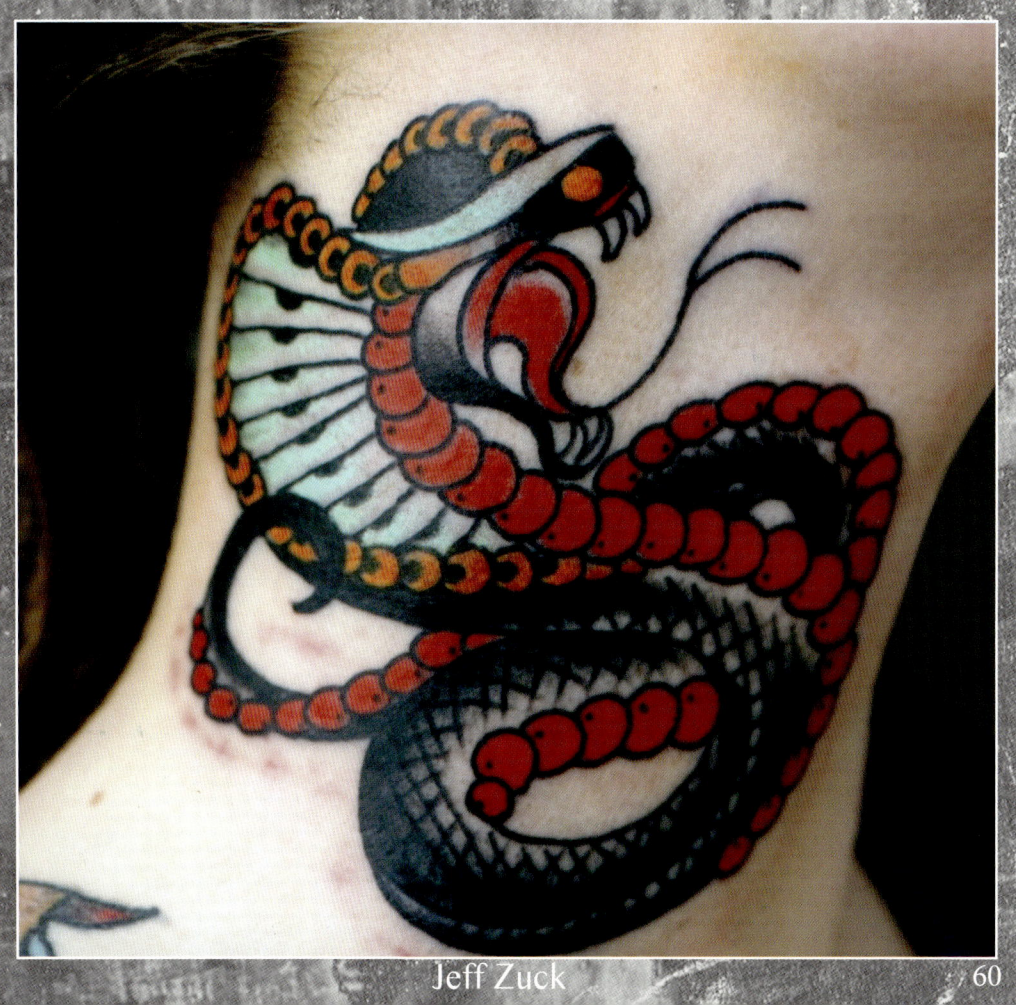

Jeff Zuck

Jeff Zuck

Jeff Zuck

Timmy B

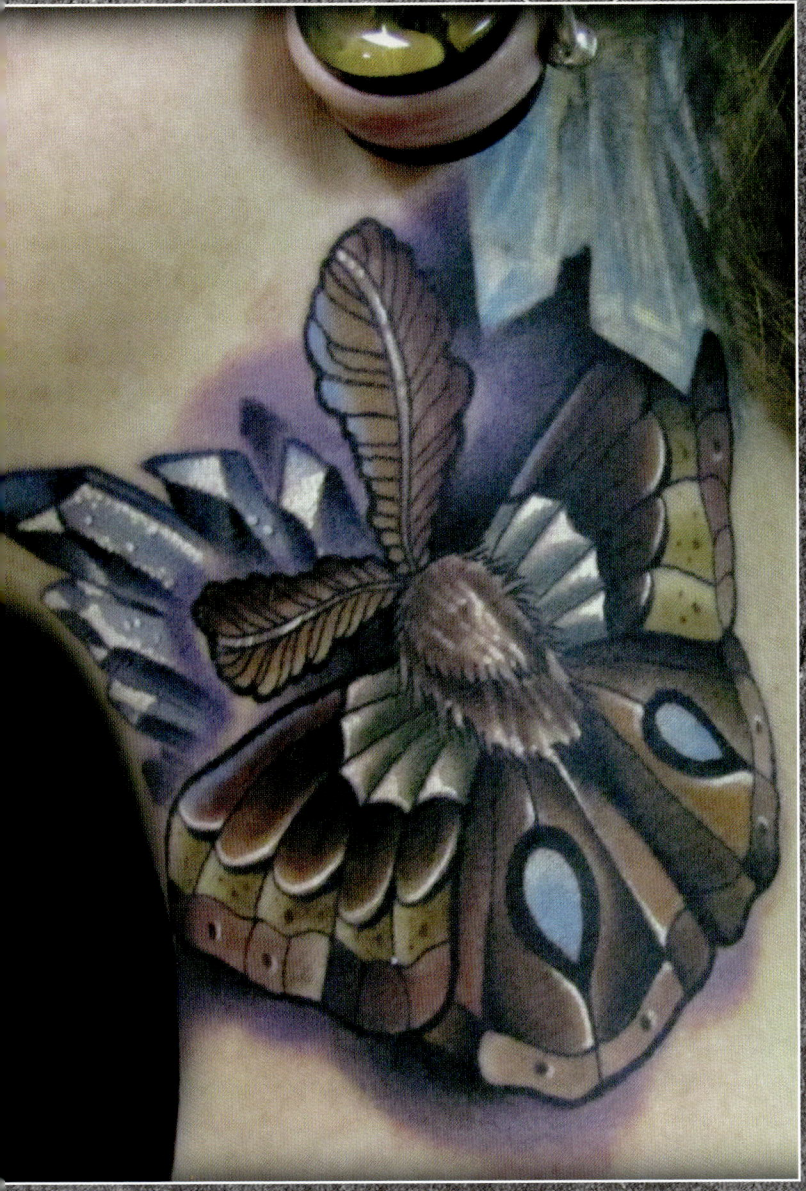

Timmy B

Timmy B

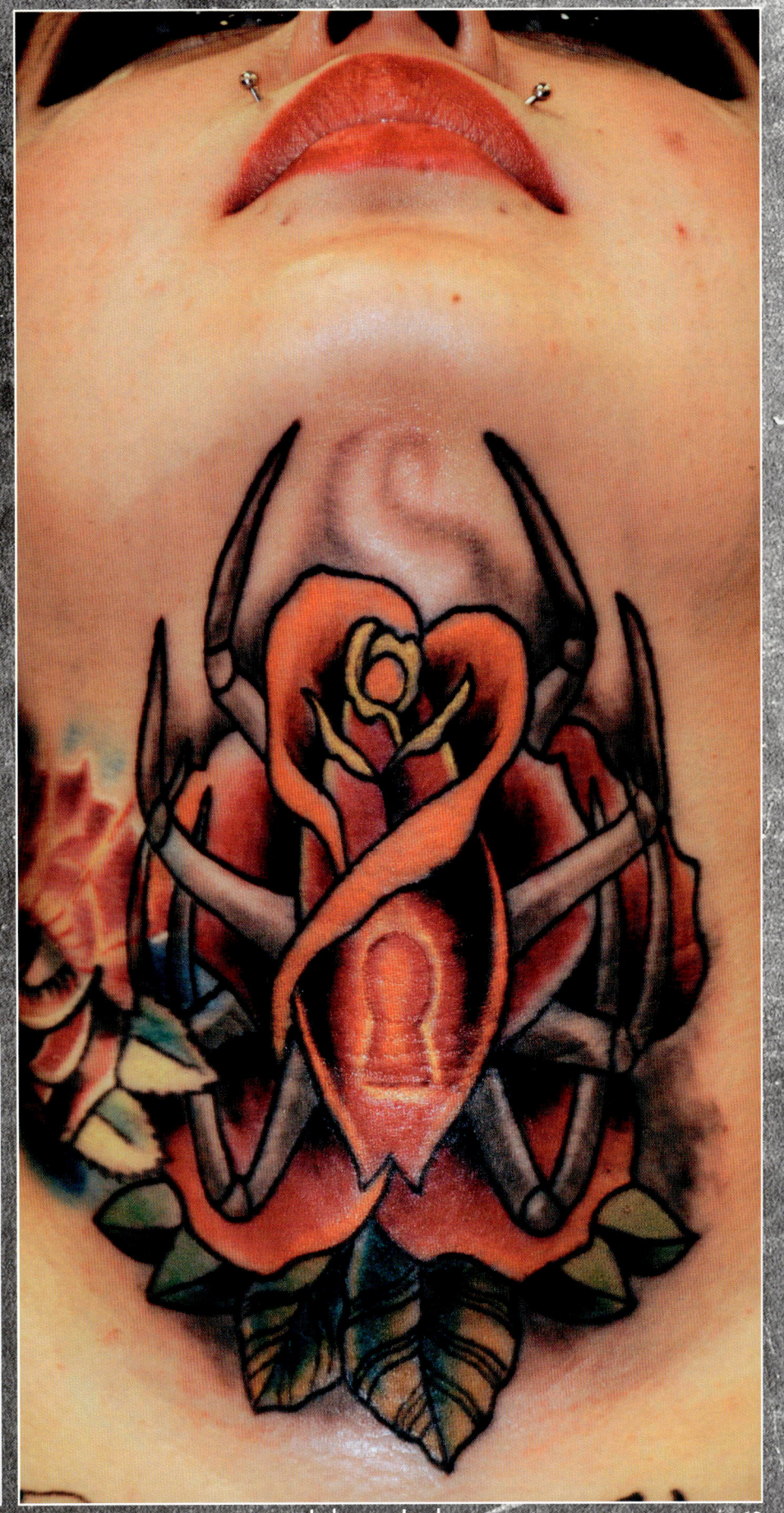

Johnny Jackson

Roman Abrego

Roman Abrego

Roman Abrego

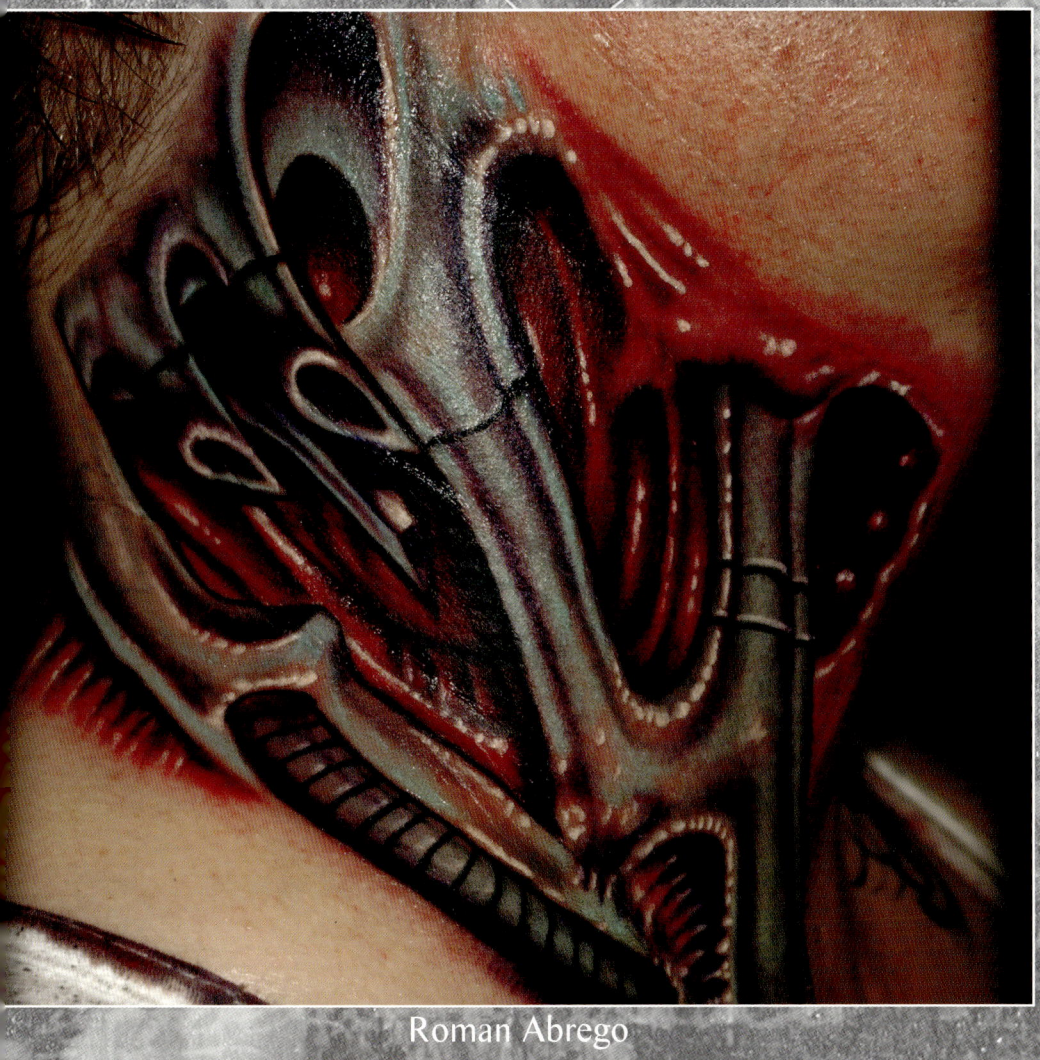

Roman Abrego

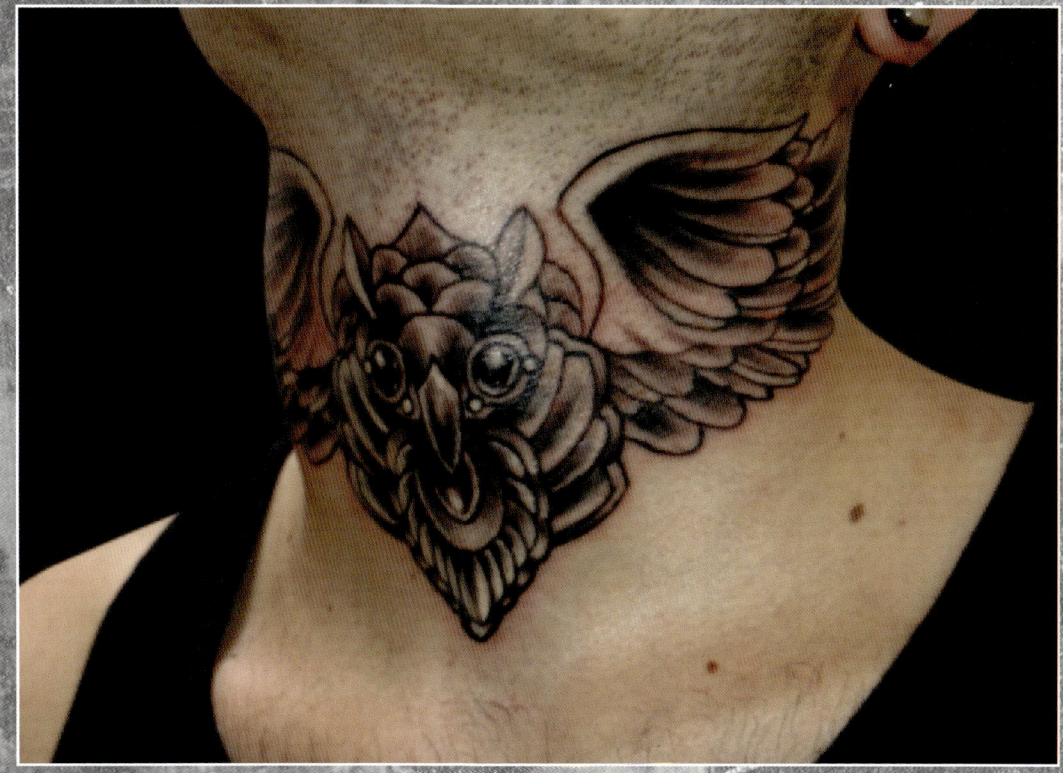

John Kosco

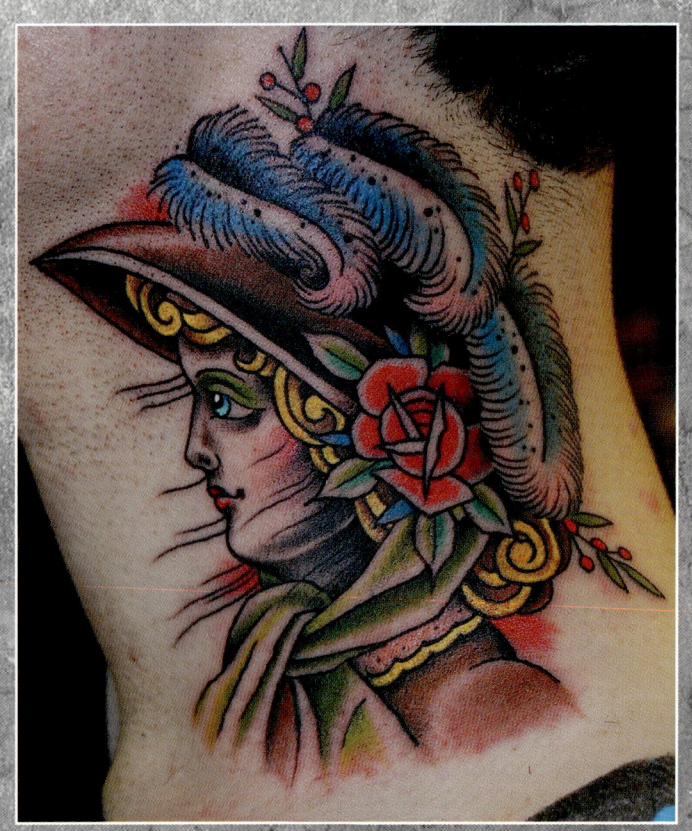

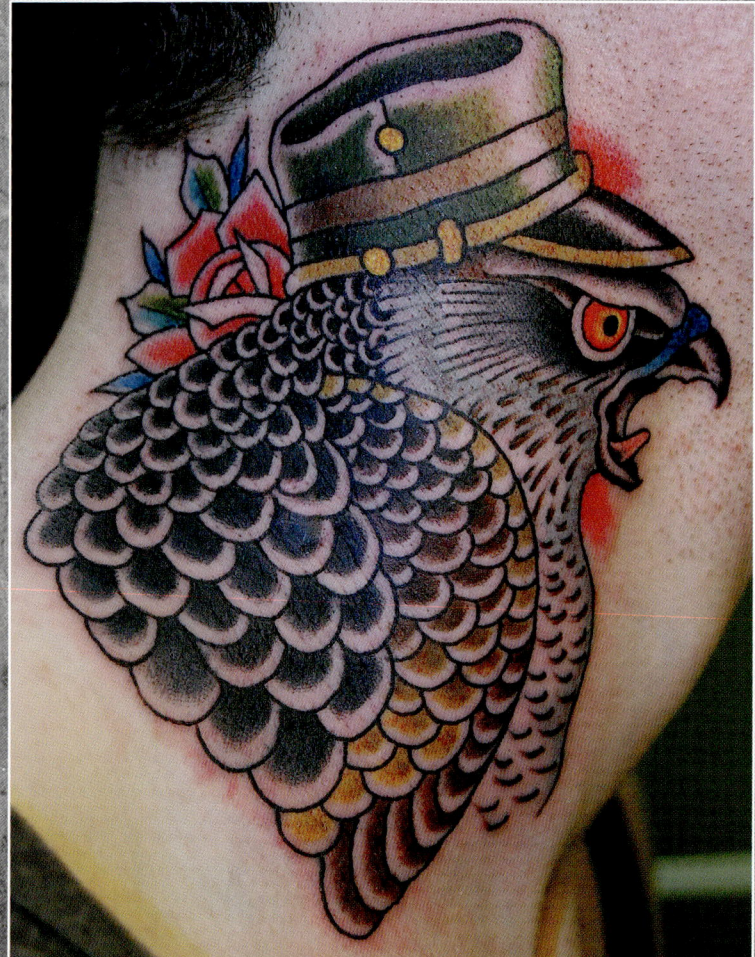

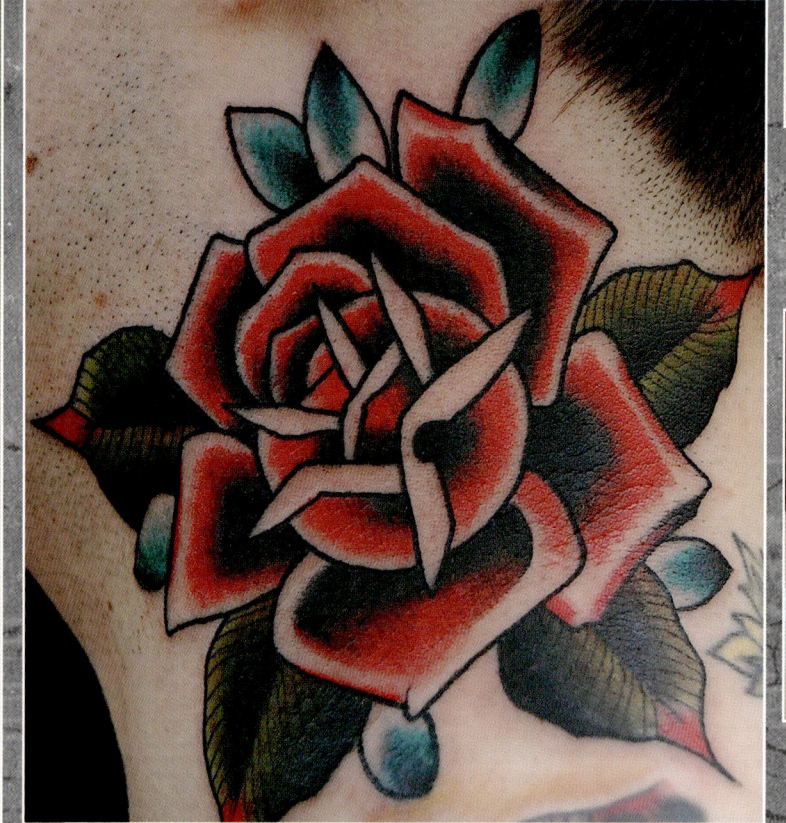

Tutti Serra

Tim Pangburn

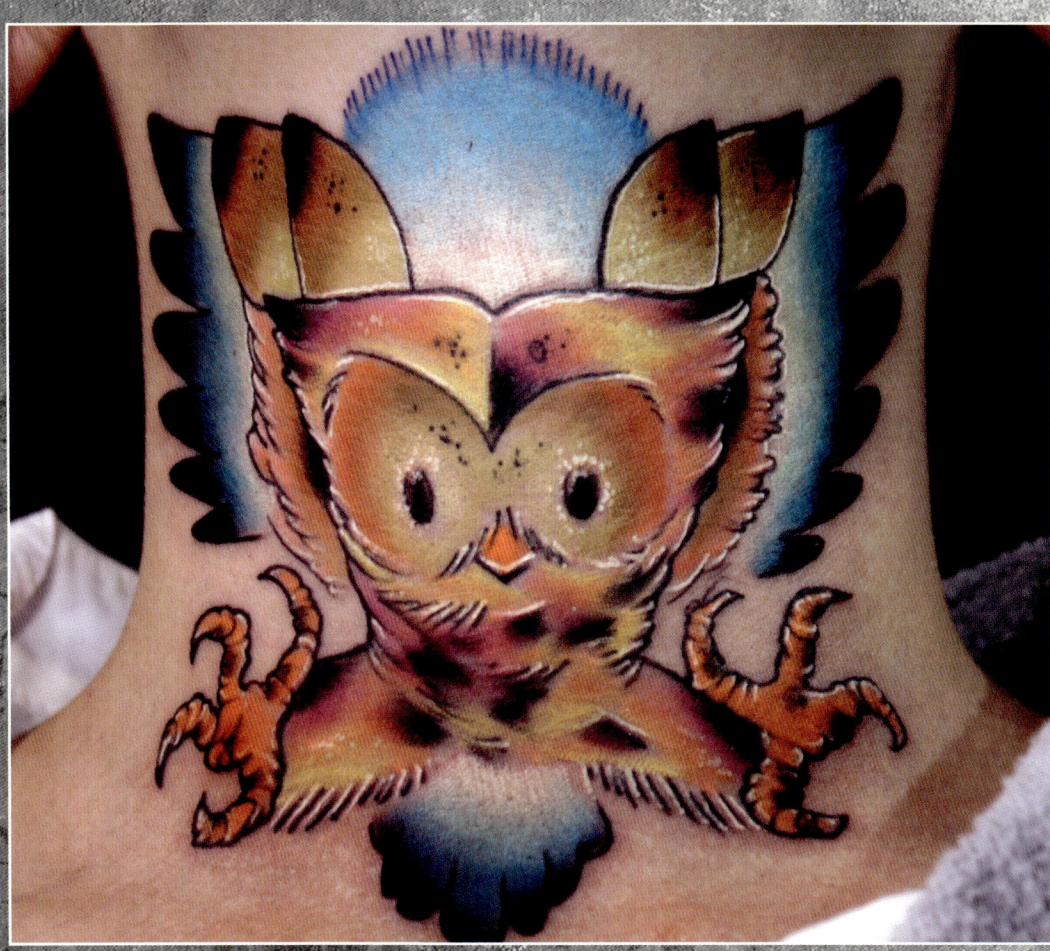

Tim Pangburn

Ron Meyers

Tim Pangburn

Indio Reyes

Indio Reyes

Josh Woods

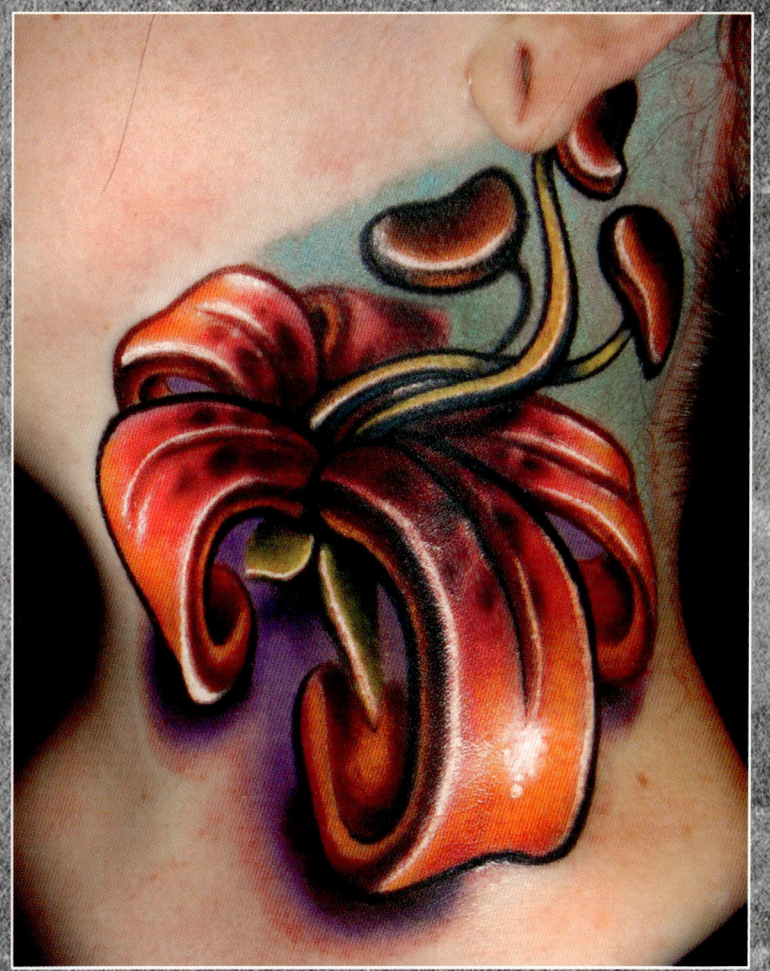

Josh Woods

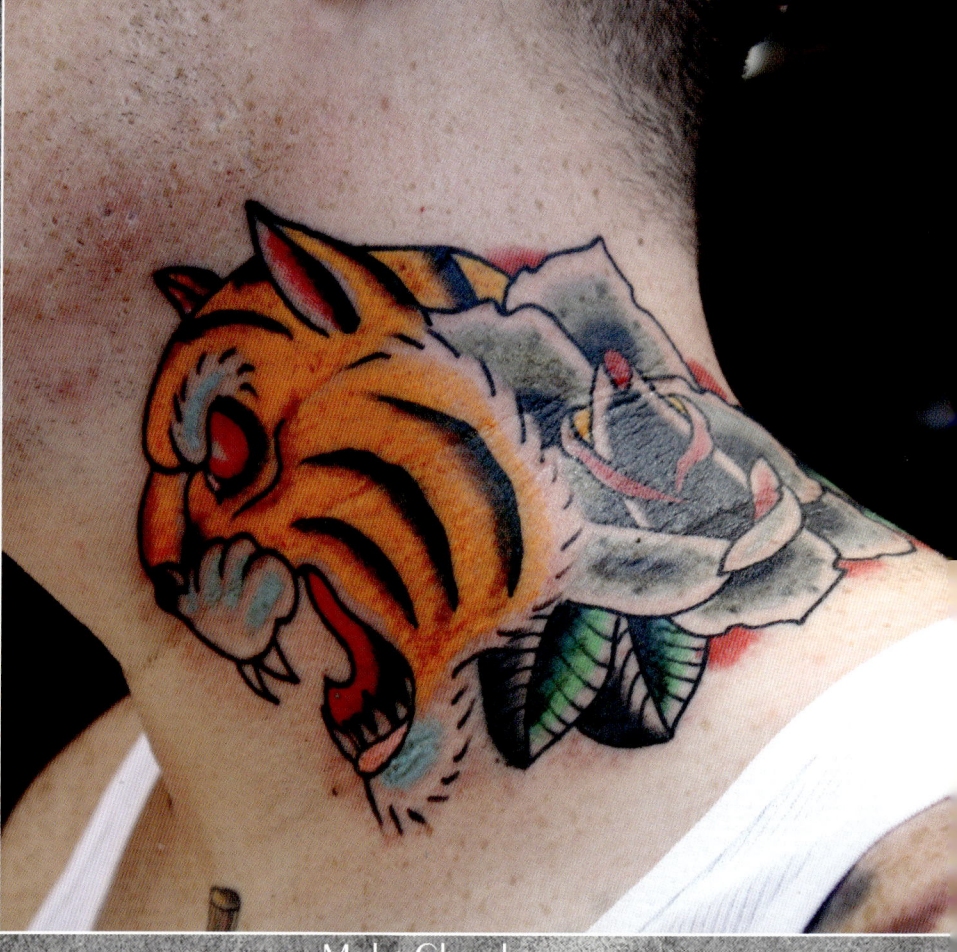

Myke Chambers

Jose Morales

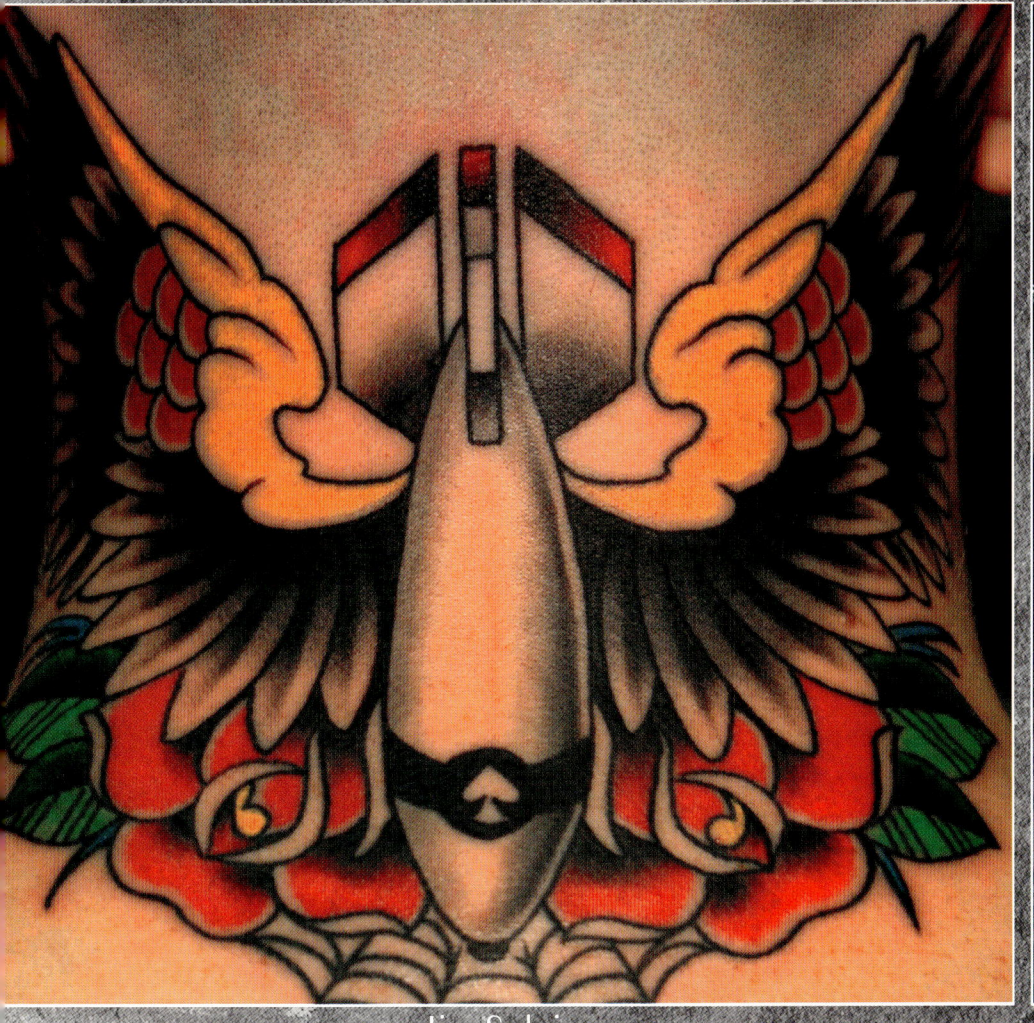

Jim Sylvia

Jim Sylvia

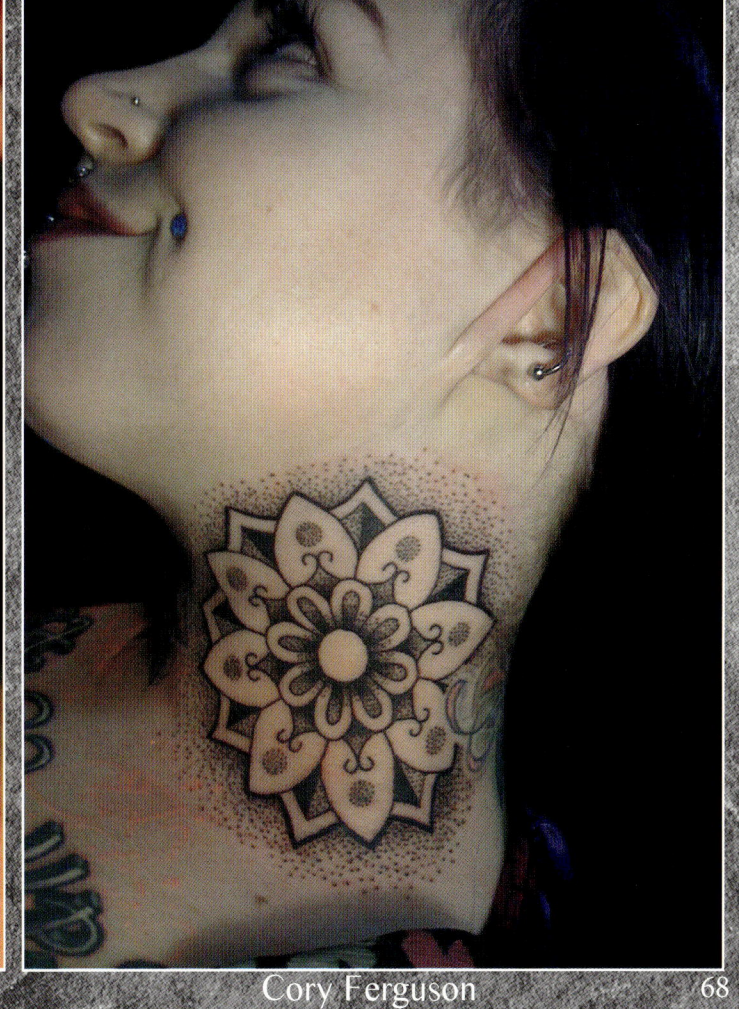

Cory Ferguson

69 Myke Chambers

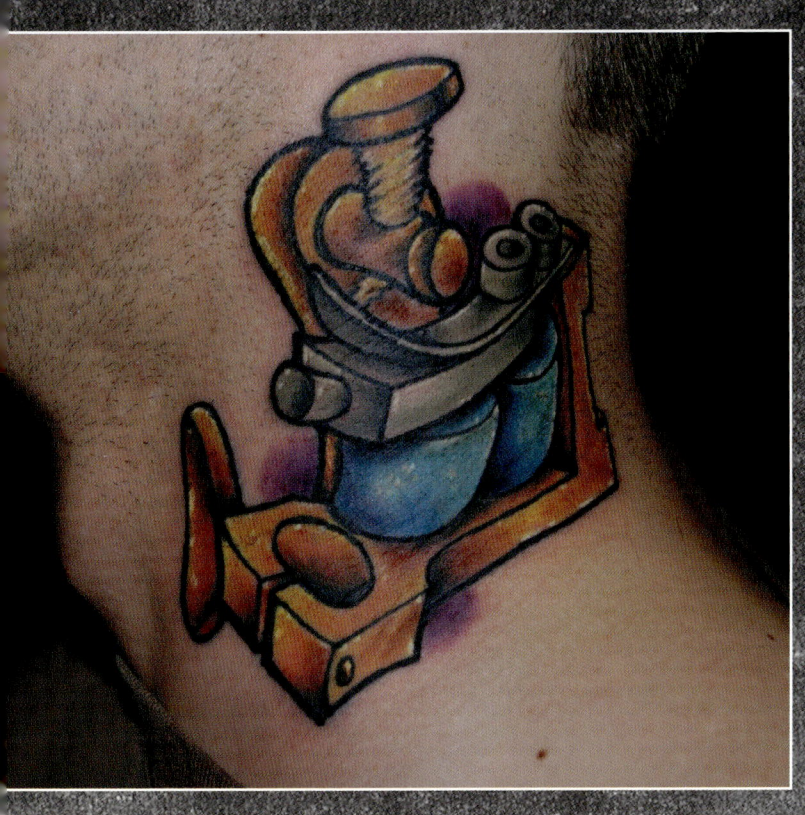

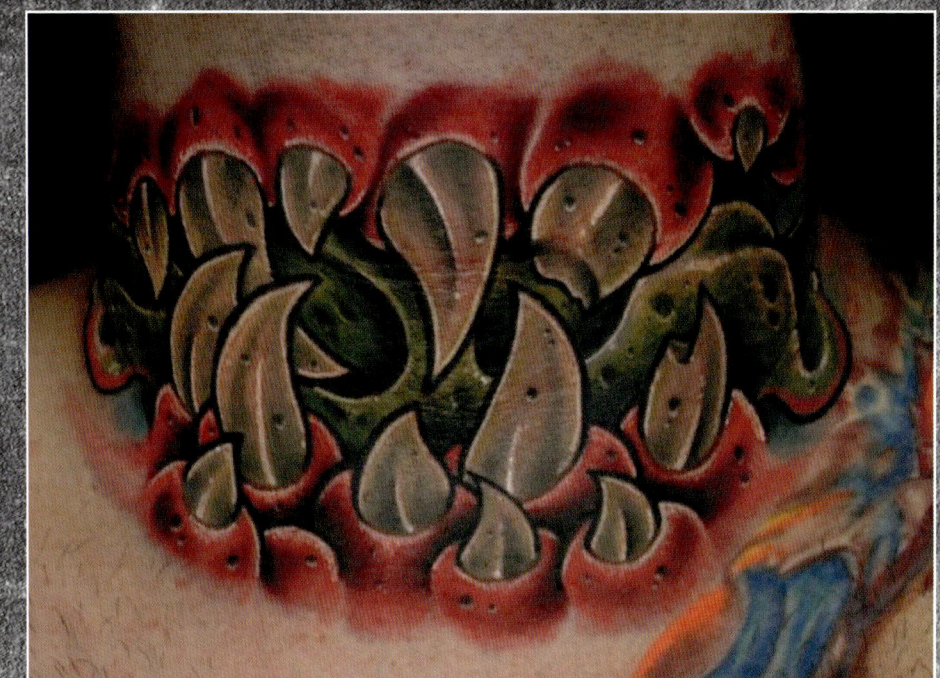

Jeremy Miller

71

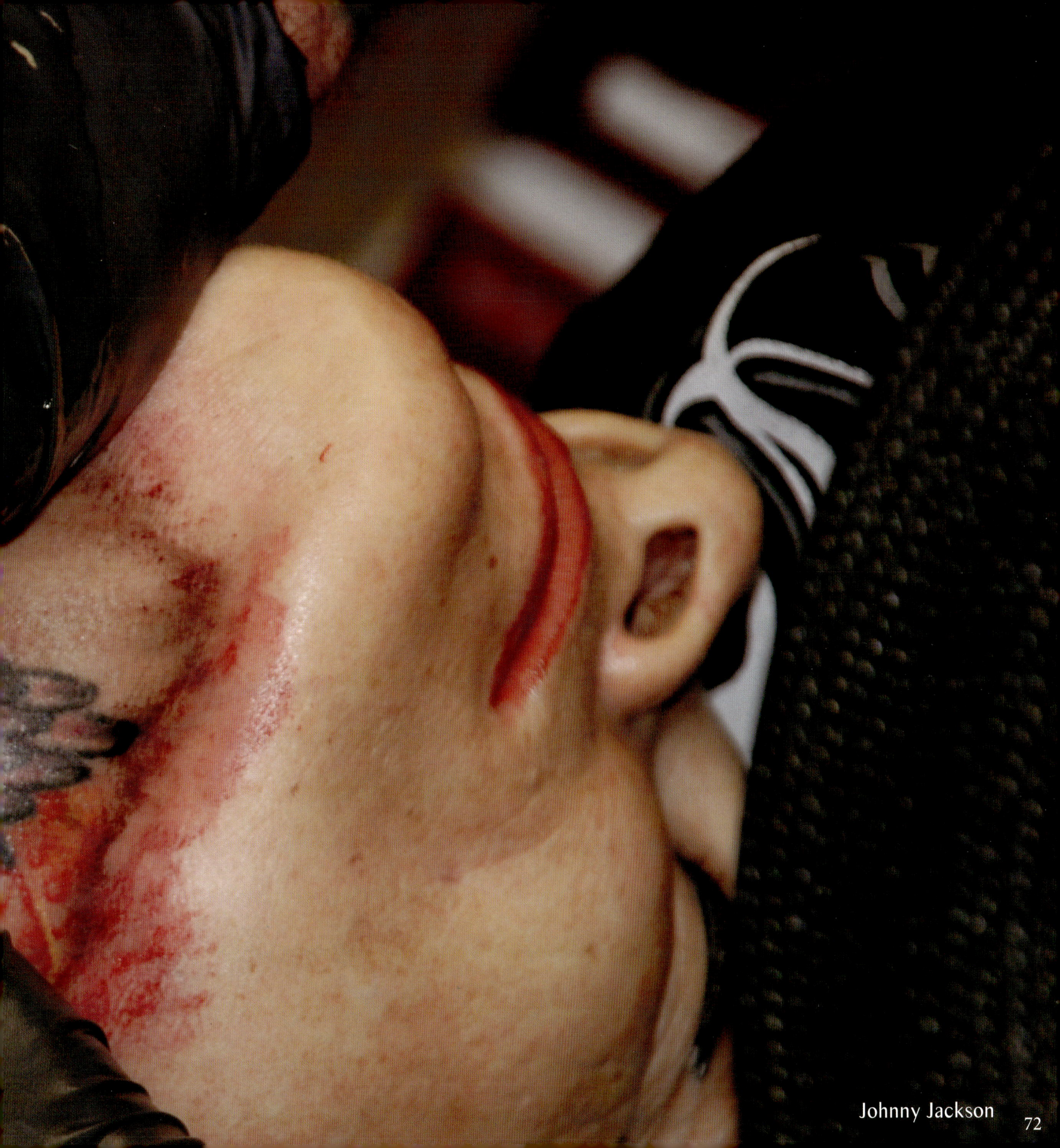

Johnny Jackson

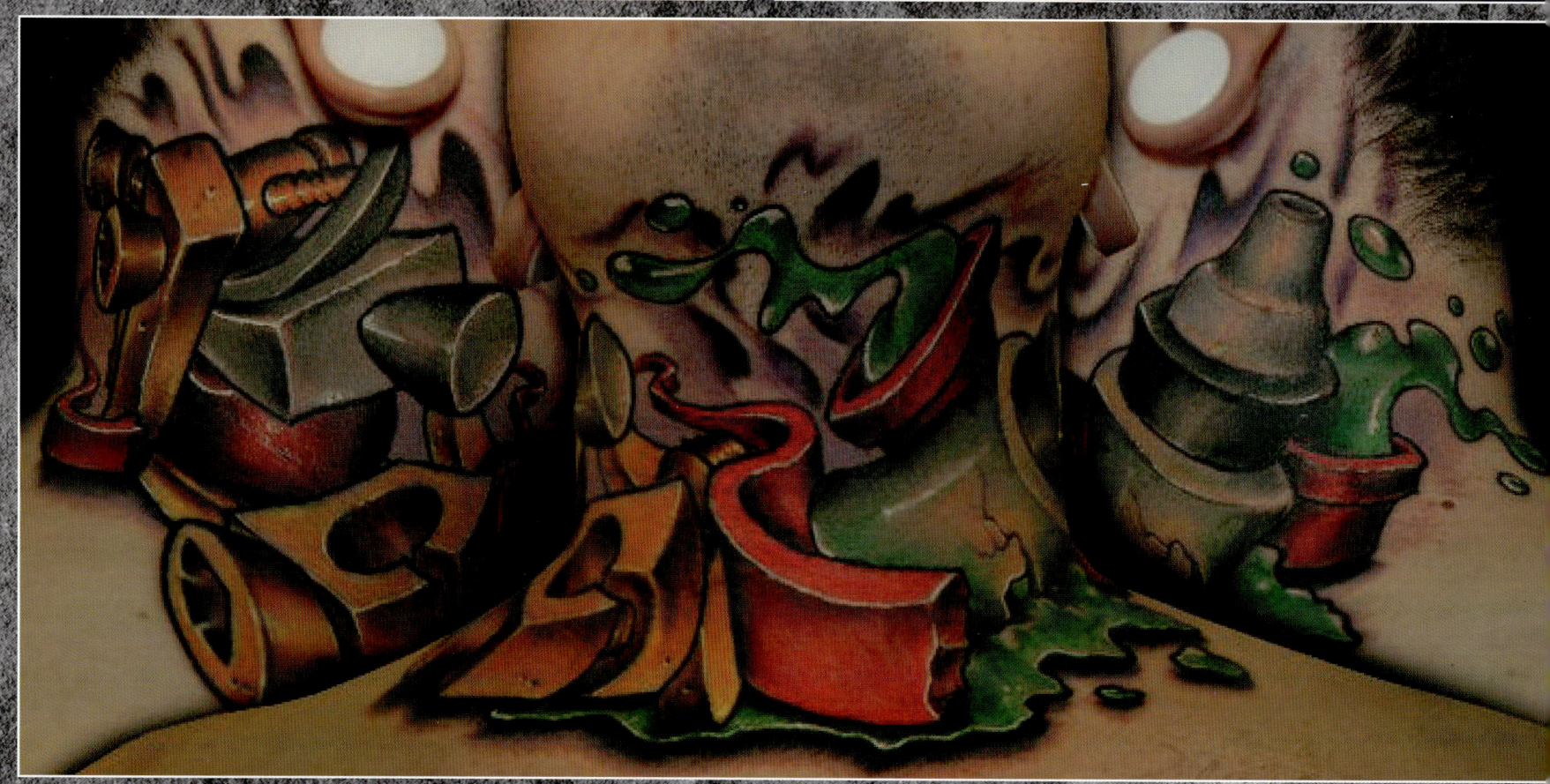

Jeremy Miller

Short

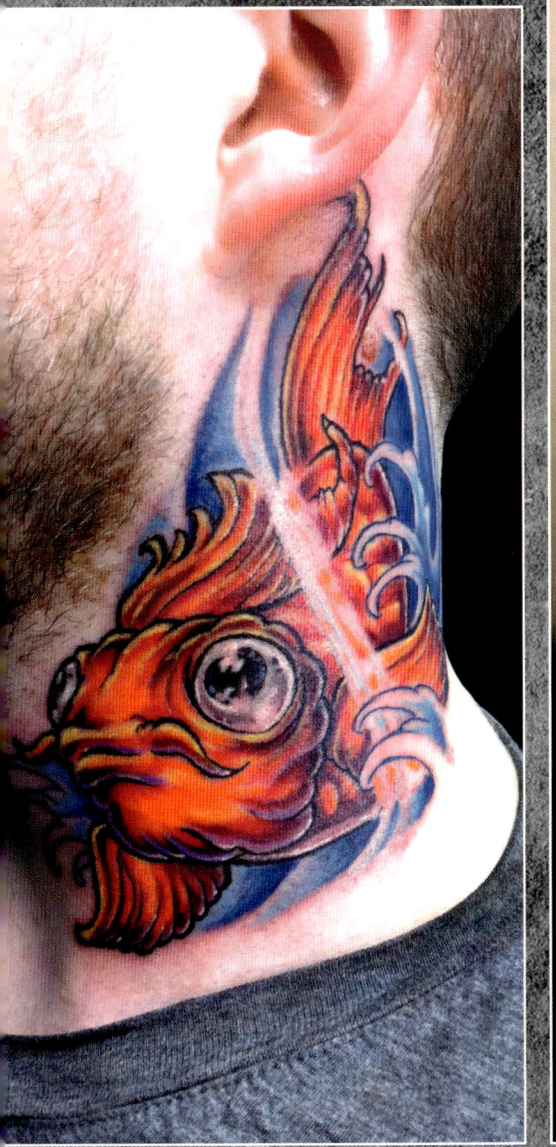

Josh Bodwell

Andy Engel

74

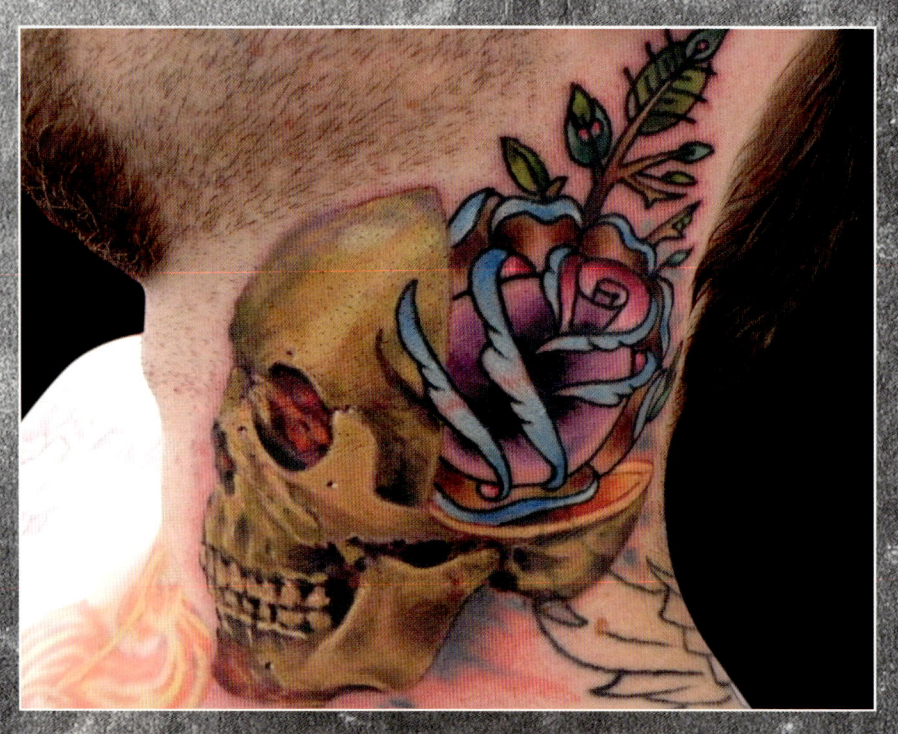

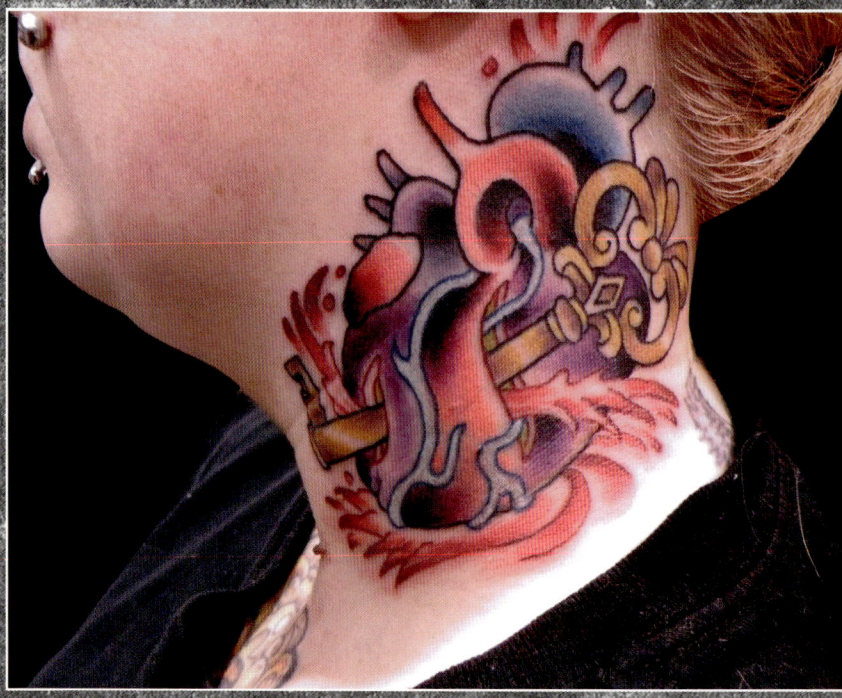

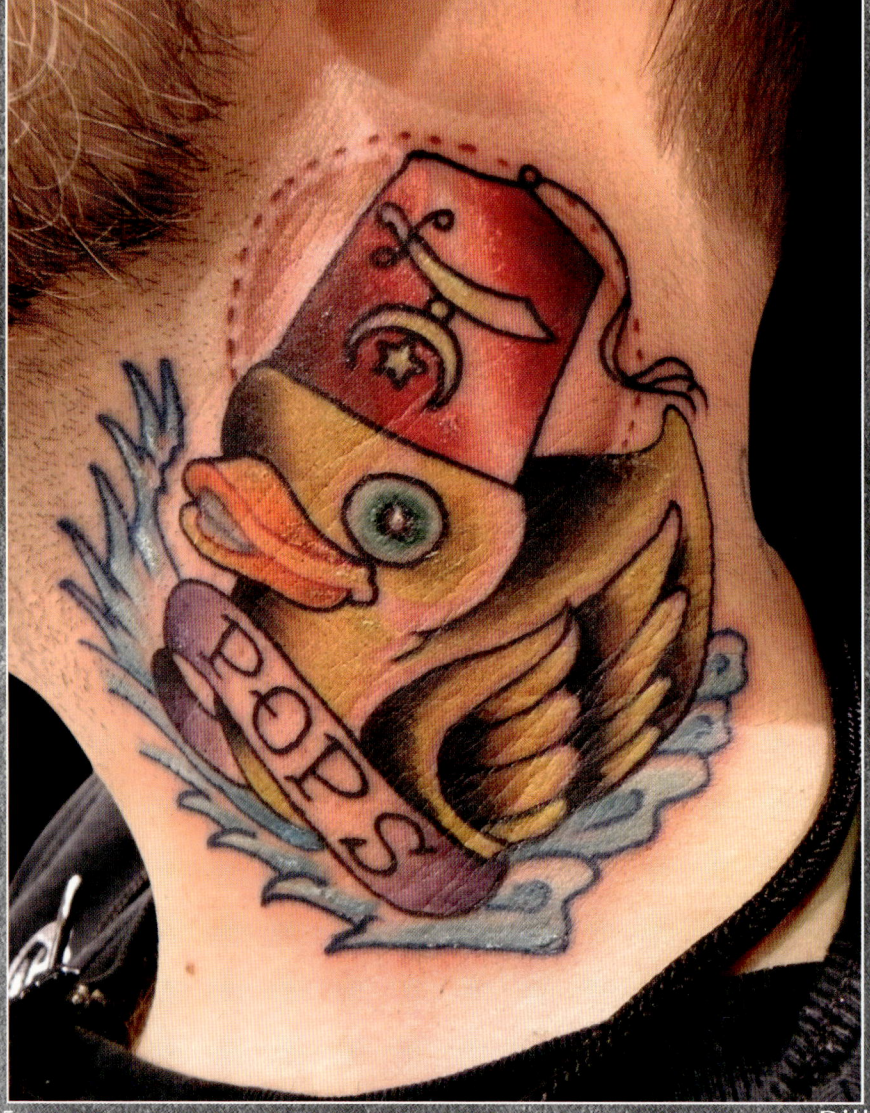

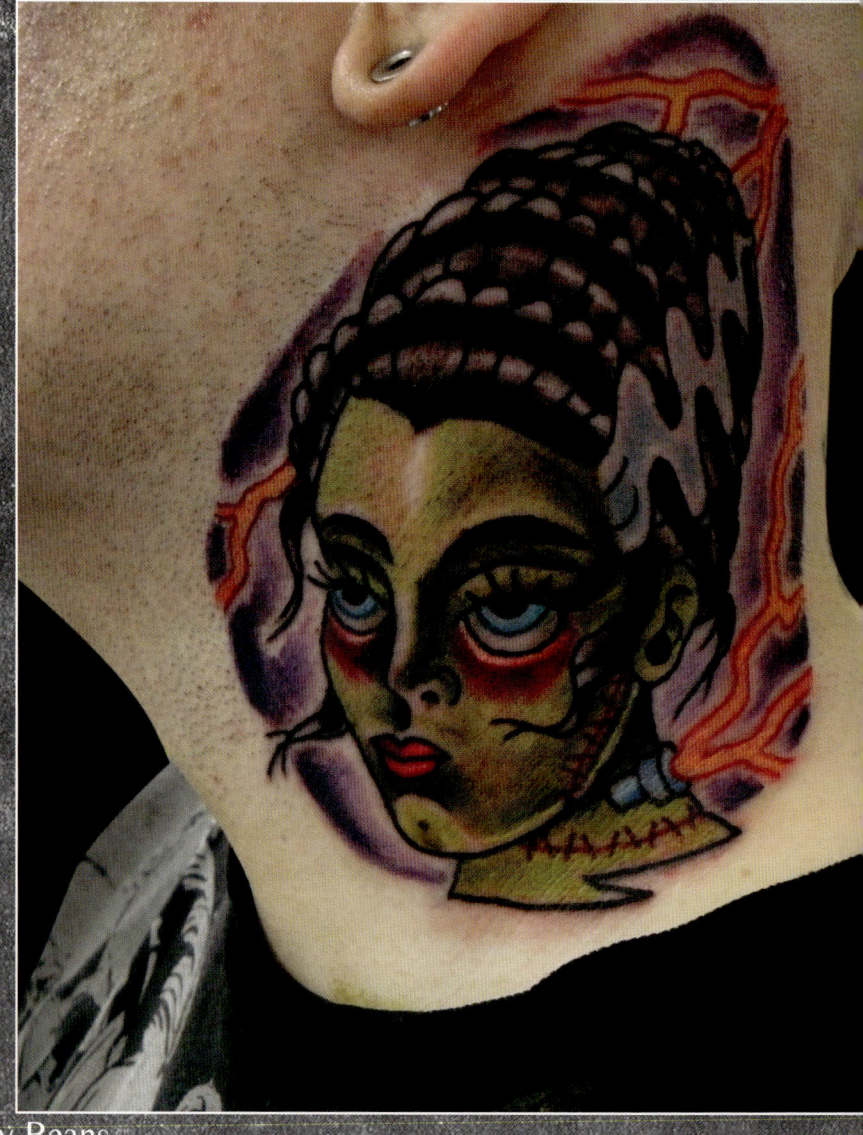

Billy Beans

Steve Byrne

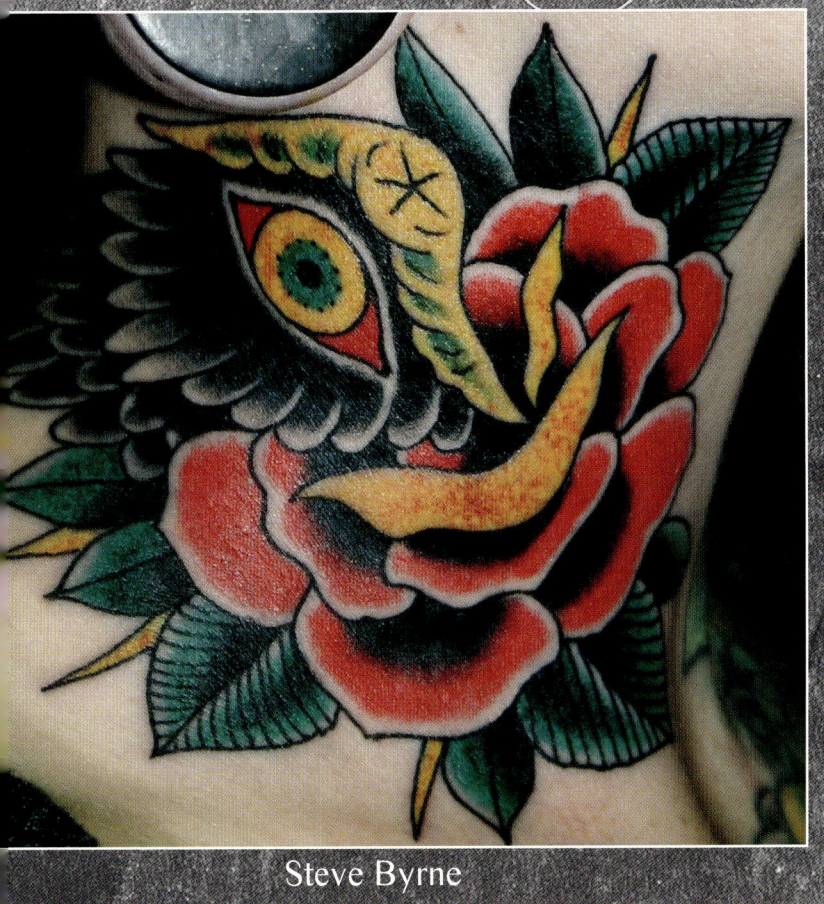

Steve Byrne

Guy Aitchison

Steve Byrne

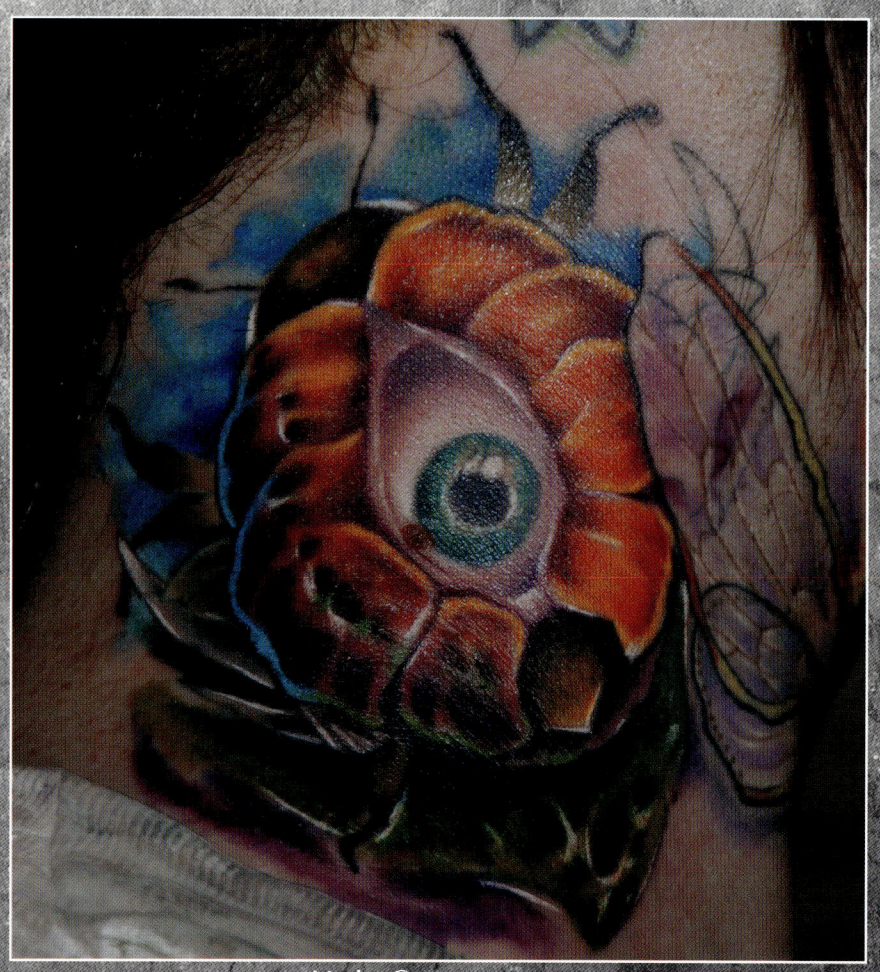

Kyle Cotterman

Daksi

Daksi

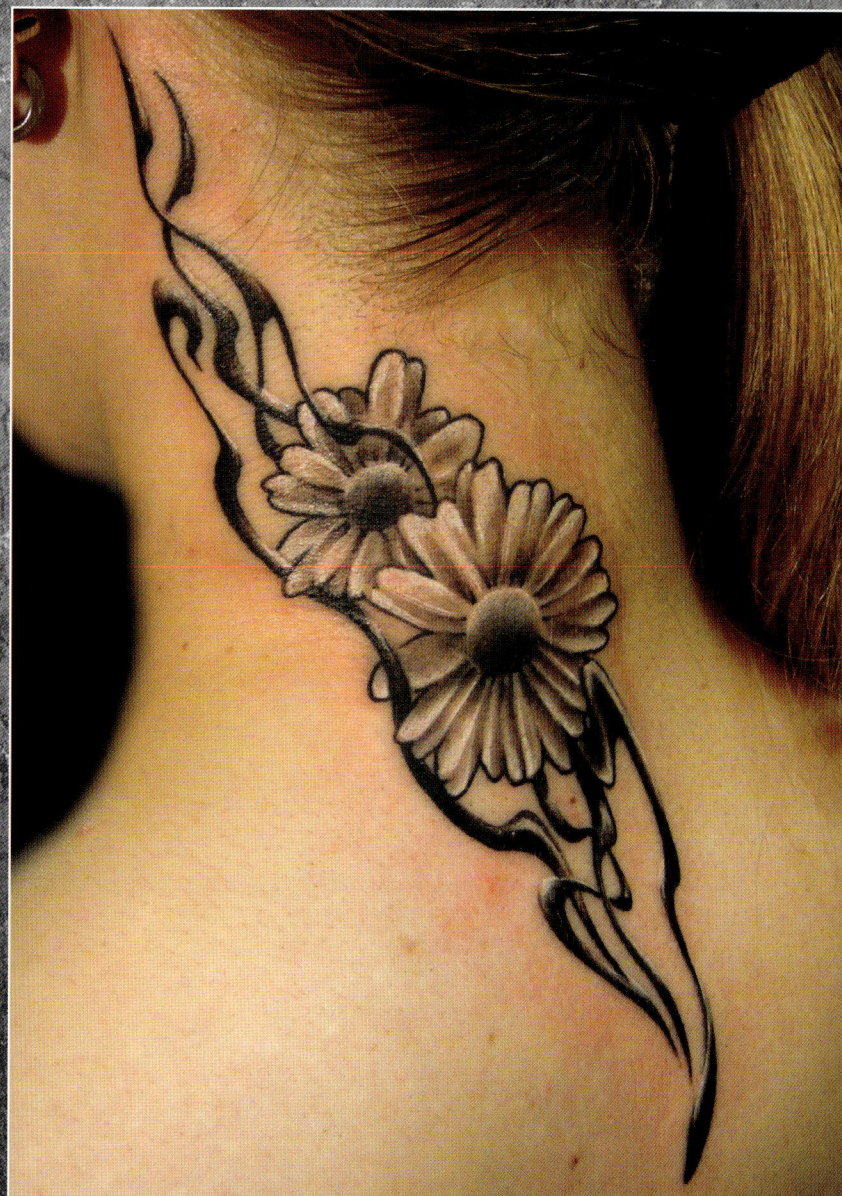

Eva Huber

Dave Tedder

Dave Tedder

Dave Tedder

Dave Tedder

Jason Vaughn

Cory Ferguson

Kurt Melancon

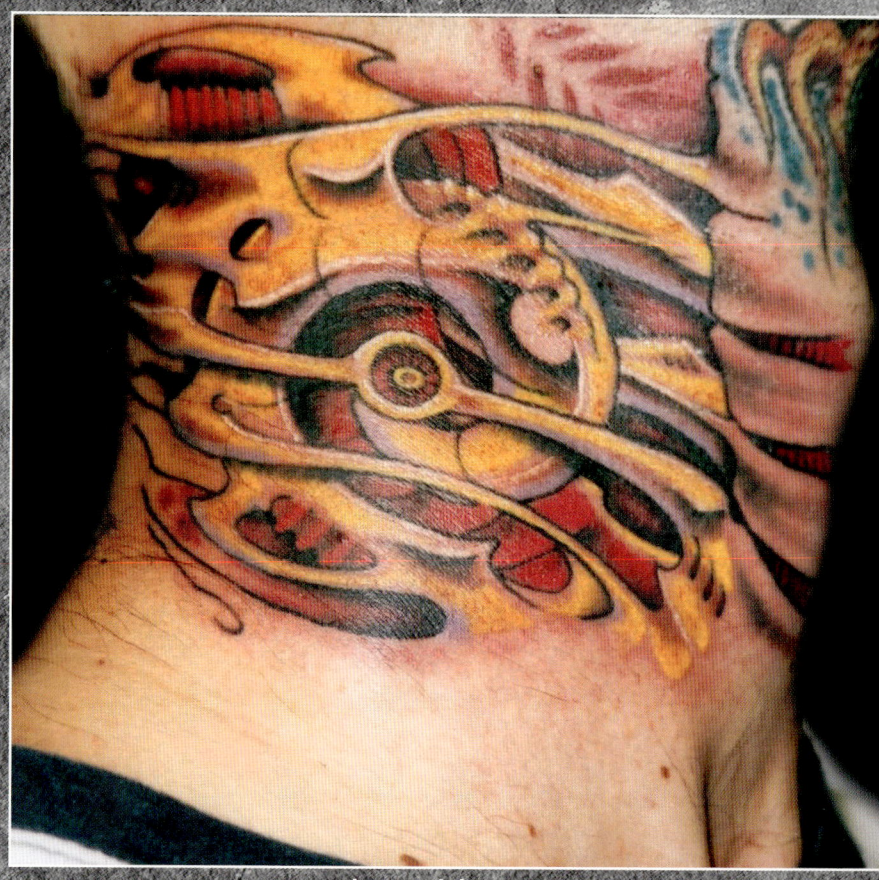

Jason Morrow

Justin Page

Justin Page

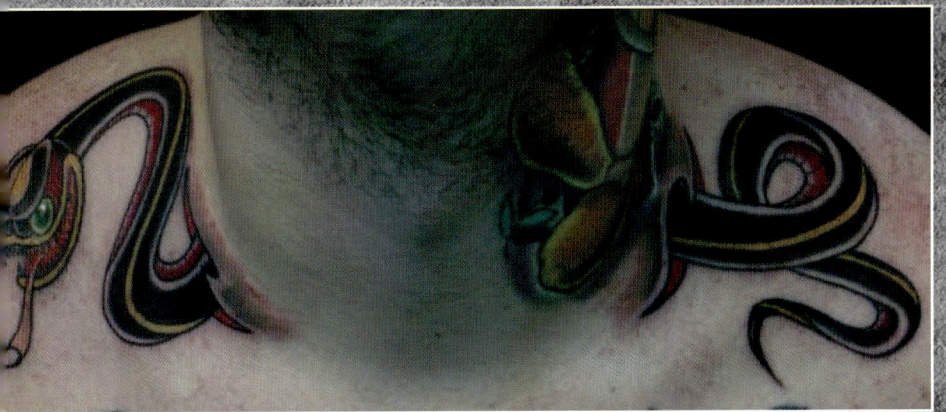

Jason Reeder

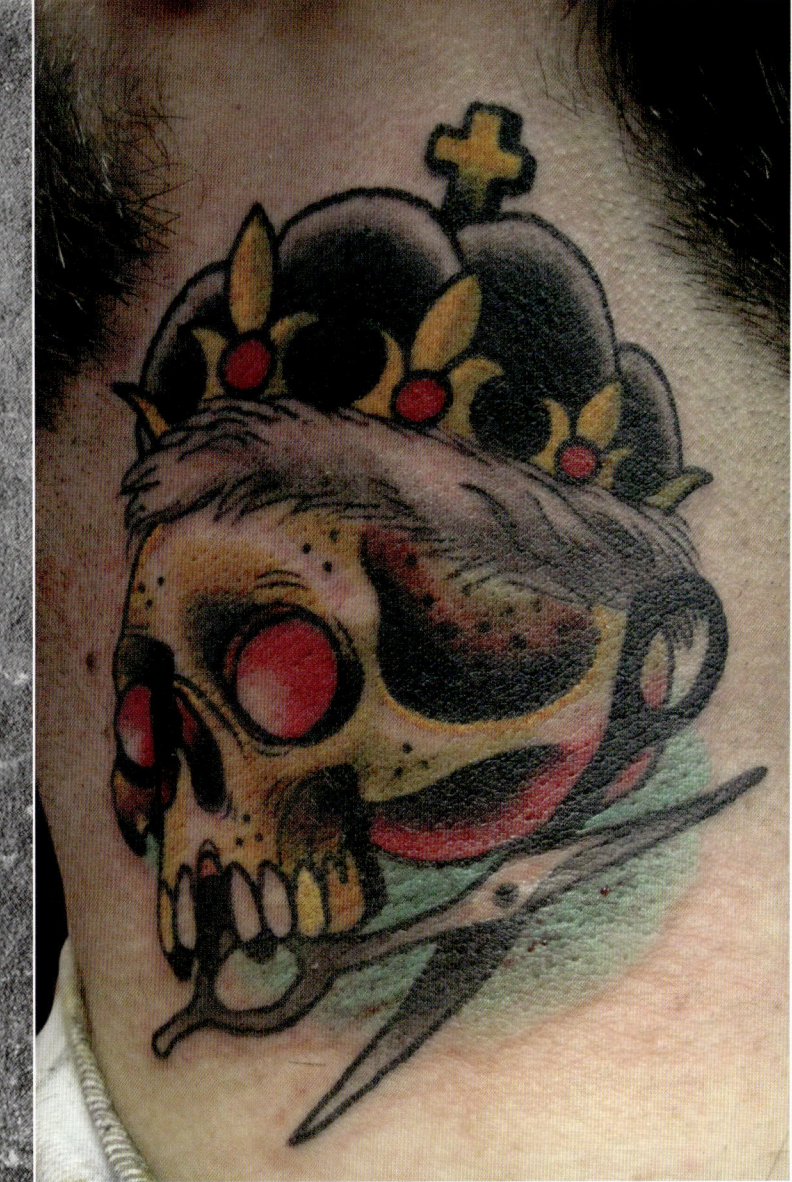

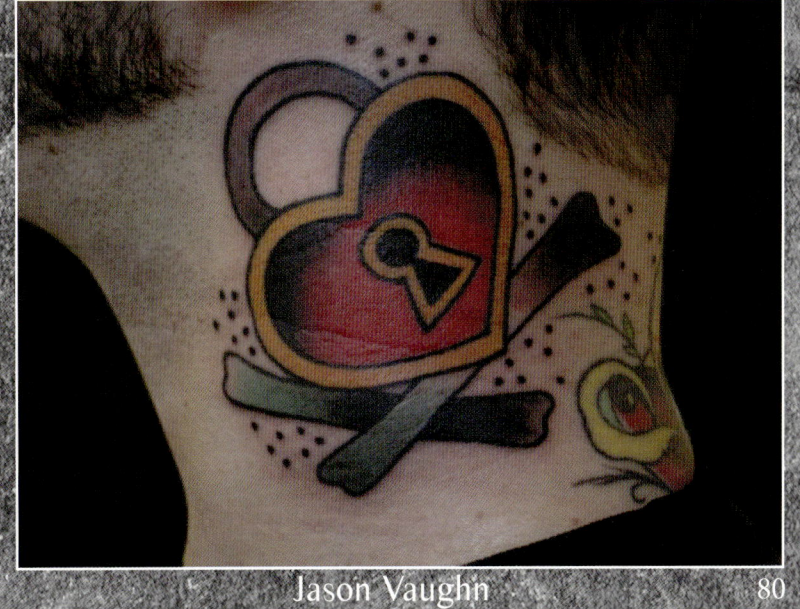

Russ Abbott

Jason Vaughn

Jason Vaughn

80

HANDS & PALMS

"Often the hands will solve a mystery that the intellect has struggled with in vain." - Carl Jung

Each day, we use our hands for a multitude of tasks, usually without even realizing how essential they are. But indeed, our hands are complex tools, with each hand containing: 29 major and minor bones, 48 nerves, 29 major joints, 34 muscles which move the fingers and thumb, and 123 named ligaments. Not only are these intricate body parts important, they are also highly visible, making them an ideal spot to place beautiful tattoo art.

Because it's difficult to conceal tattooed hands, it is definitely a big step and commitment for ink lovers to make. Unless one wears gloves or is quick to plunk them into pockets, it's pretty likely that most of the world will see the creative works of art that are inked upon the hands. The shape of the hands makes tattoo designs fun for artists to develop, with the possibilities of what the work entails, endless.

By using the fingers, thumbs, wrist, and connection to the arm, the hand can become a marionette of ingenuity, with the fingers as the strings. Seen as a symbol of protection since ancient times, the hands provide an imaginative canvas for tattooists to work with. This chapter features an array of outstanding tattooed hand works from artists far and wide.

John Anderton

Guy Aitchison

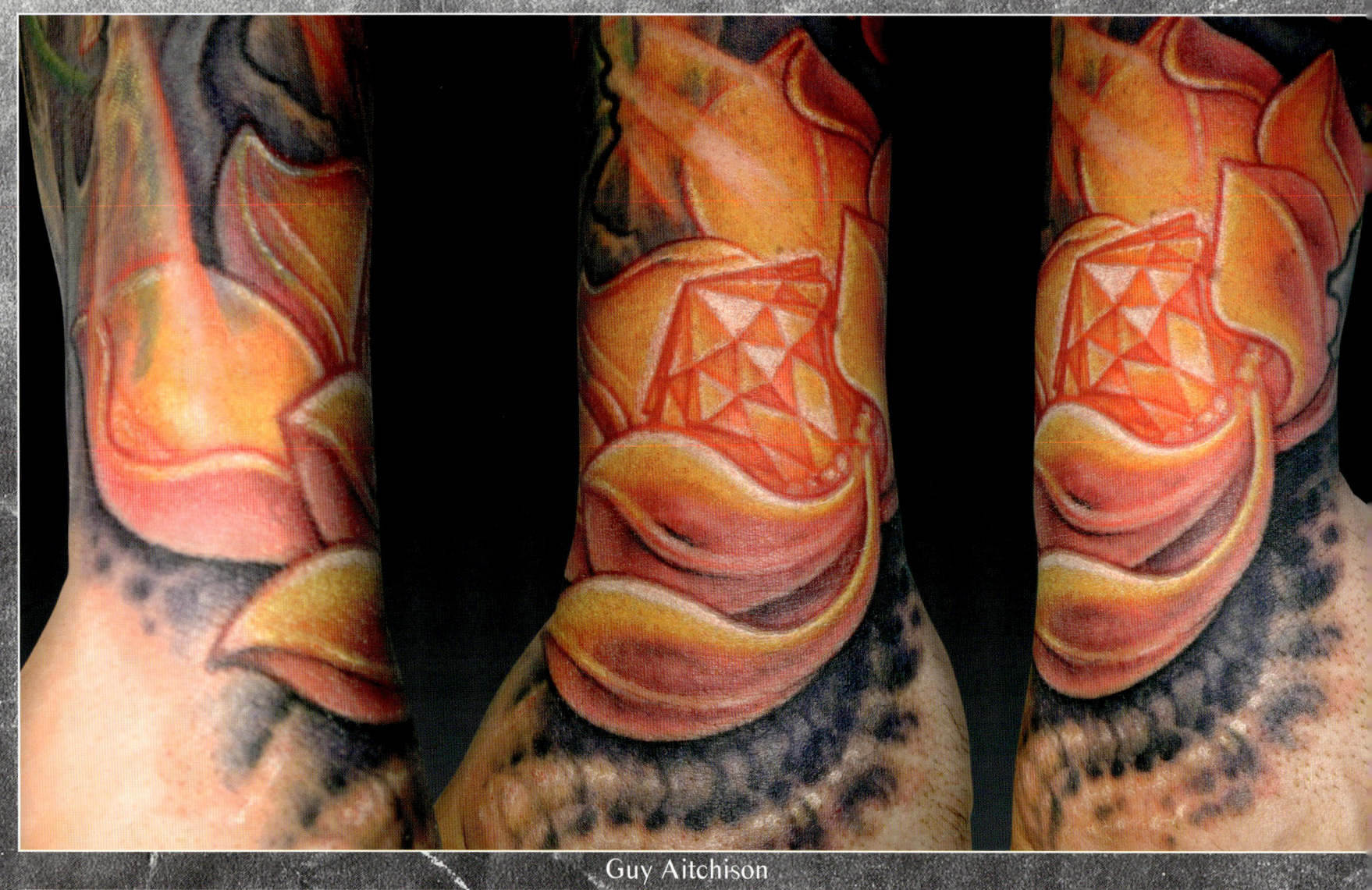

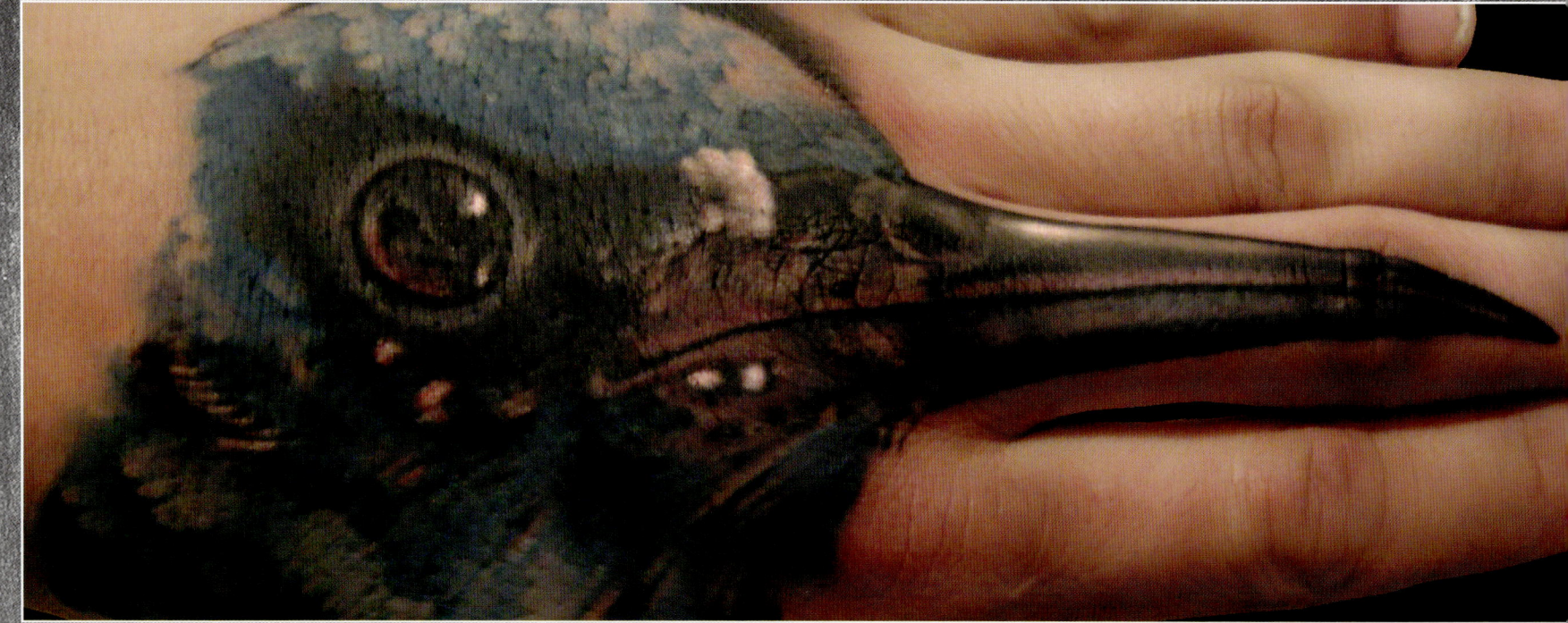

Nicklas Westin

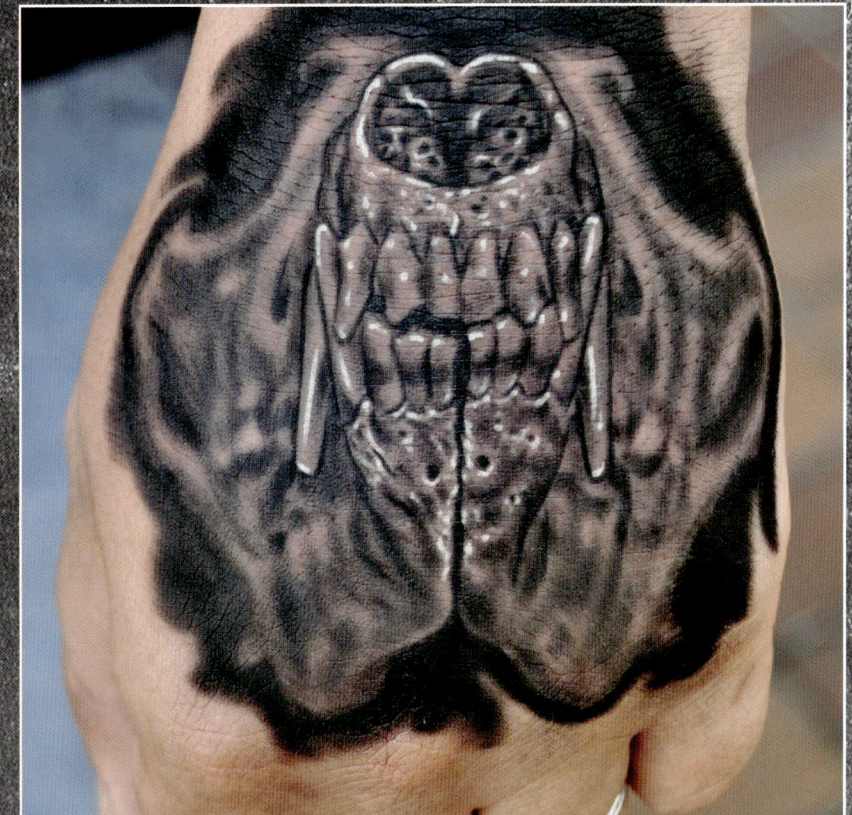

Mike DeVries

Mike DeVries

Nick Baxter

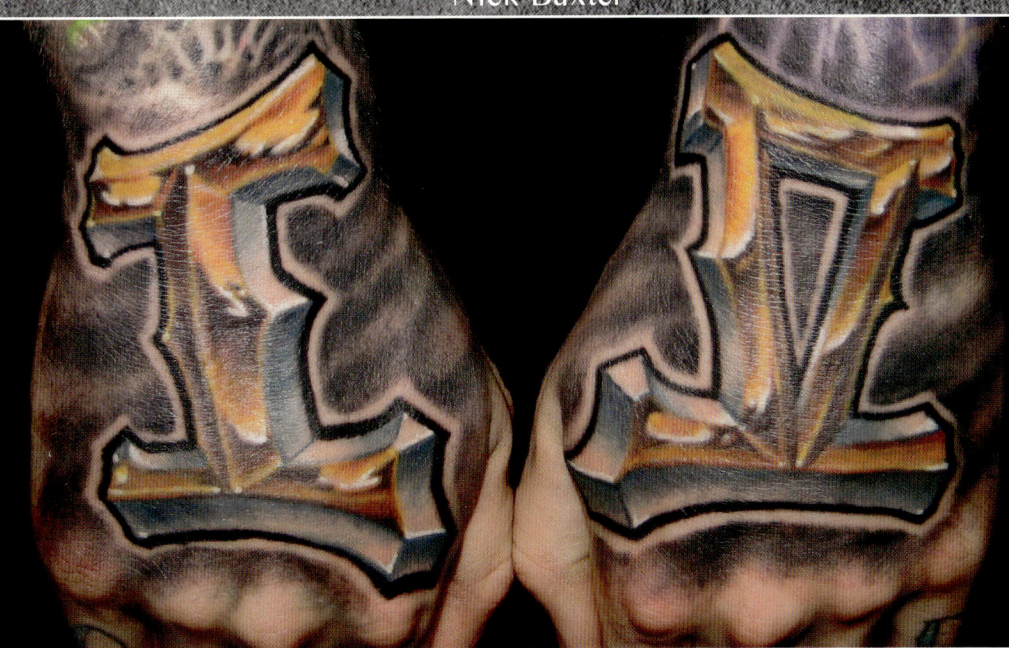

Dan Hazelton

86

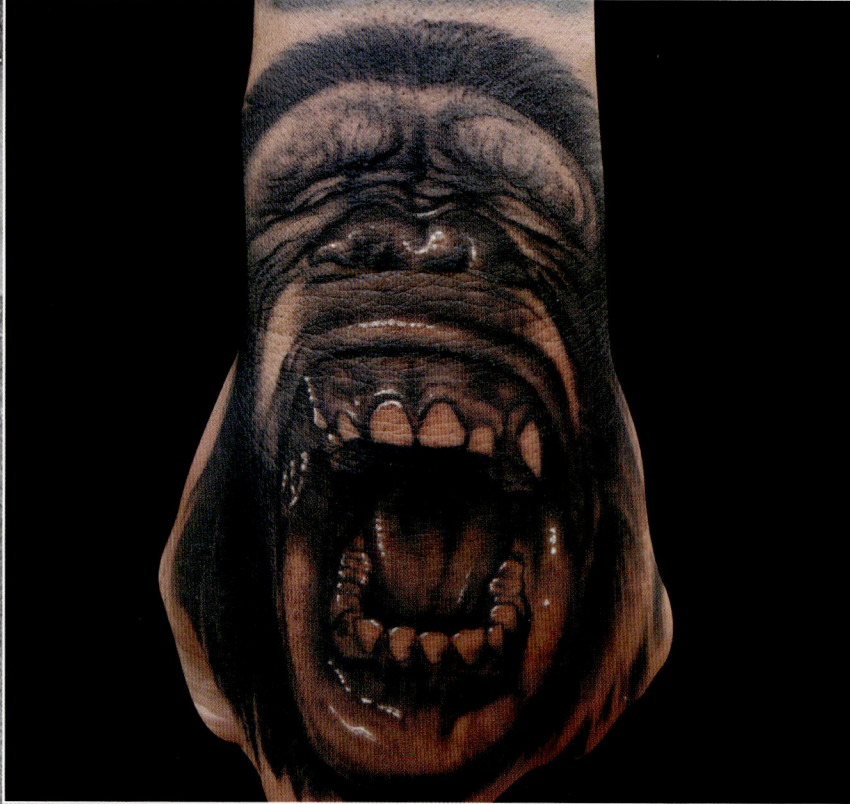

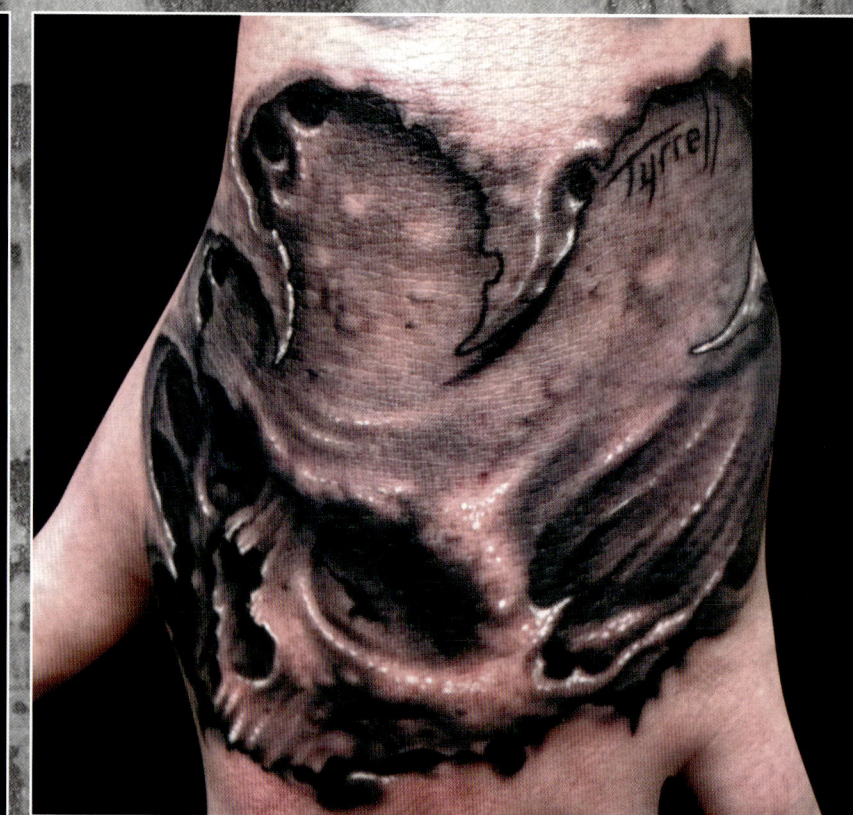

Bob Tyrrell

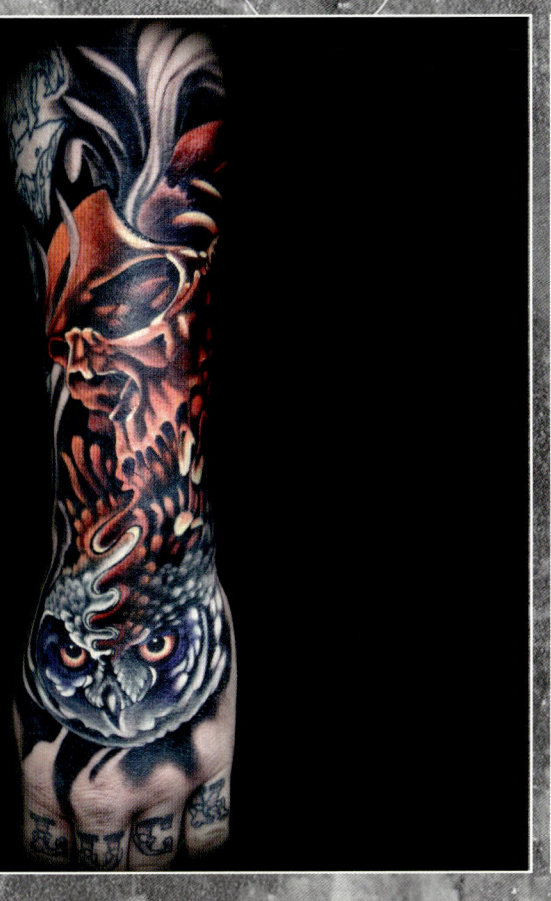

Fabz

Josh Duffy

Paris Pierides

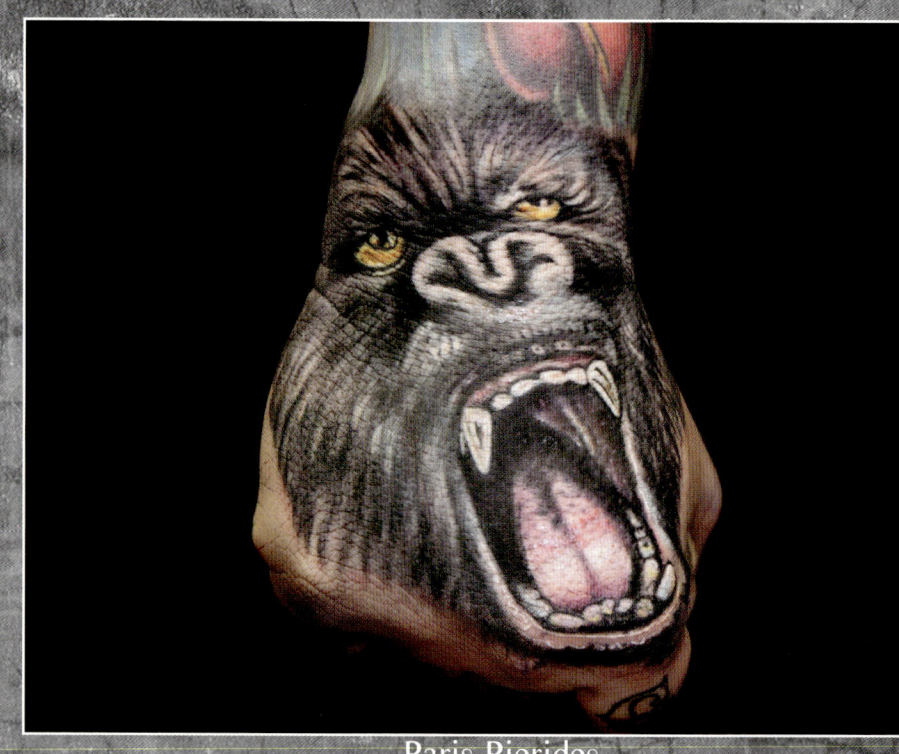

Paris Pierides

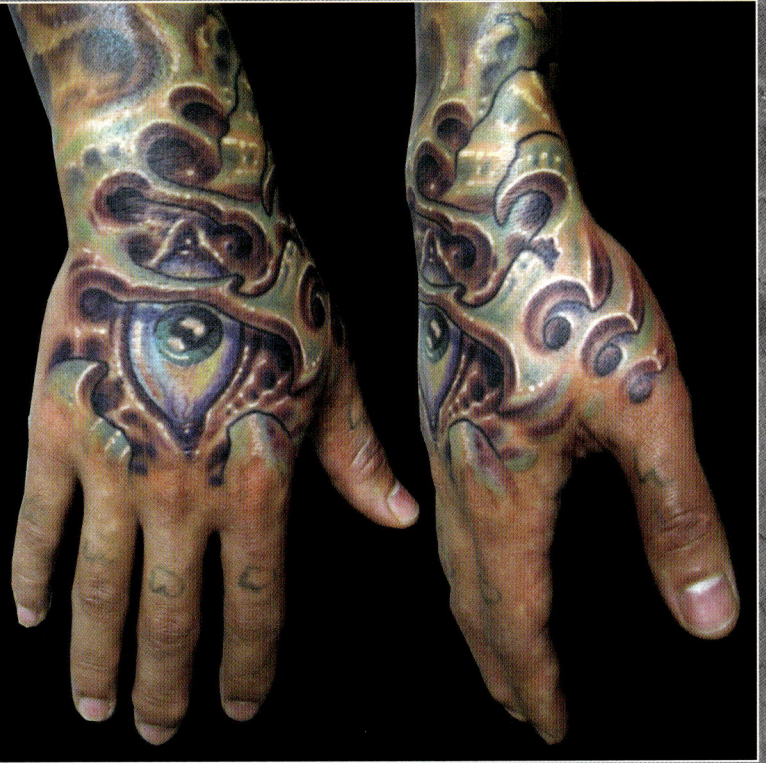

Christian Perez

90

Jason Adkins

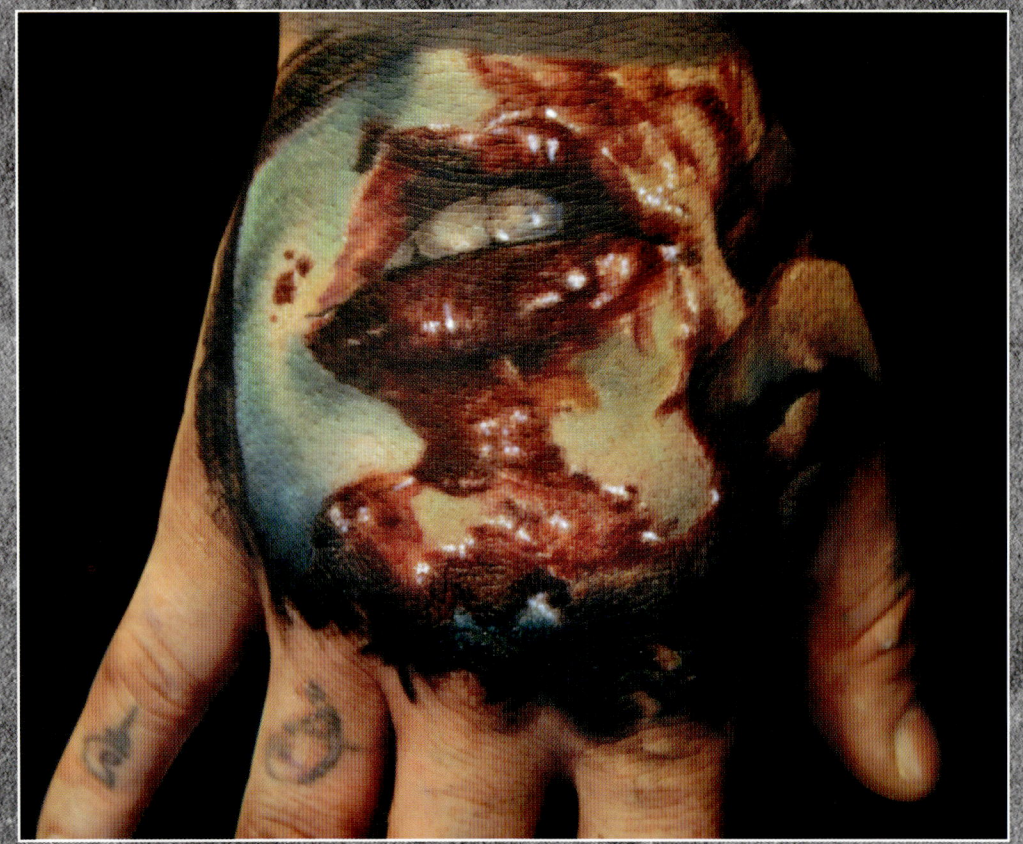

Florian Karg

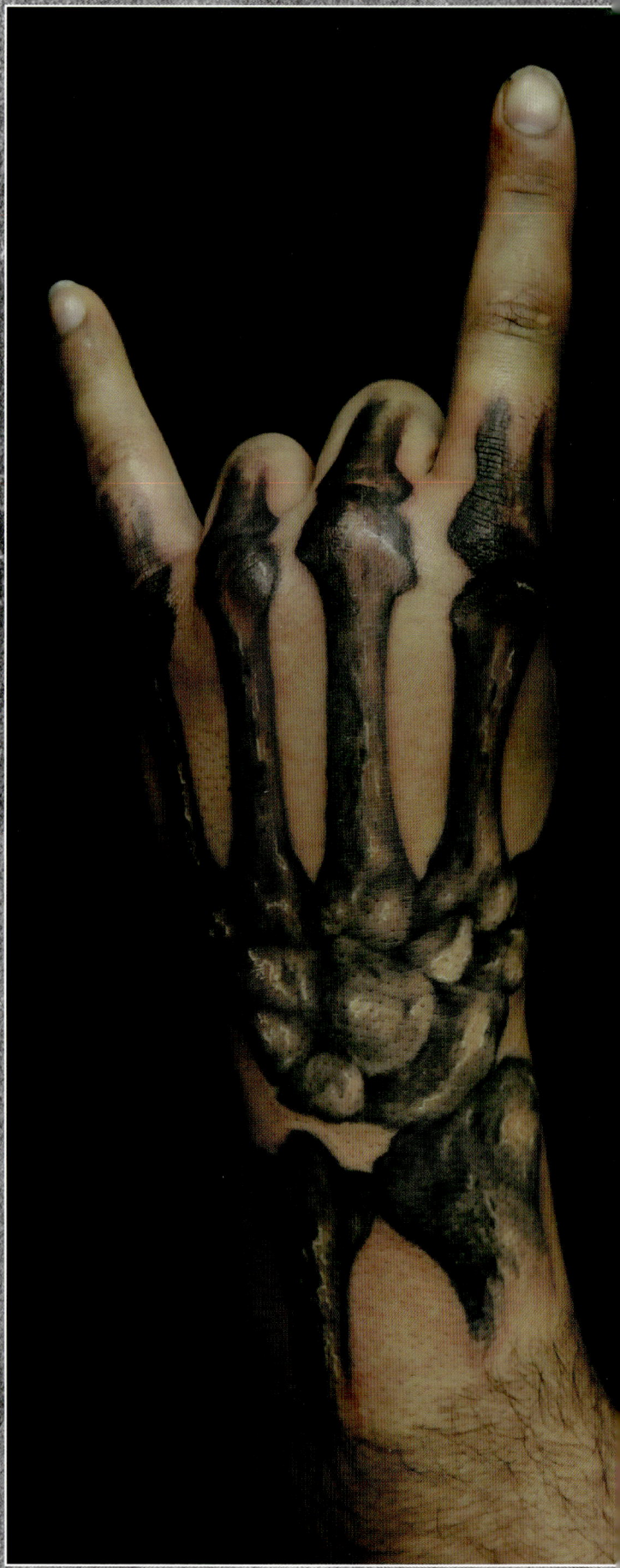

Florian Karg

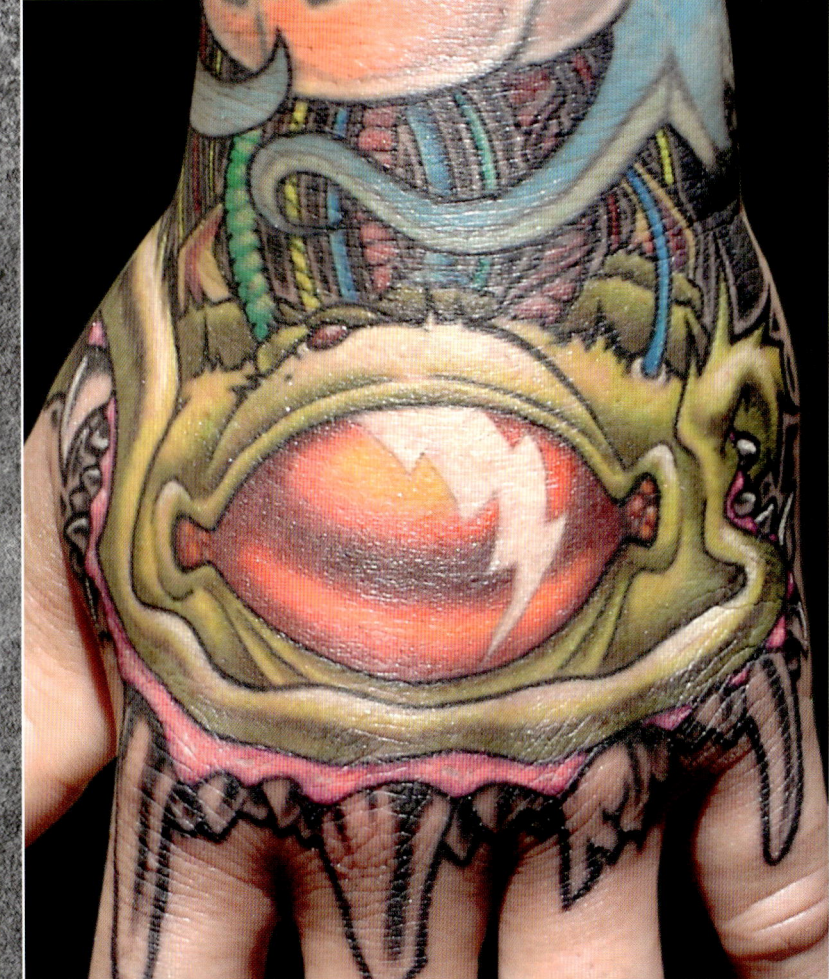

Jesse Smith

Tim Senecal

Tim Senecal

Shawn Barber

Shawn Barber

Shawn Barber

Uncle Allan

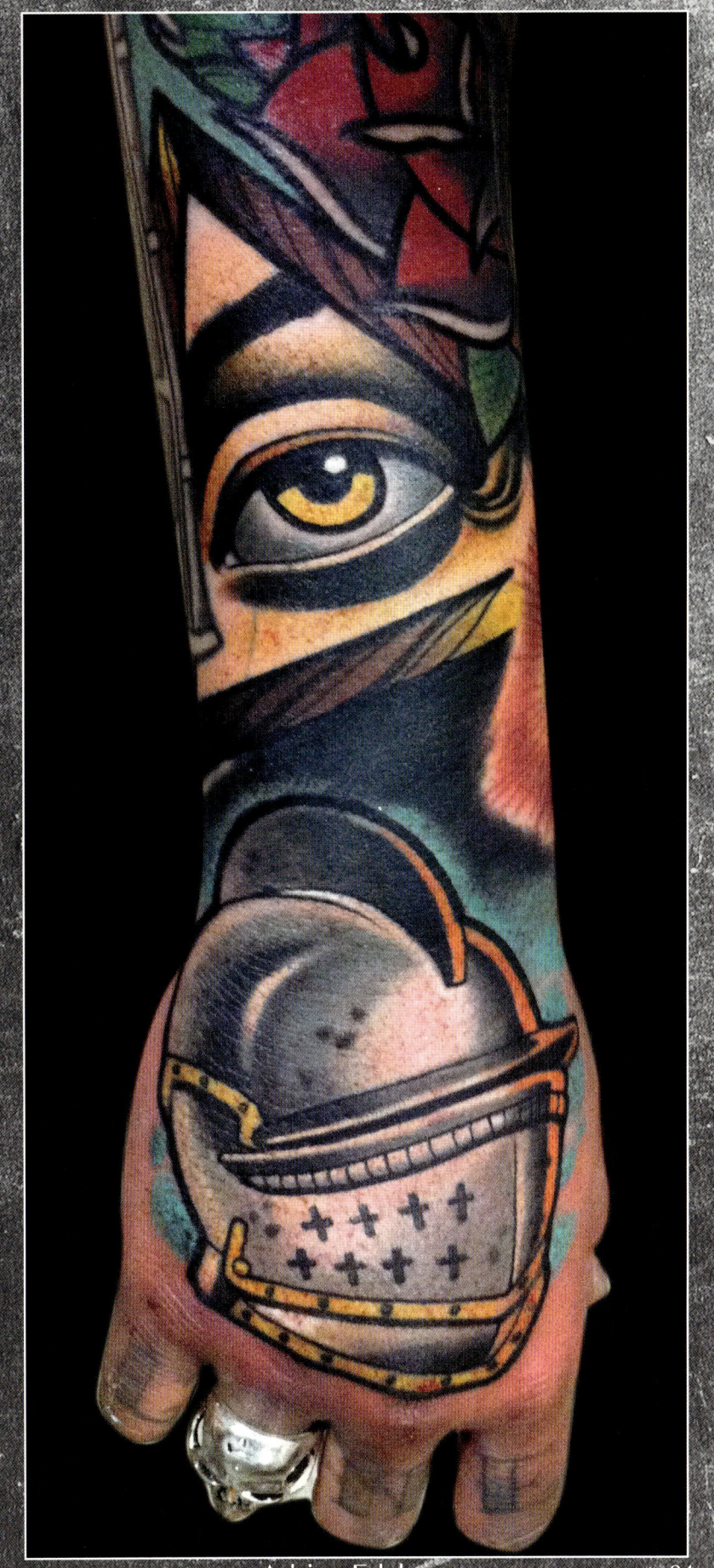

Adrian Edek

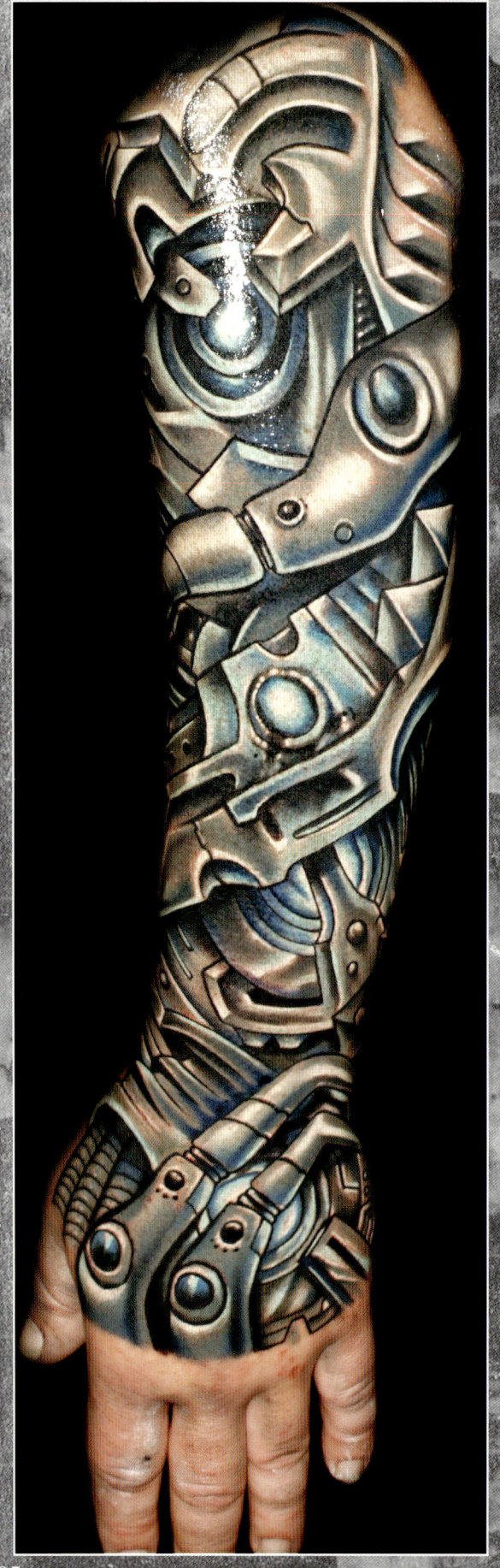

Roman Abrego

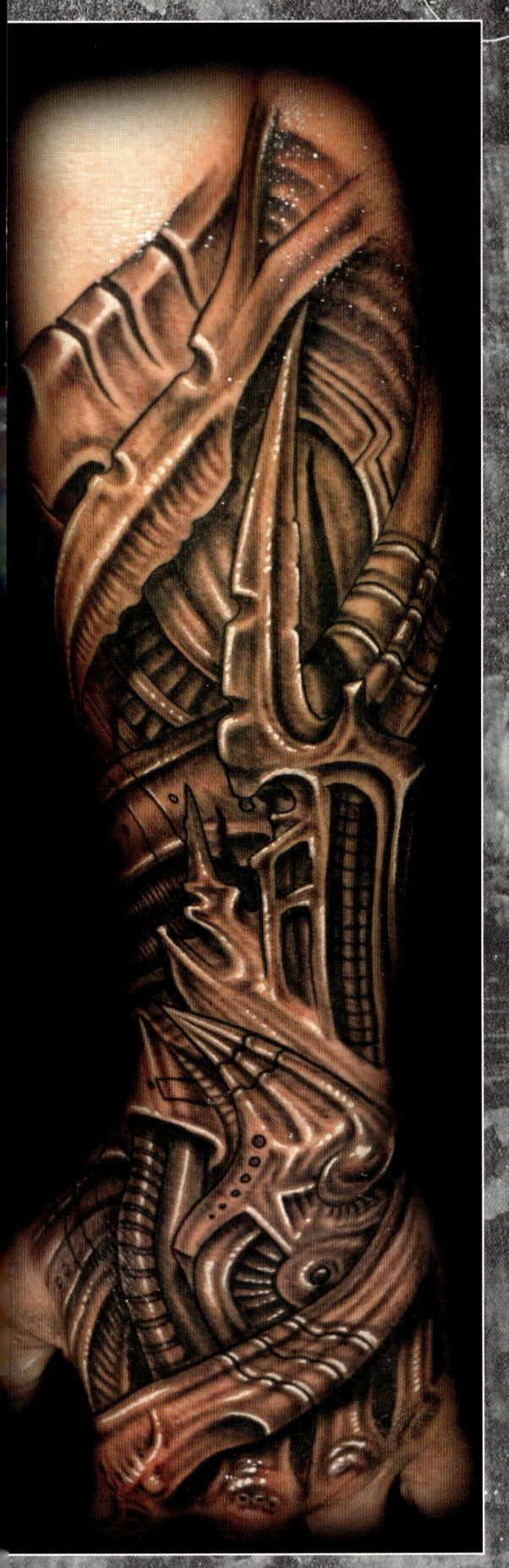

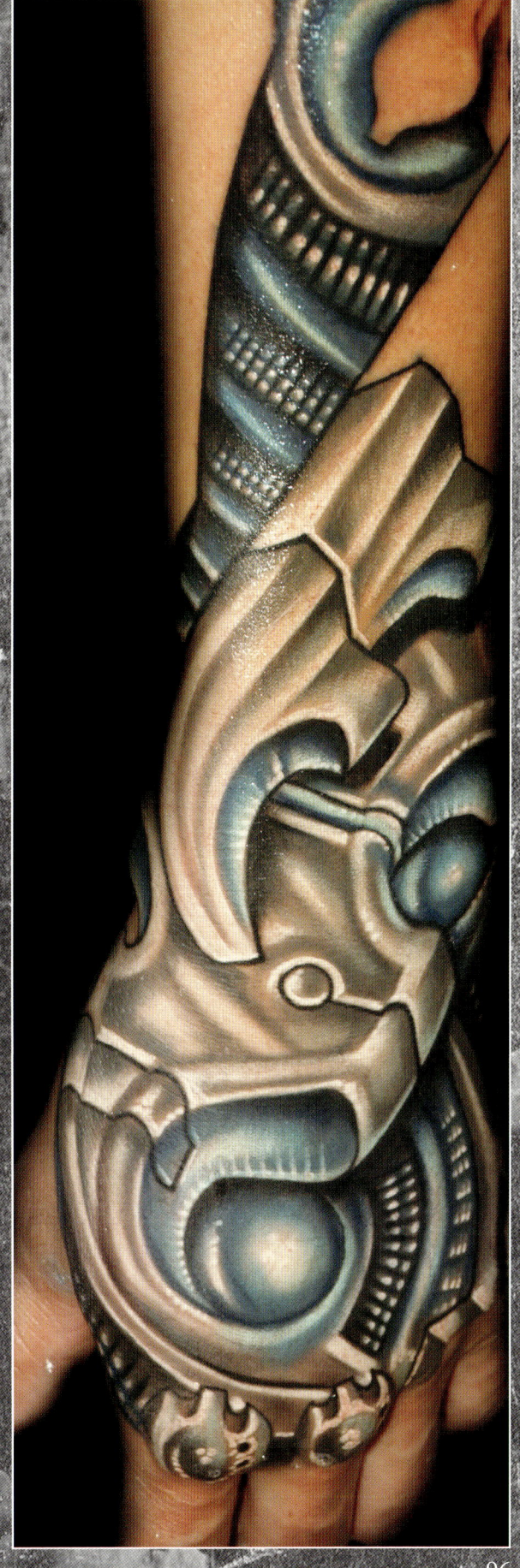

Roman Abrego

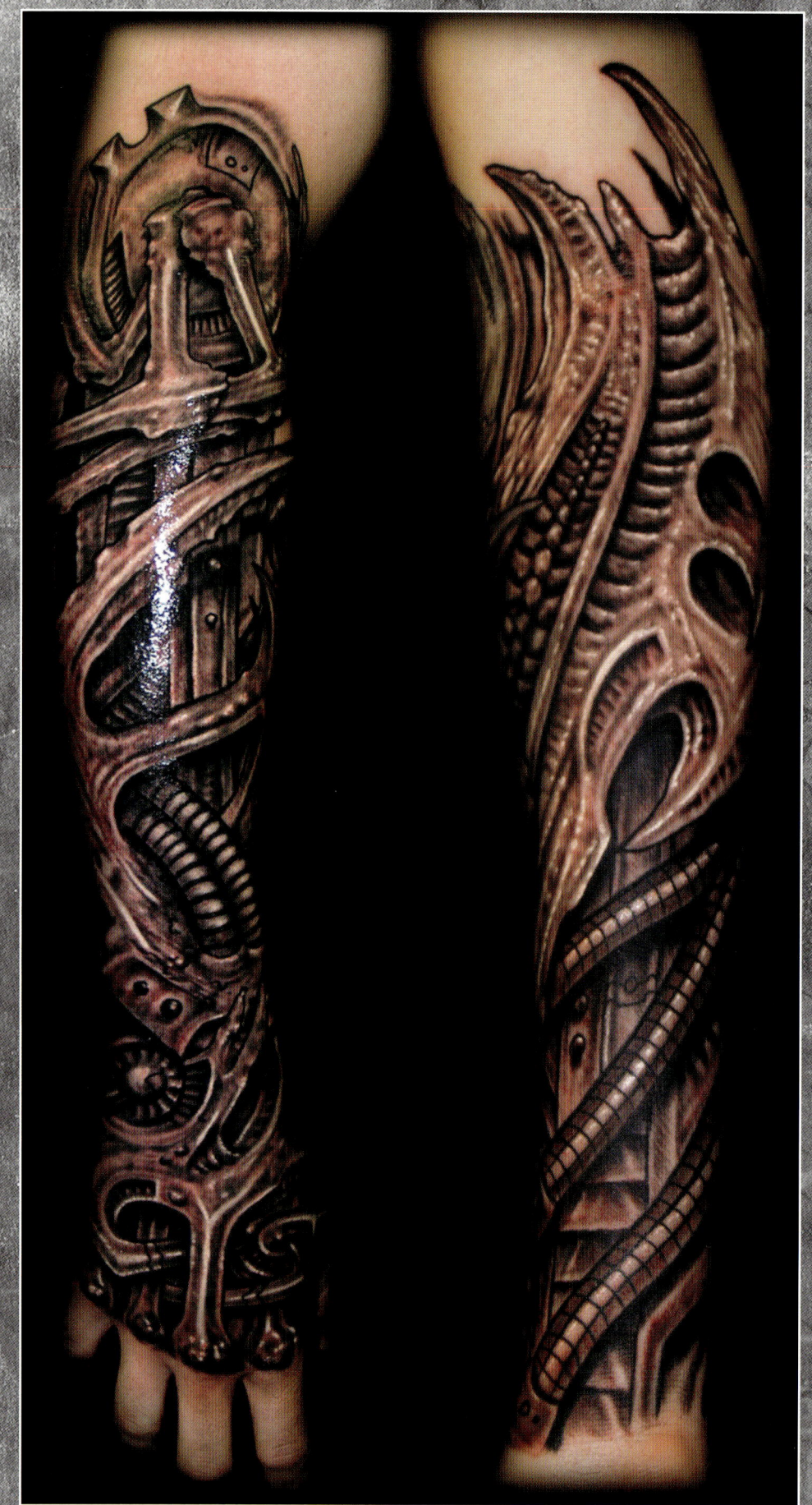

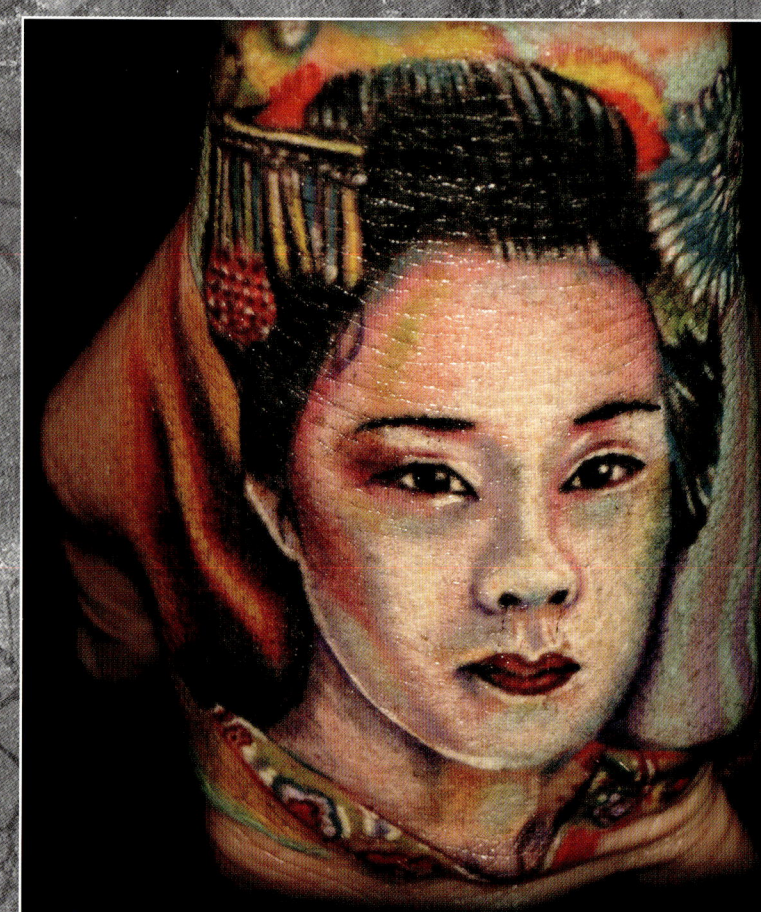

Roman Abrego

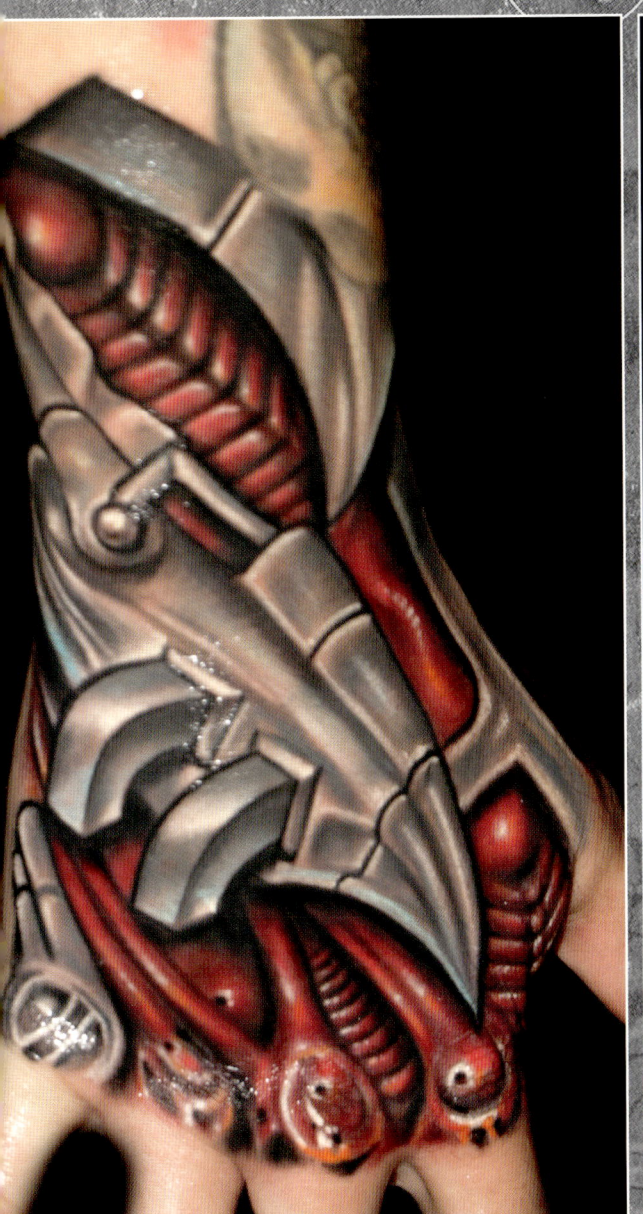

Roman Abrego

Roman Abrego

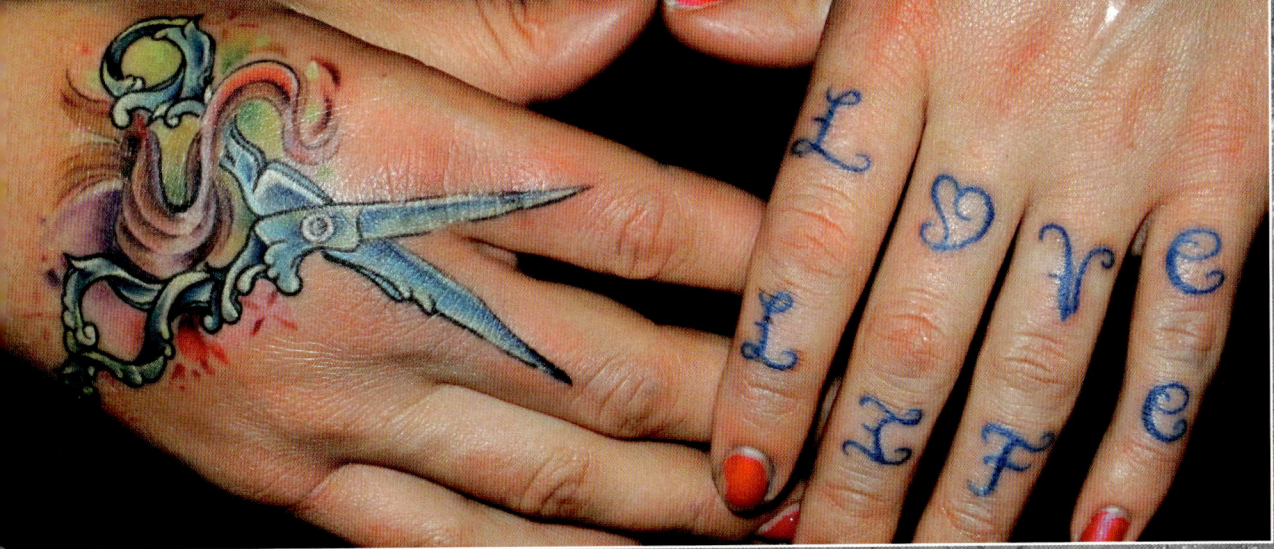

Melissa Fusco

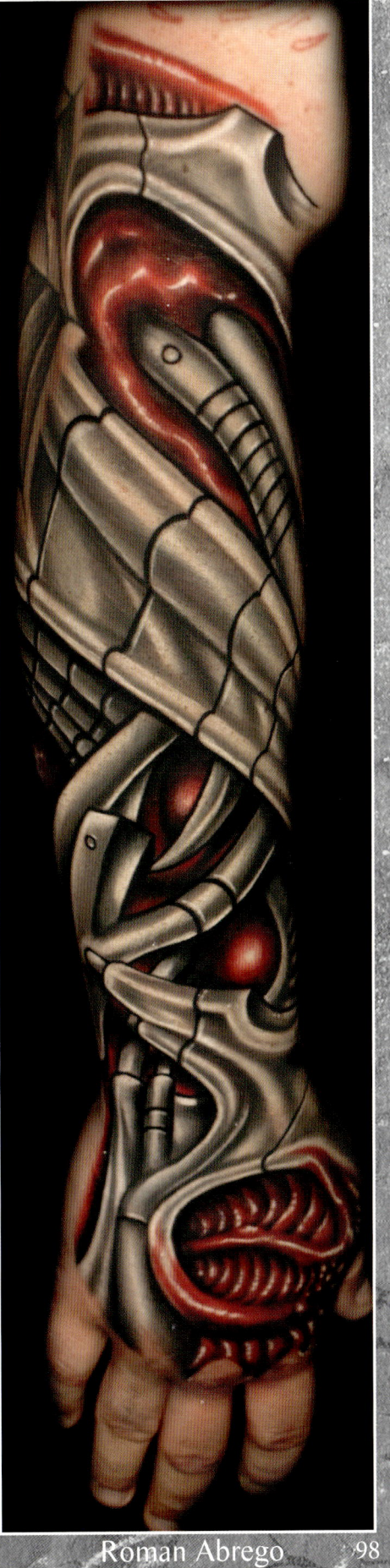

Roman Abrego

98

Andy Barrett

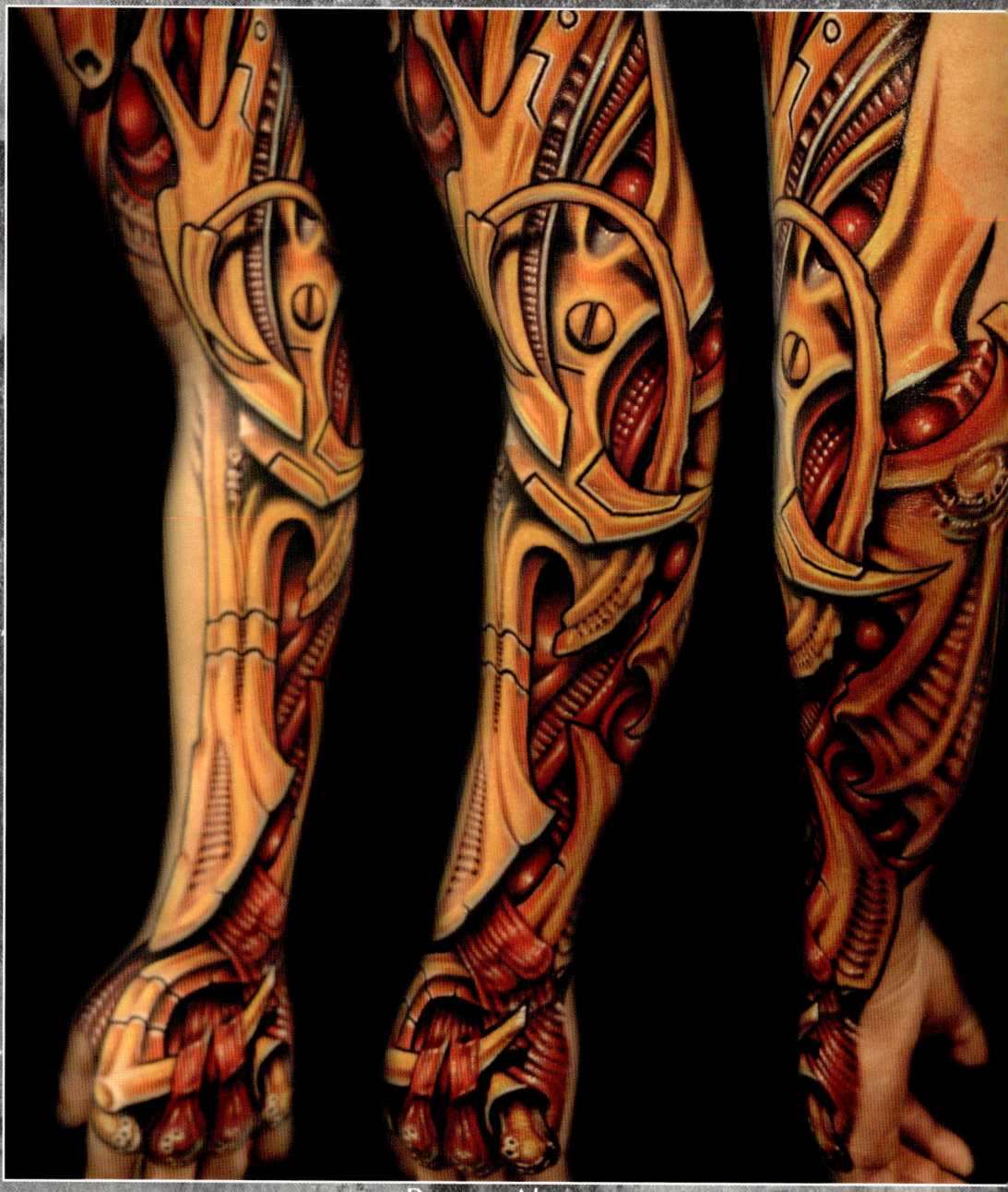

Roman Abrego

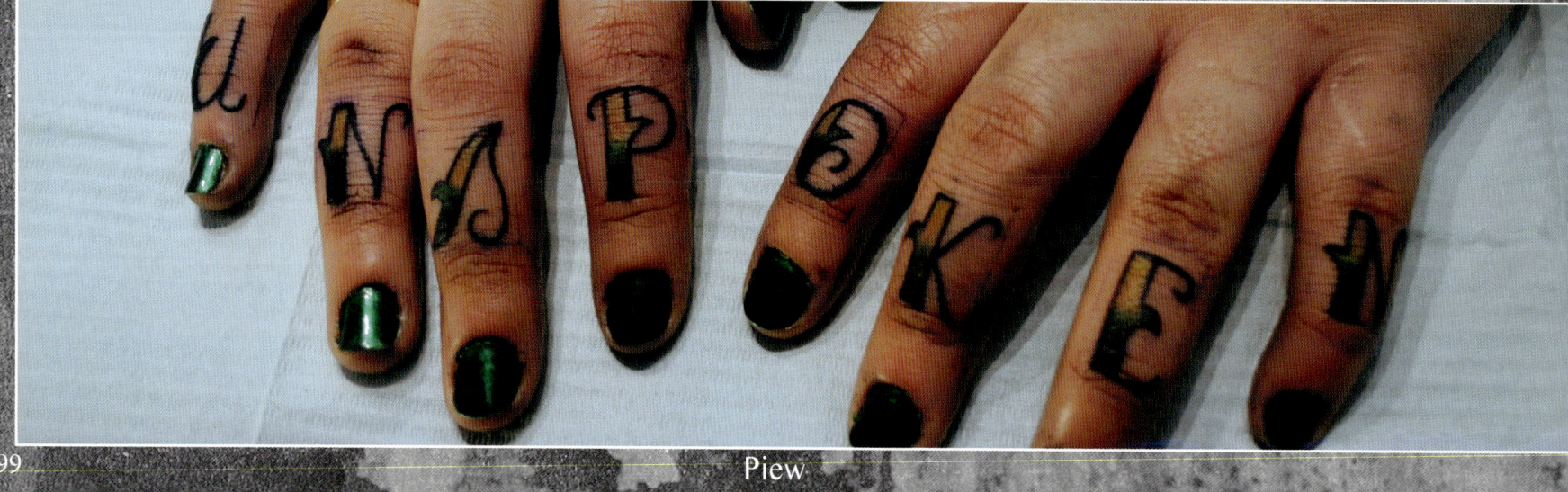

Piew

Holly Azzara

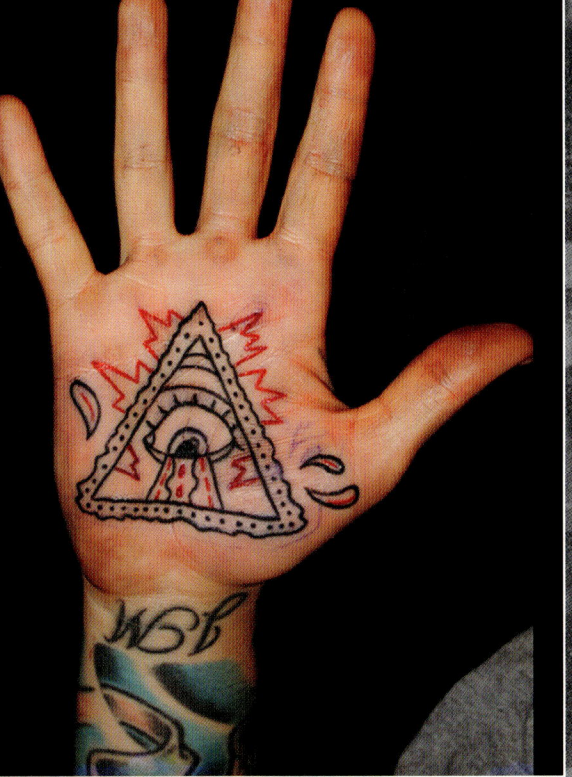

Jim Sylvia

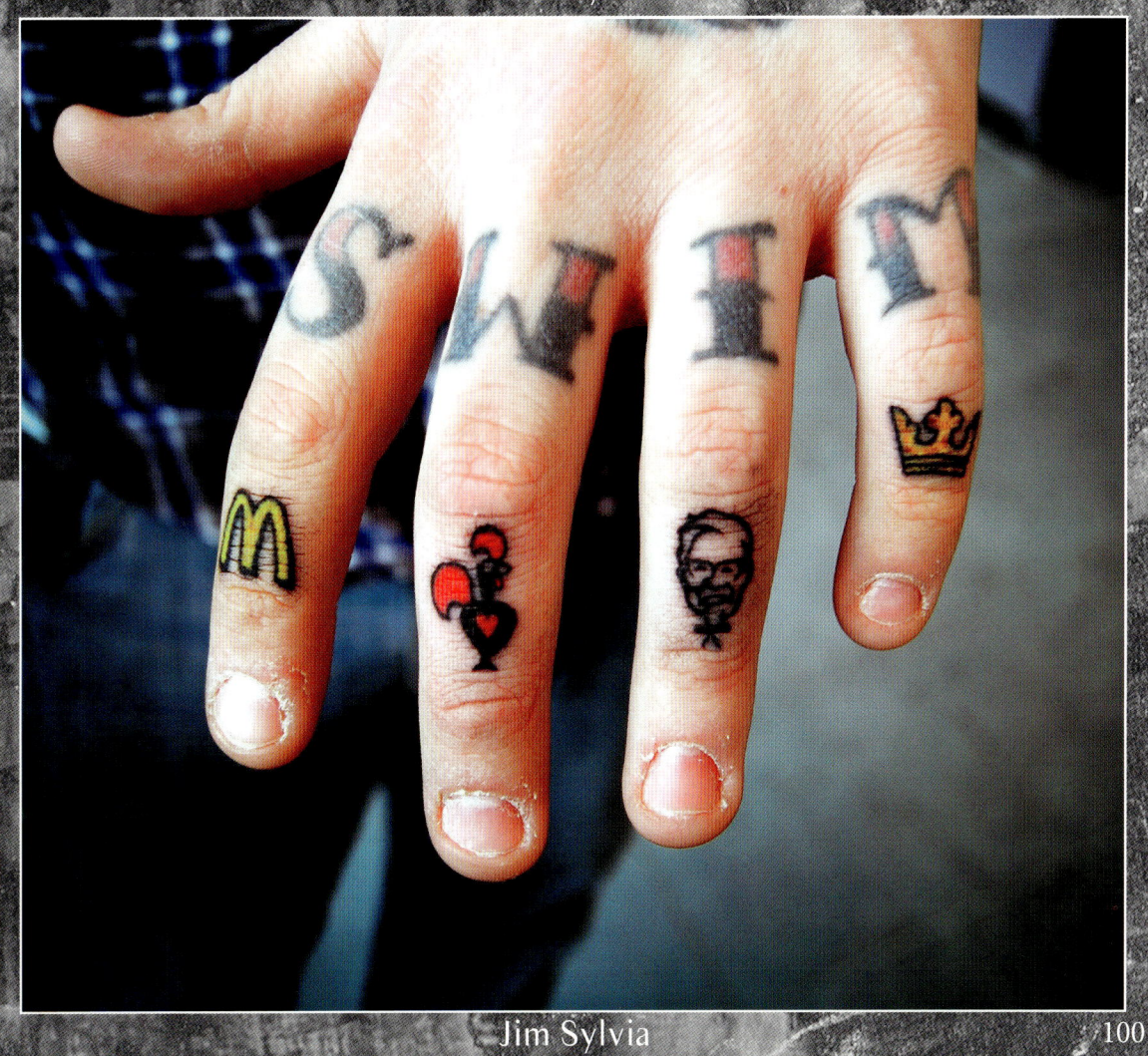

Piew

Jim Sylvia

100

Franco Vescovi

Franco Vescovi

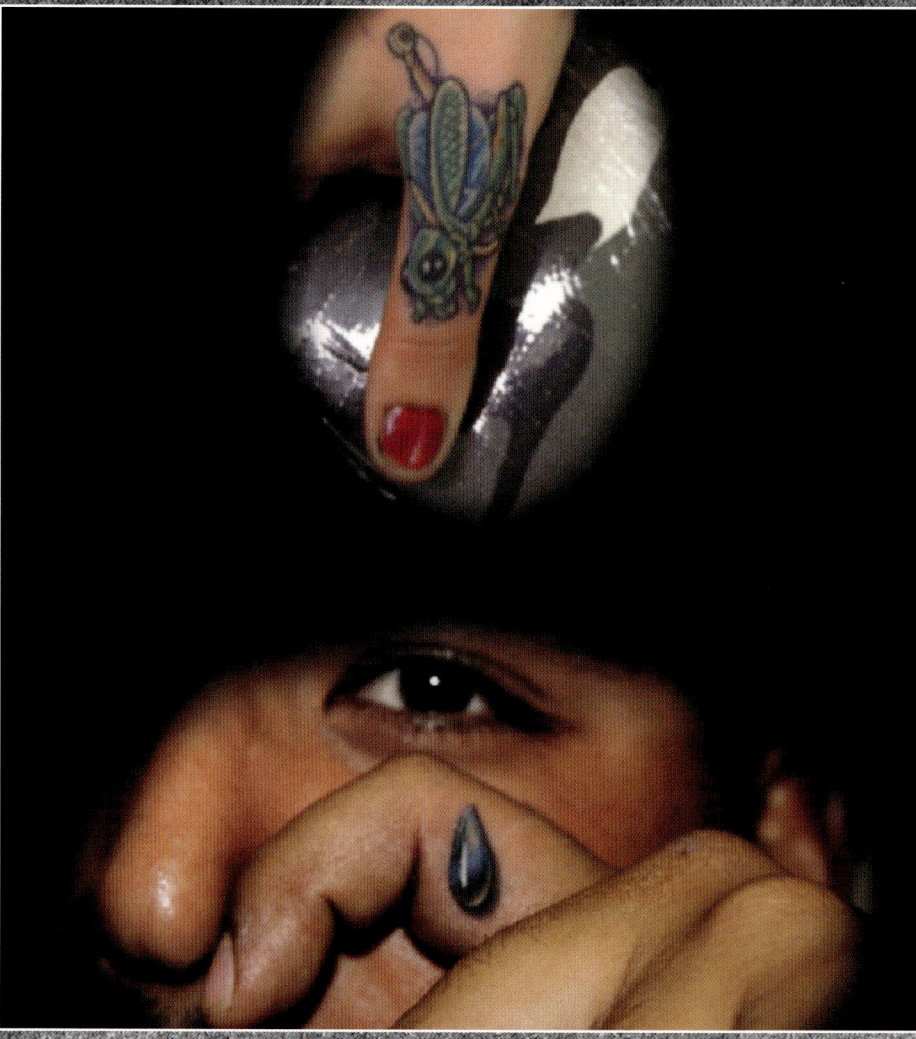

Eva Huber

Billy Weigler

Cory Ferguson

Ben Corn

Steve Byrne

102

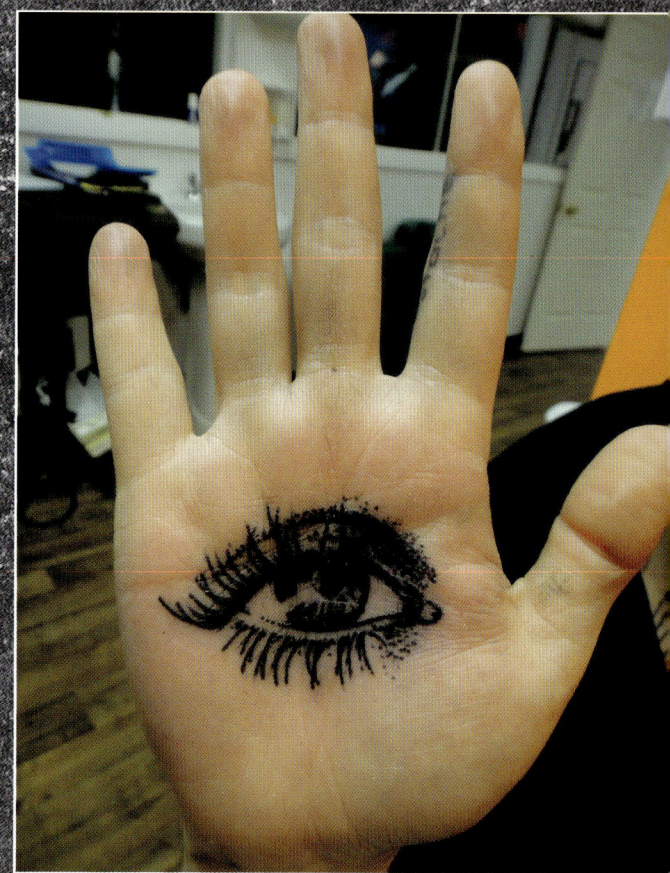

Steve Byrne

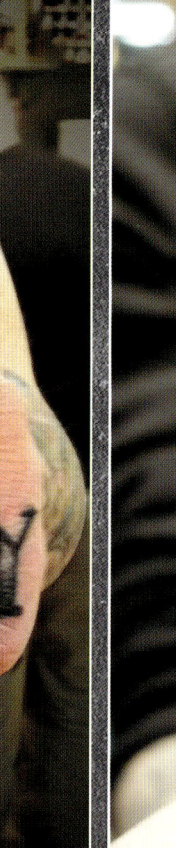

Loïc Lavenu

Miro Tomas

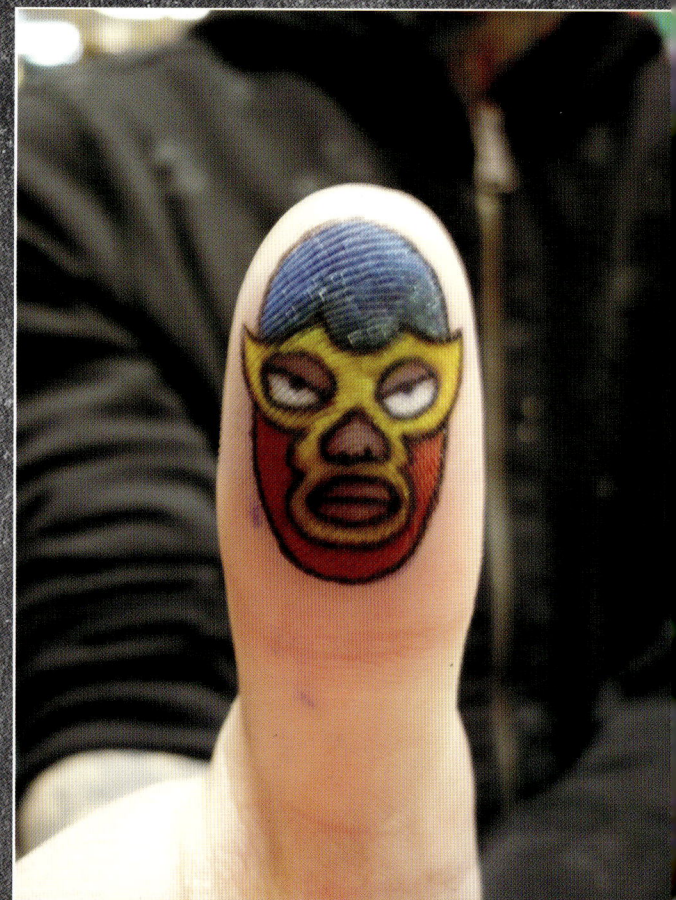

John Lloyd

Miro Tomas

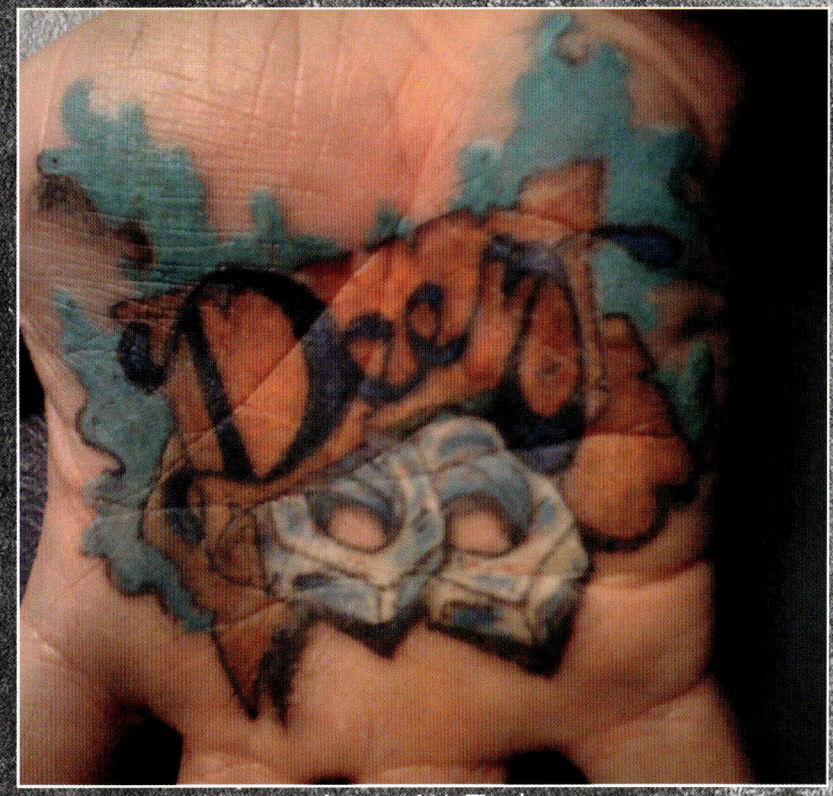

James W. Taylor

Brady Willmott

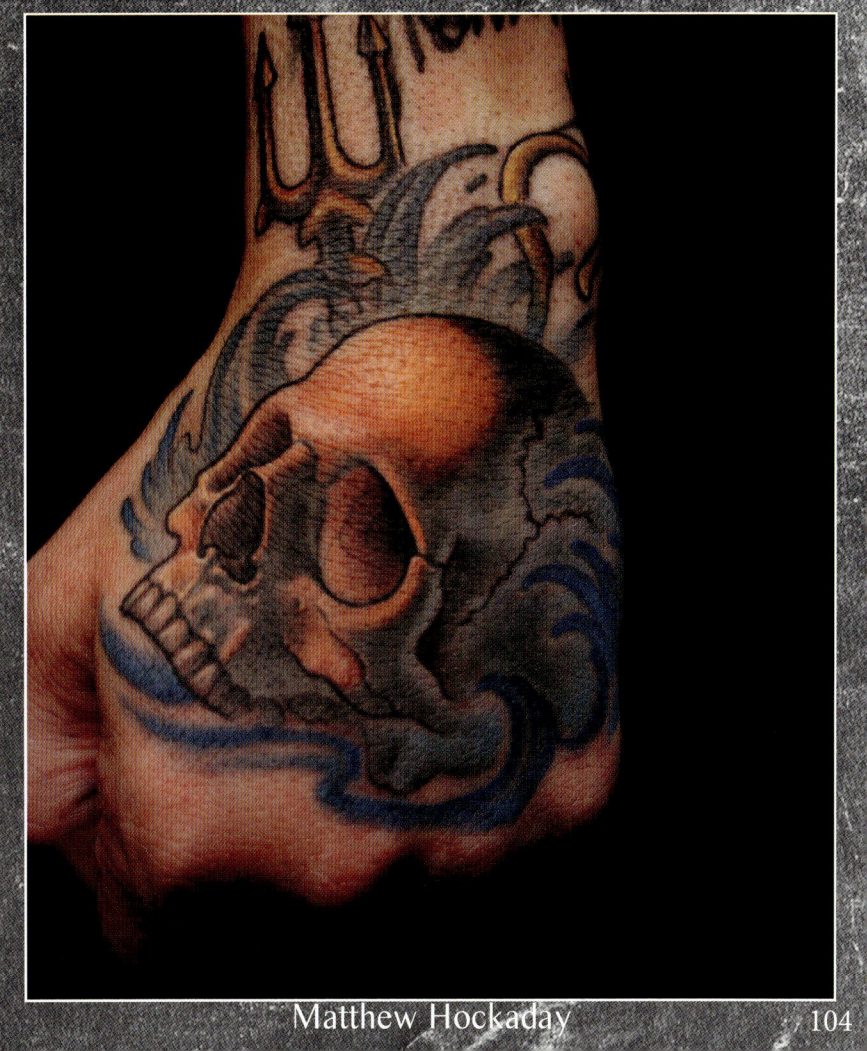

Matthew Hockaday

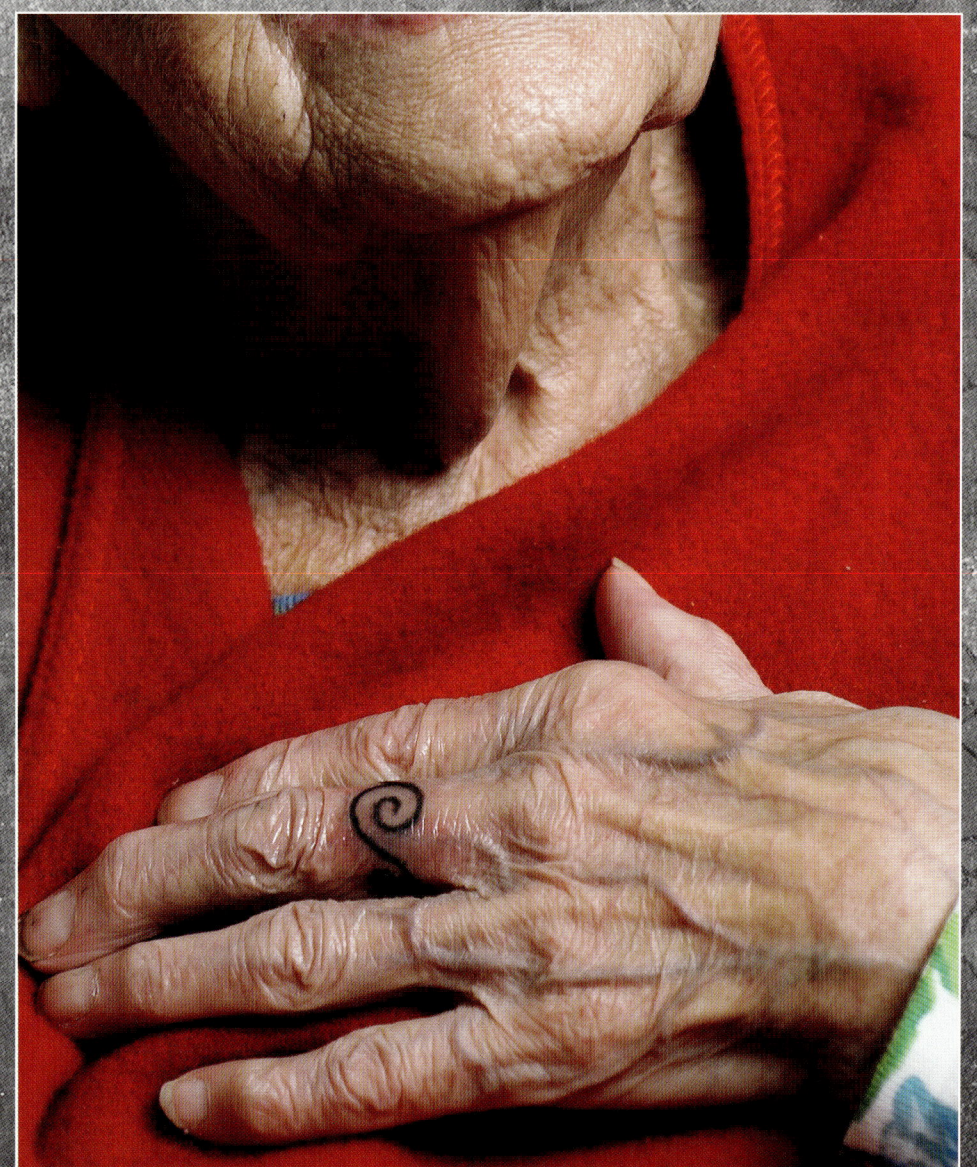

Colin Dale

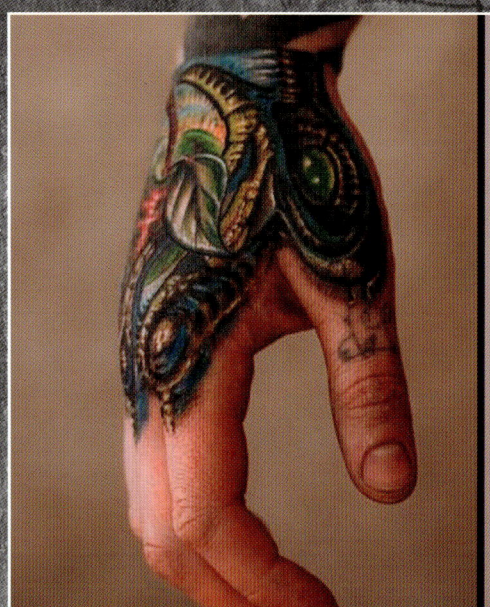

Dave Tedder

Vova Mult

Vova Mult

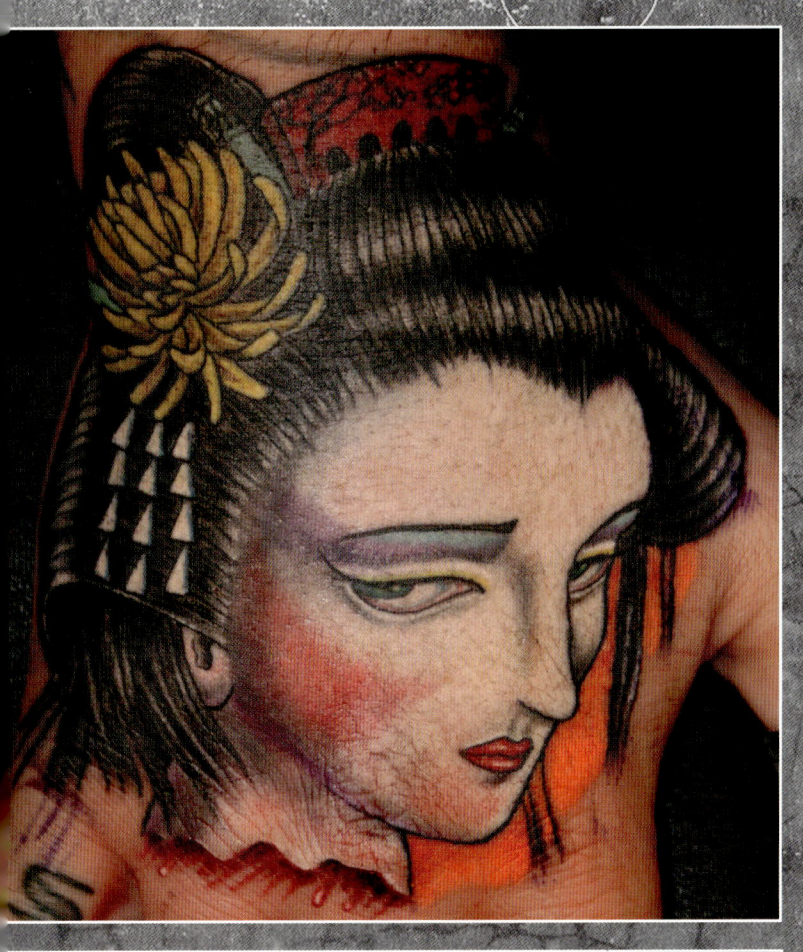

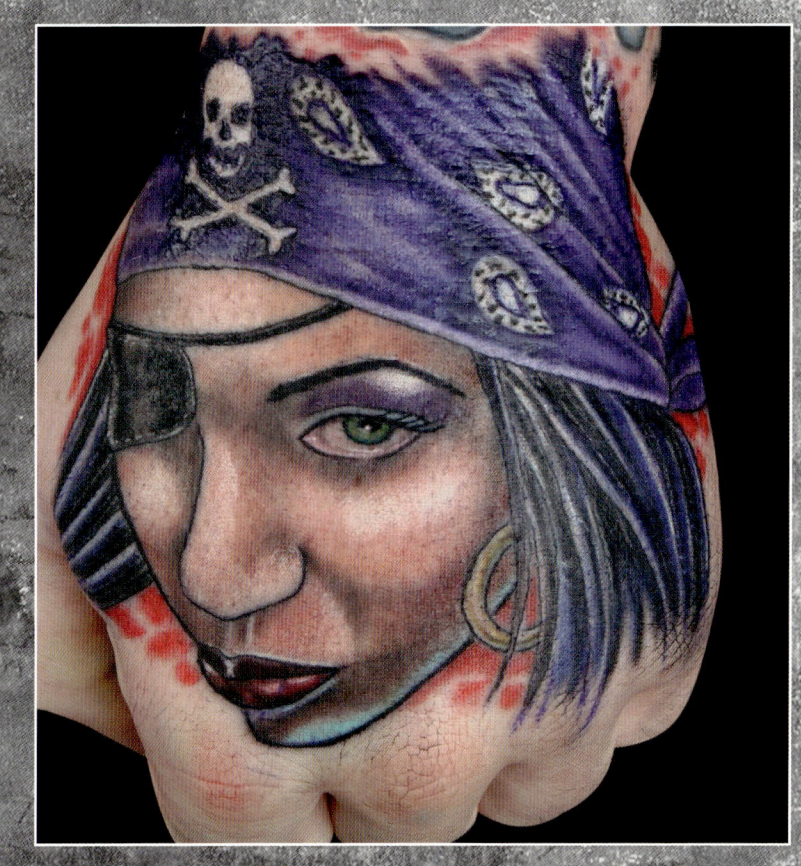

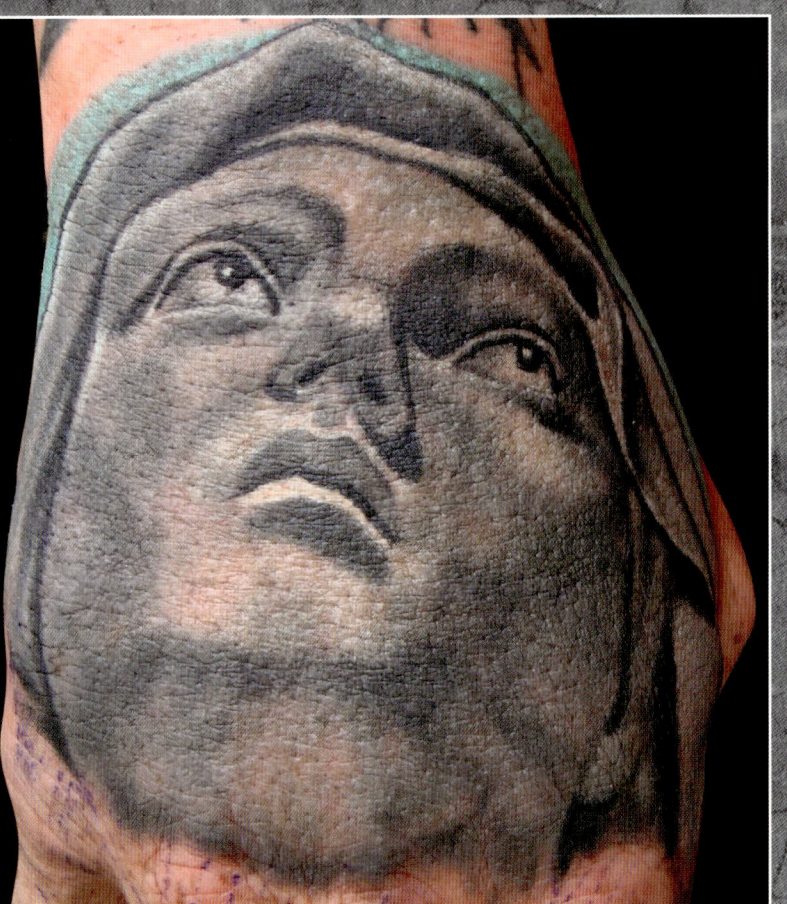

Larry Brogan

Larry Brogan

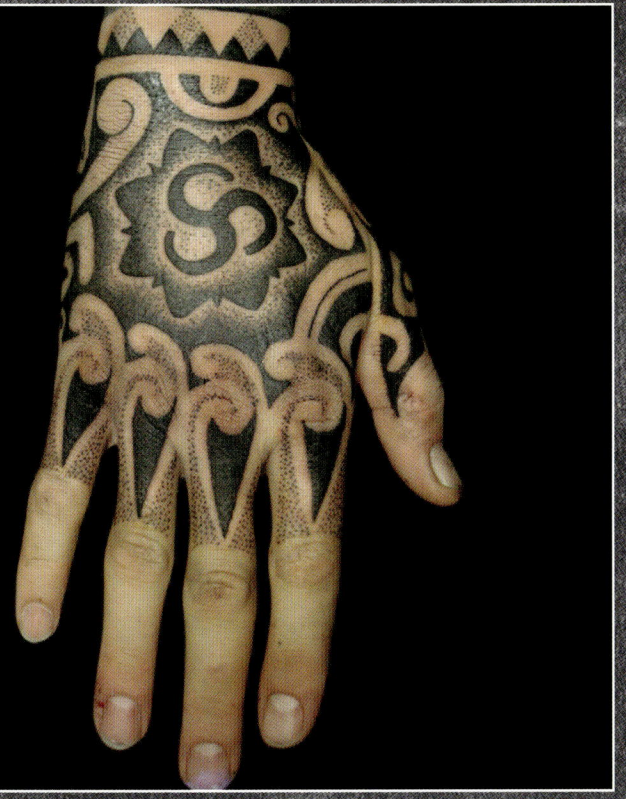

Cory Ferguson

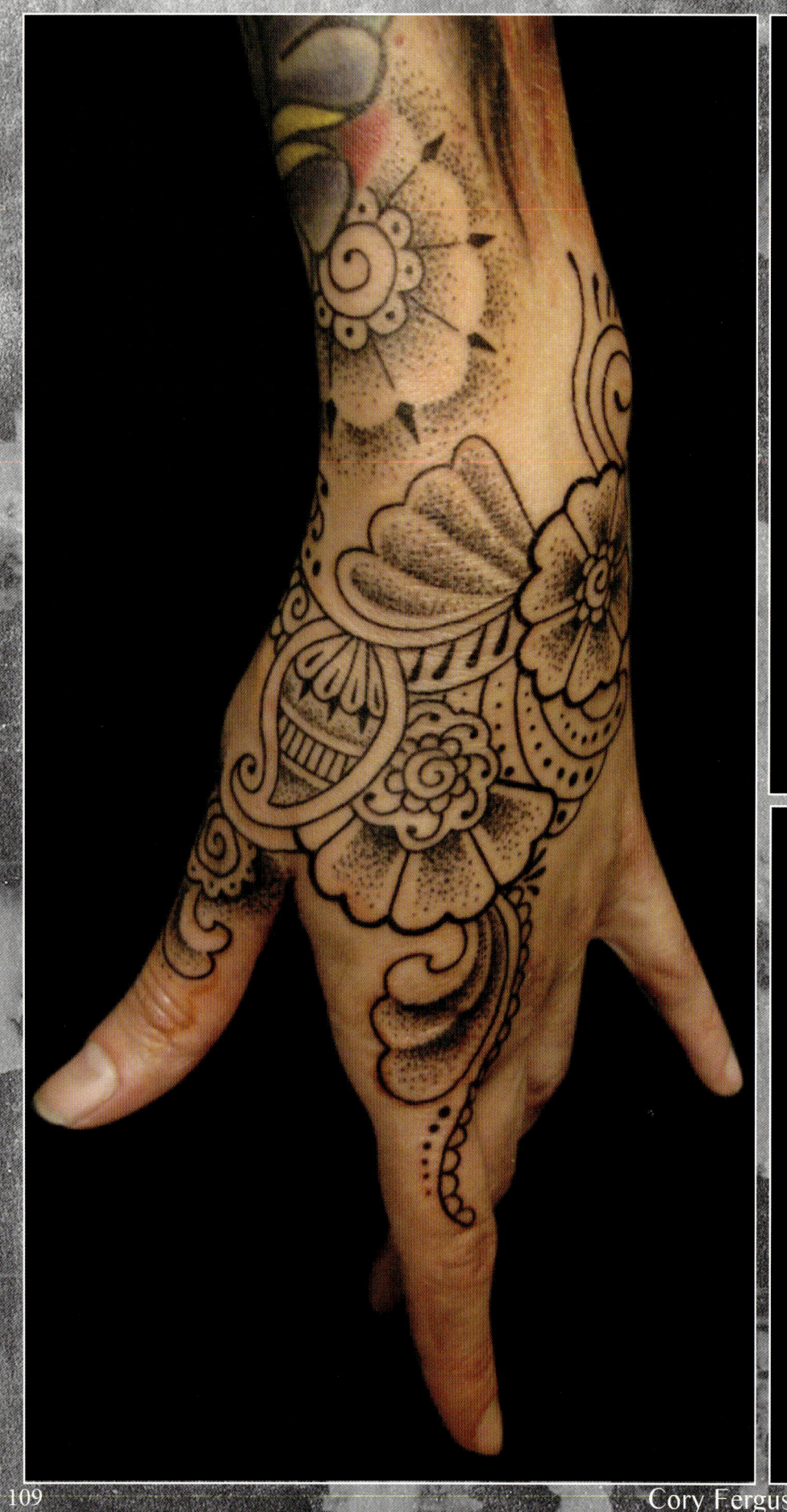

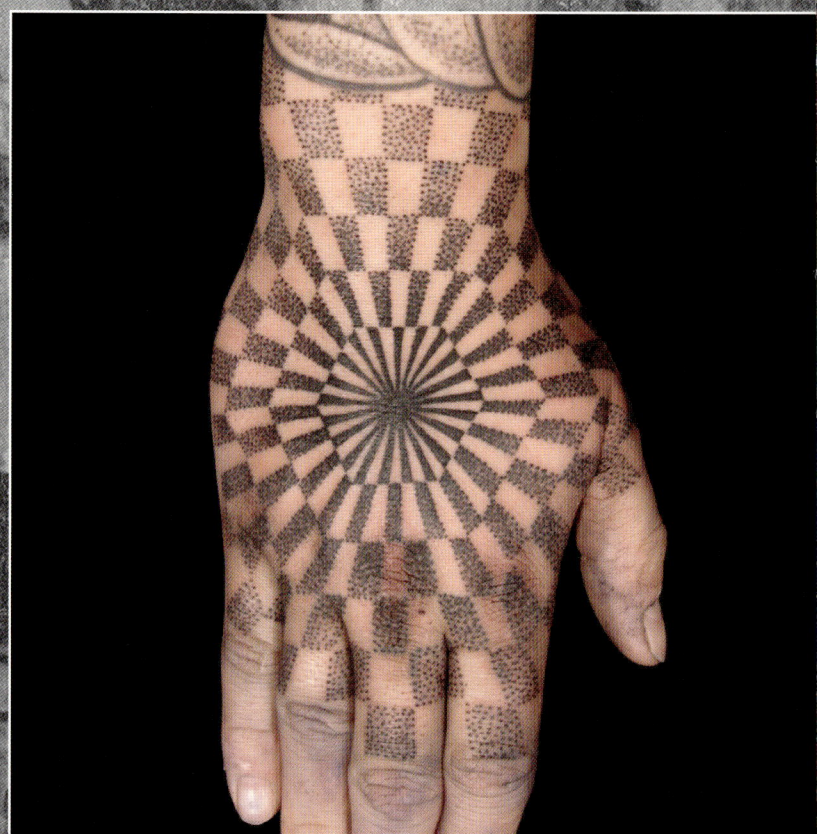

Cory Ferguson

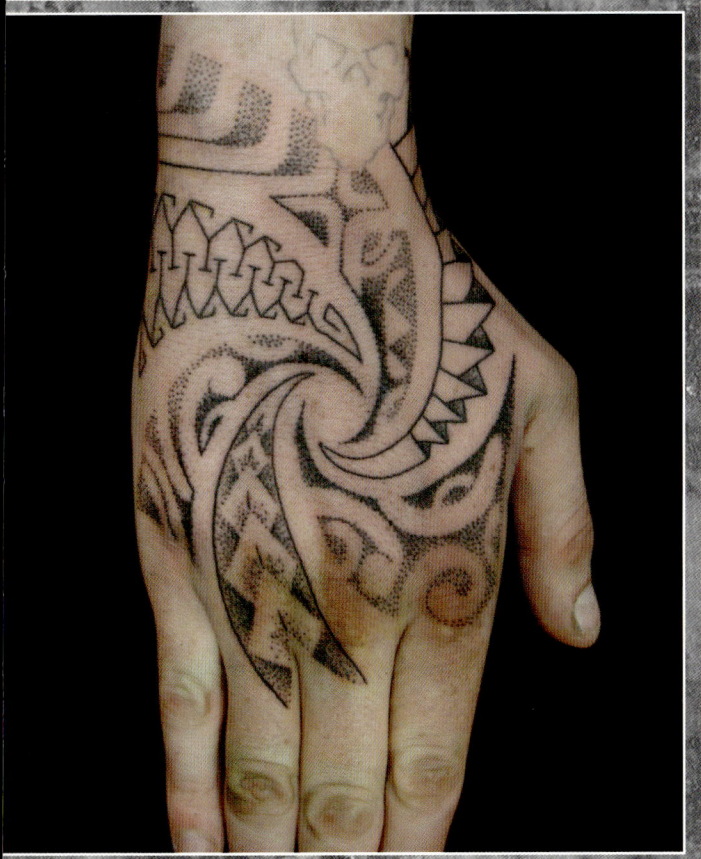

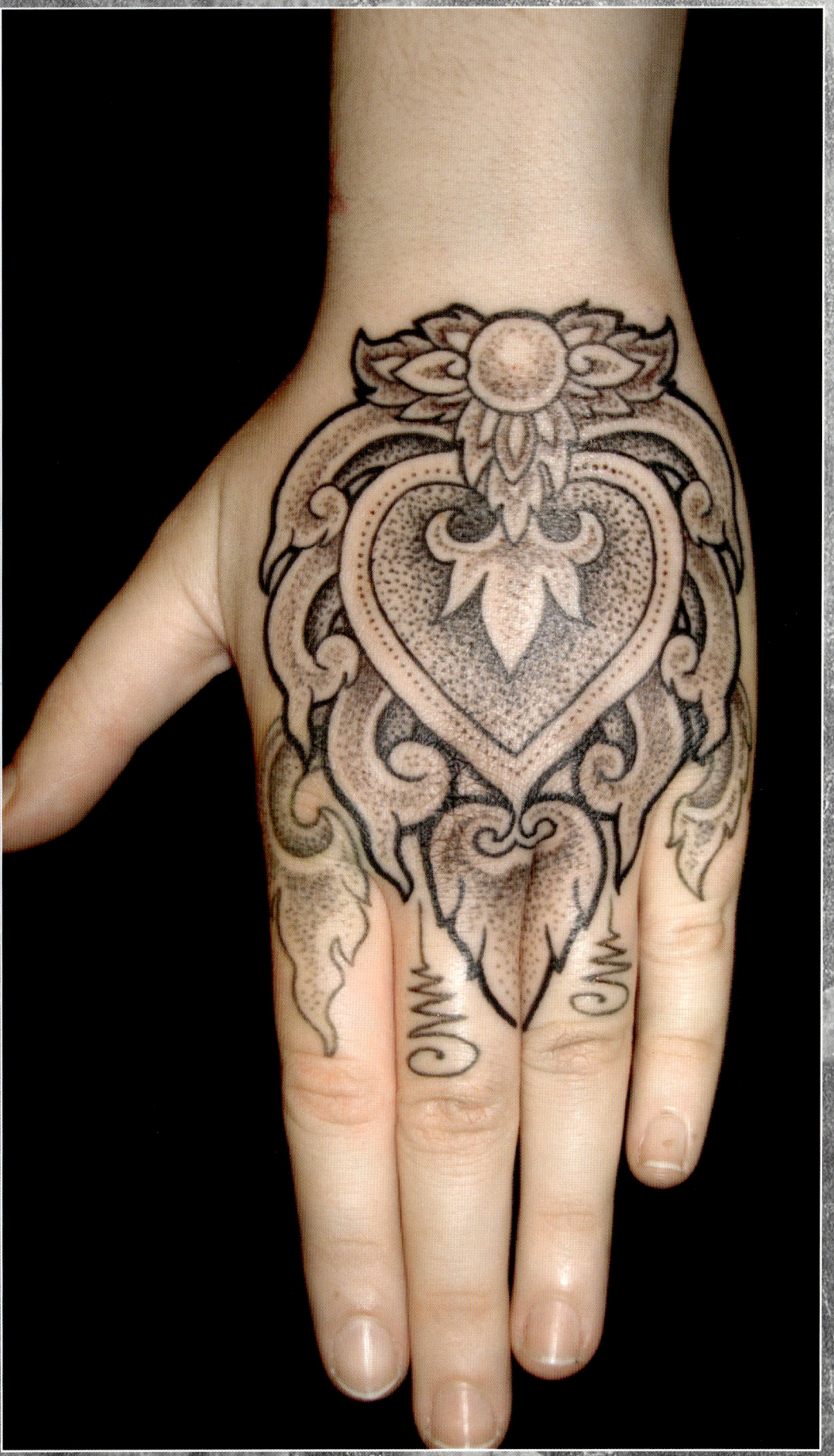

Cory Ferguson

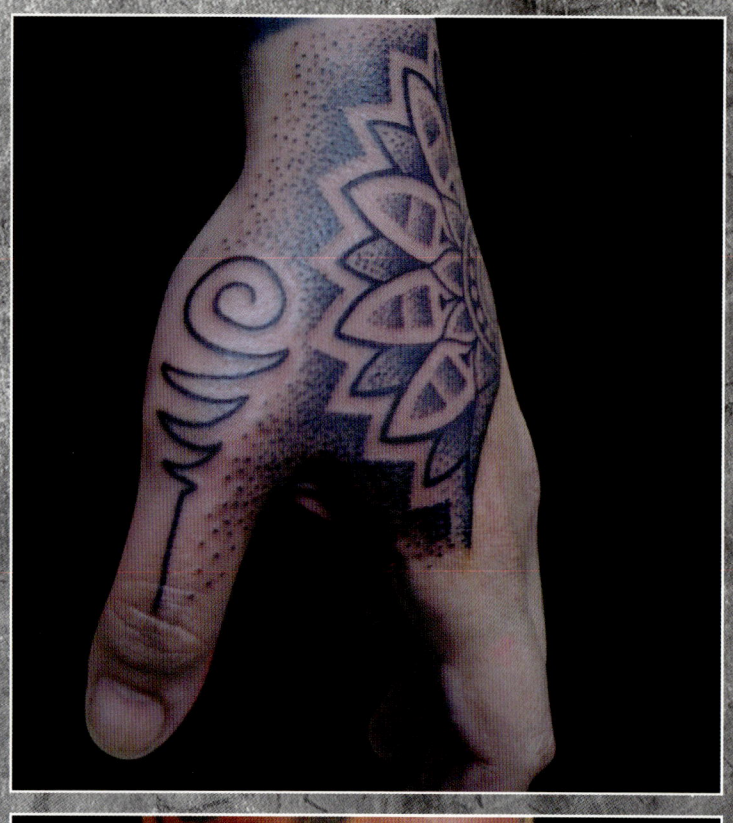

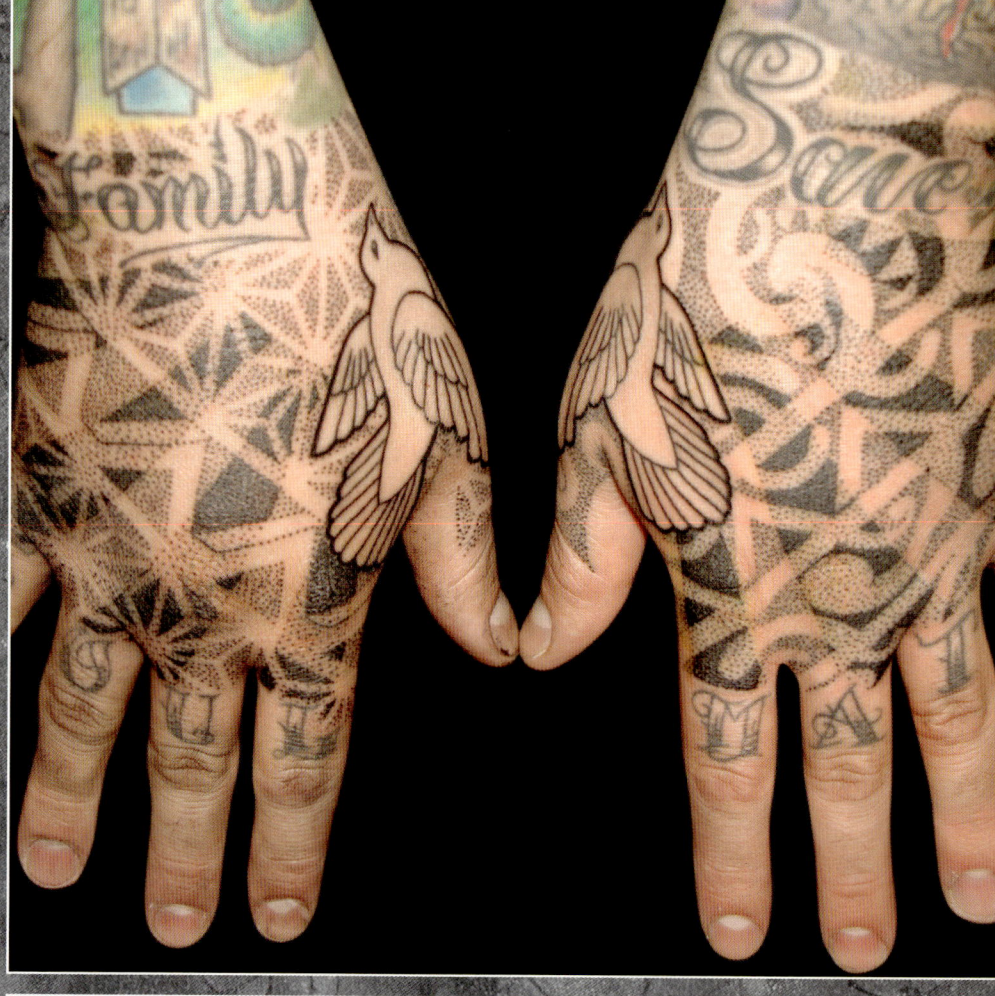

Cory Ferguson

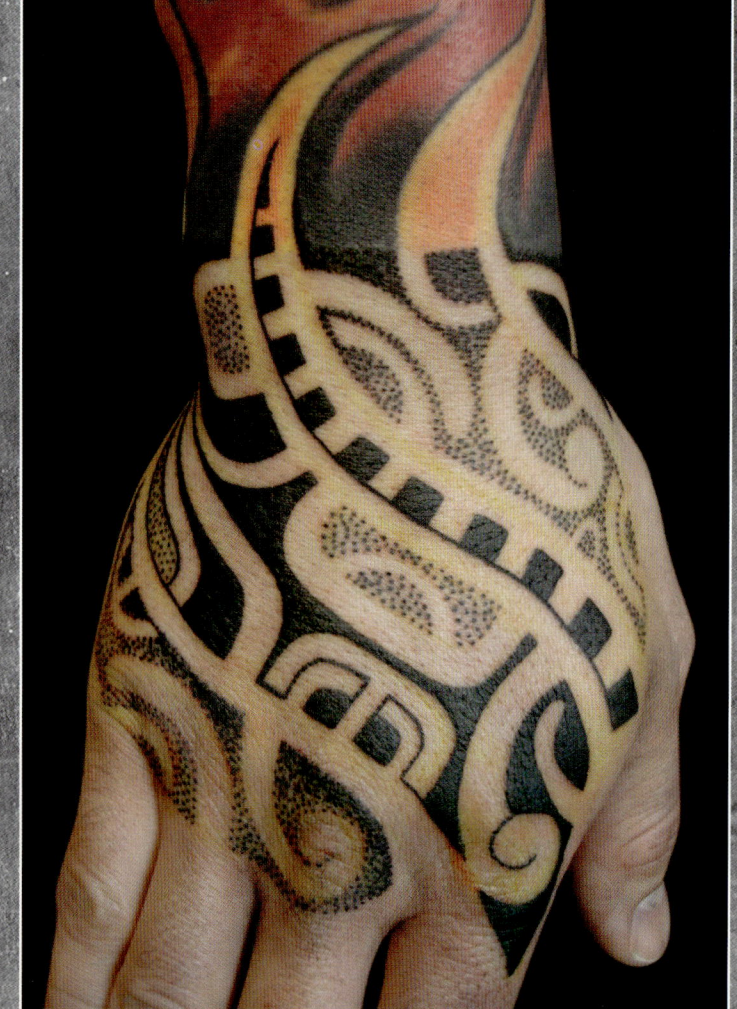

Cory Ferguson

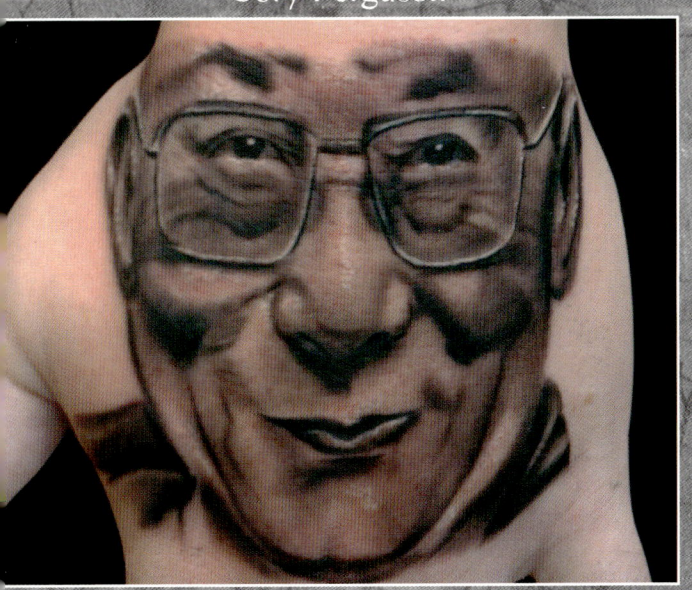

Colt Brown

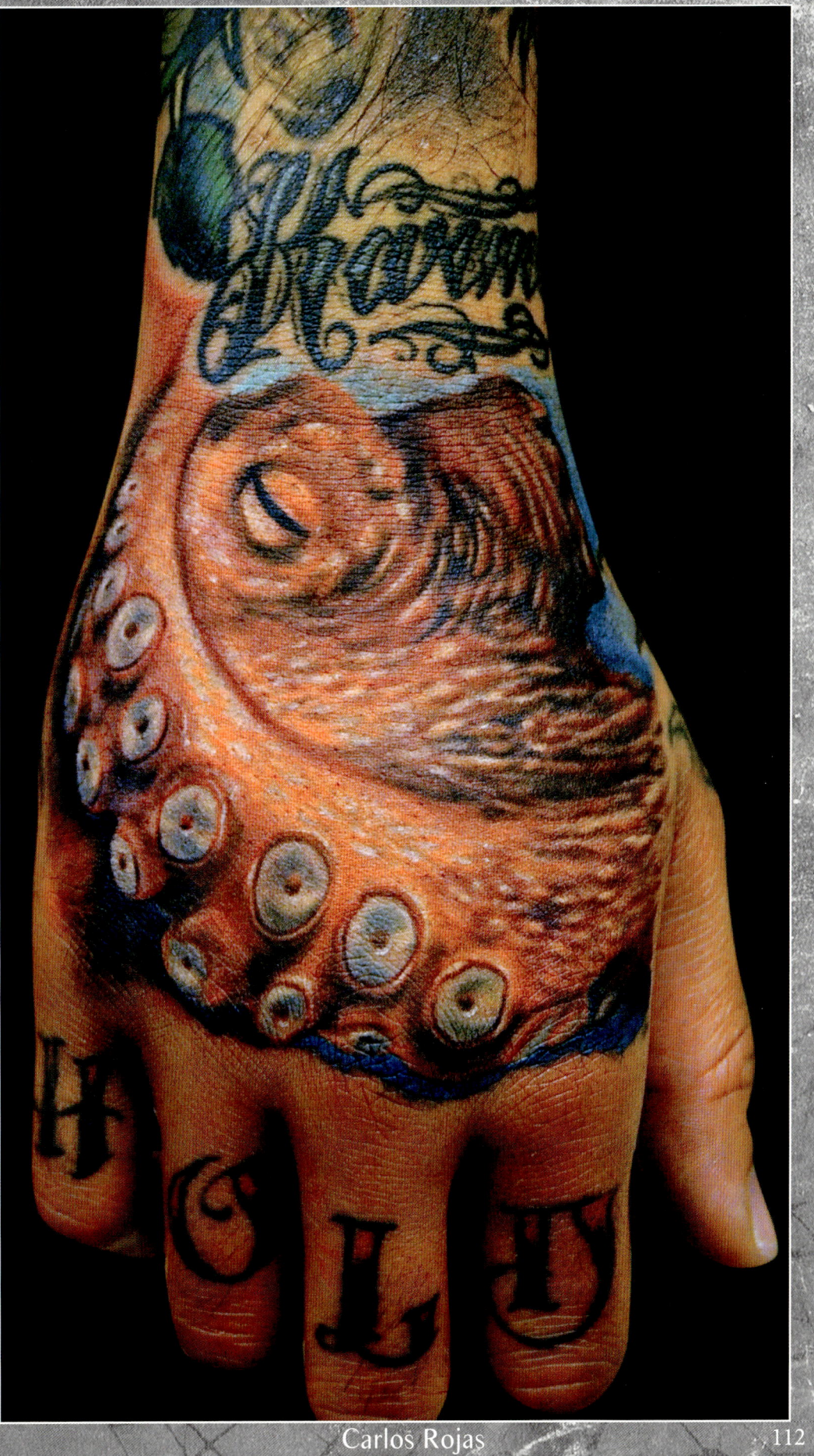

Carlos Rojas

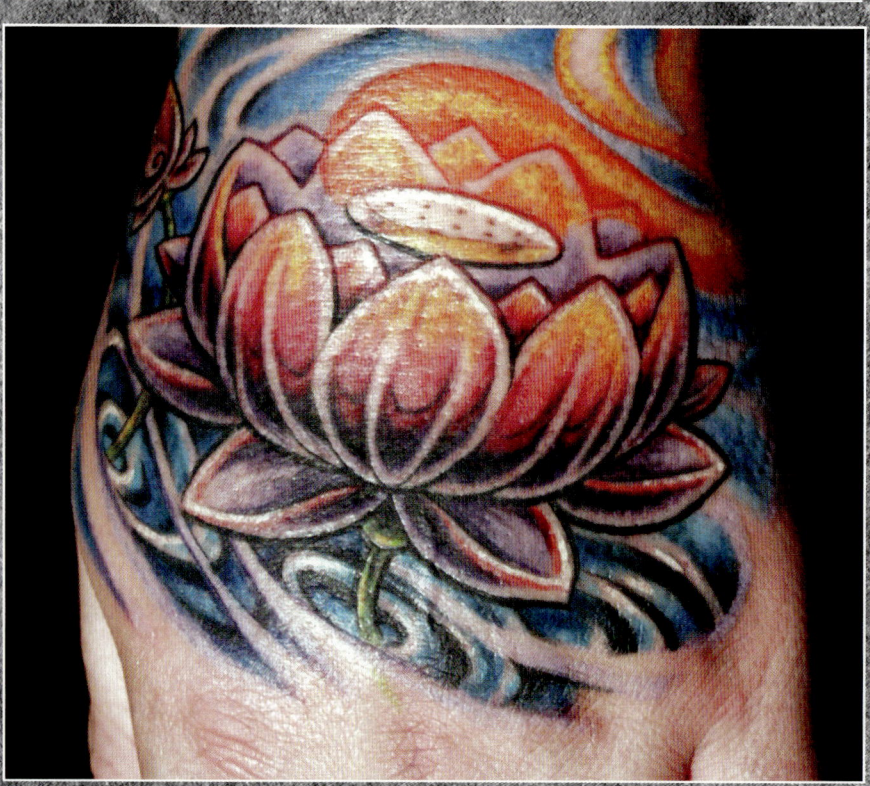

James Kern

James Kern

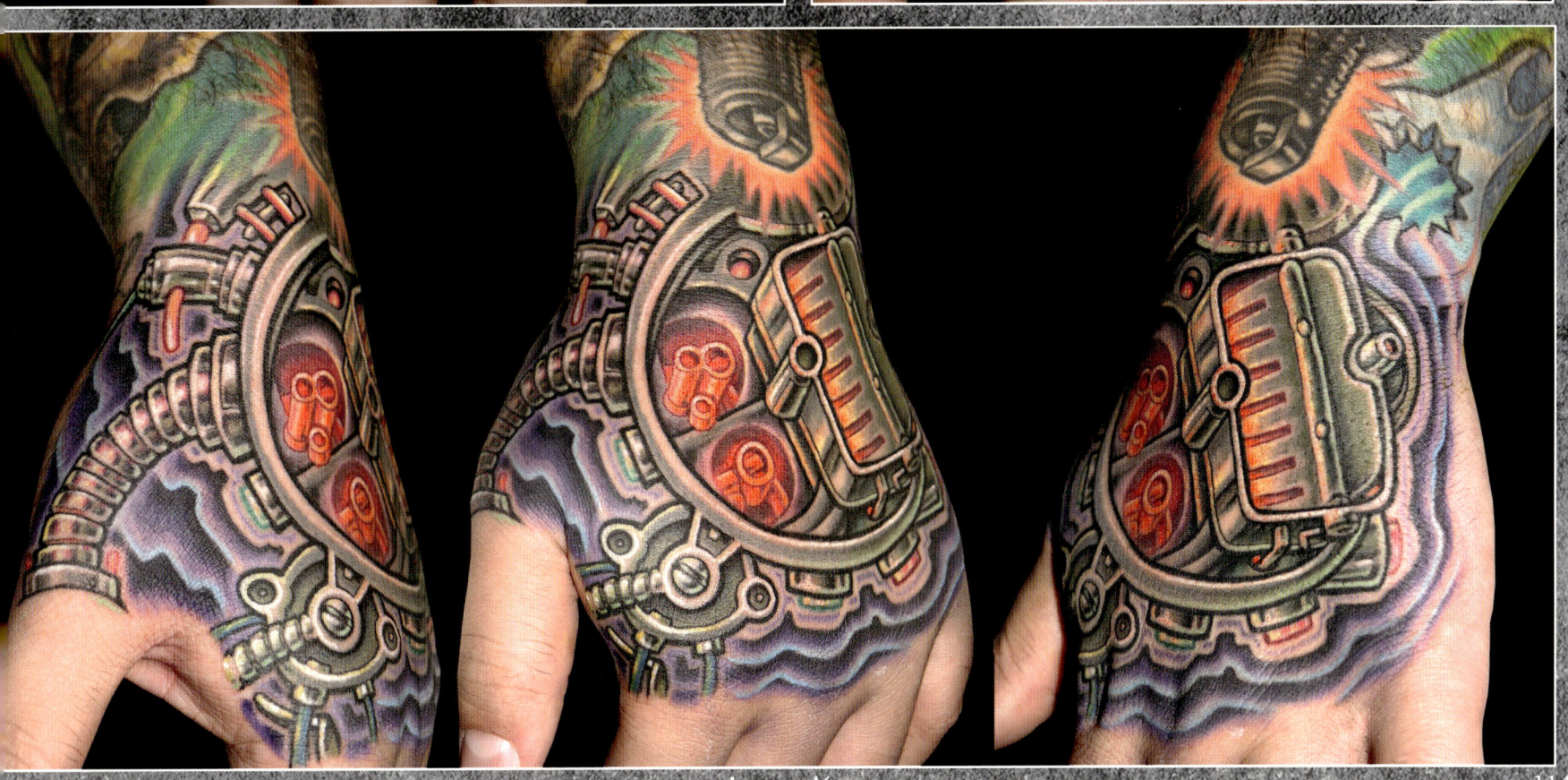

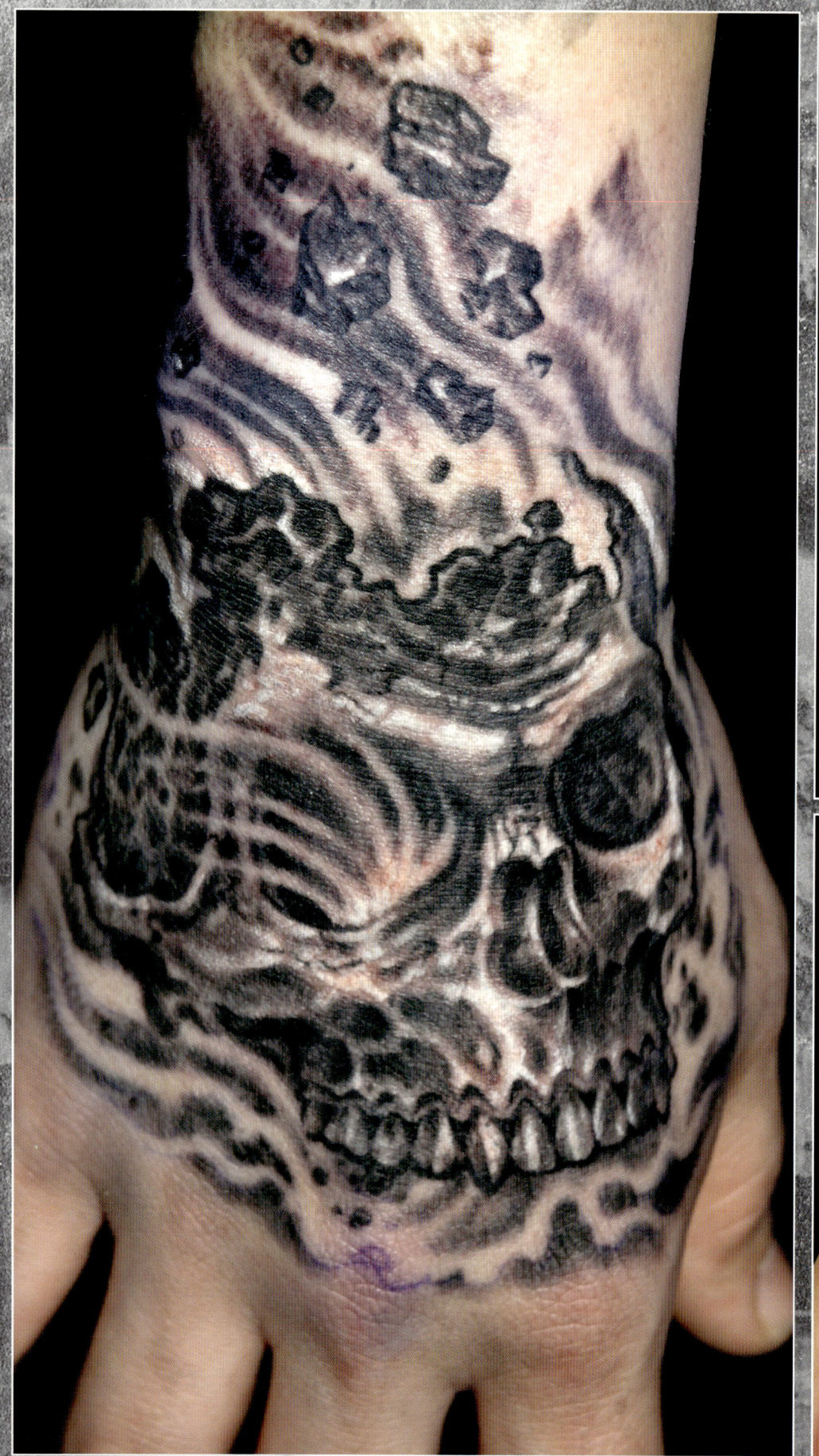

James Kern

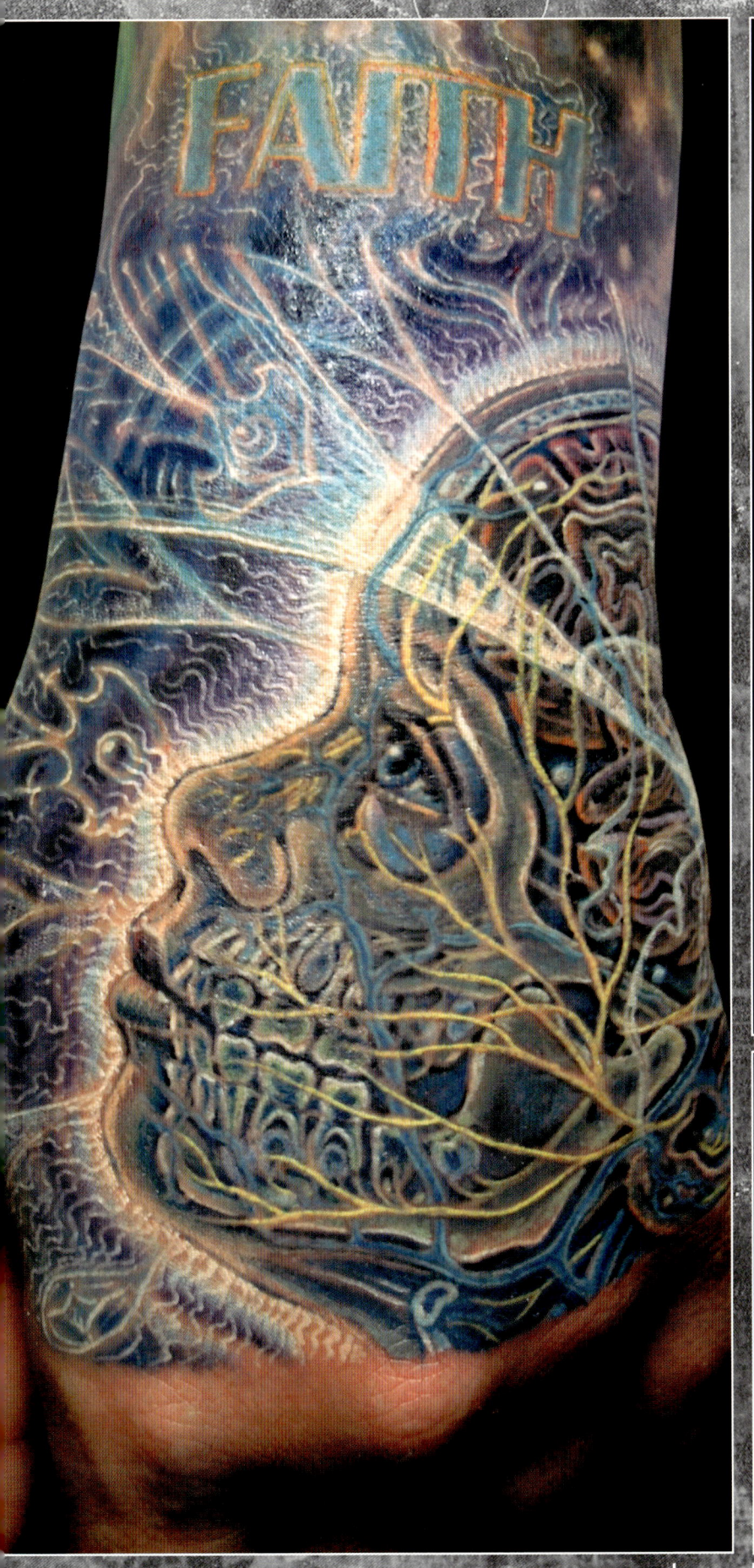

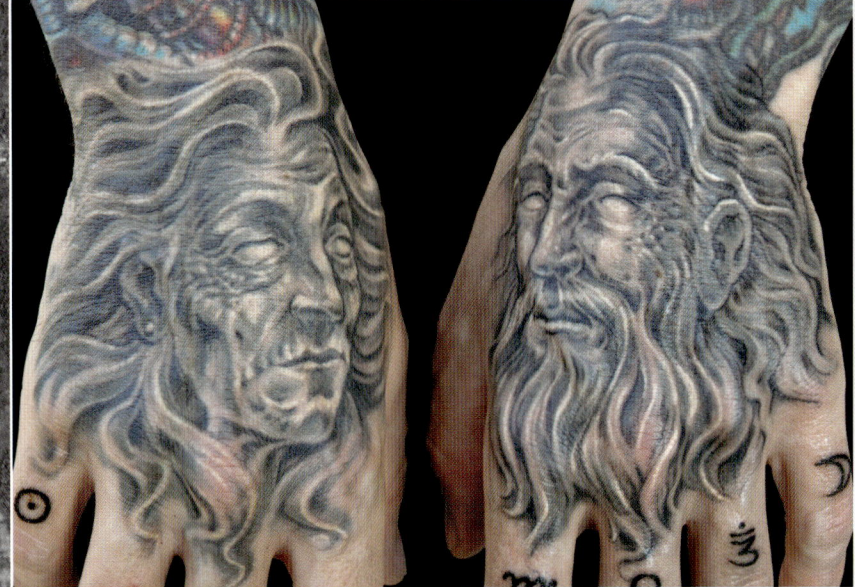

James Kern

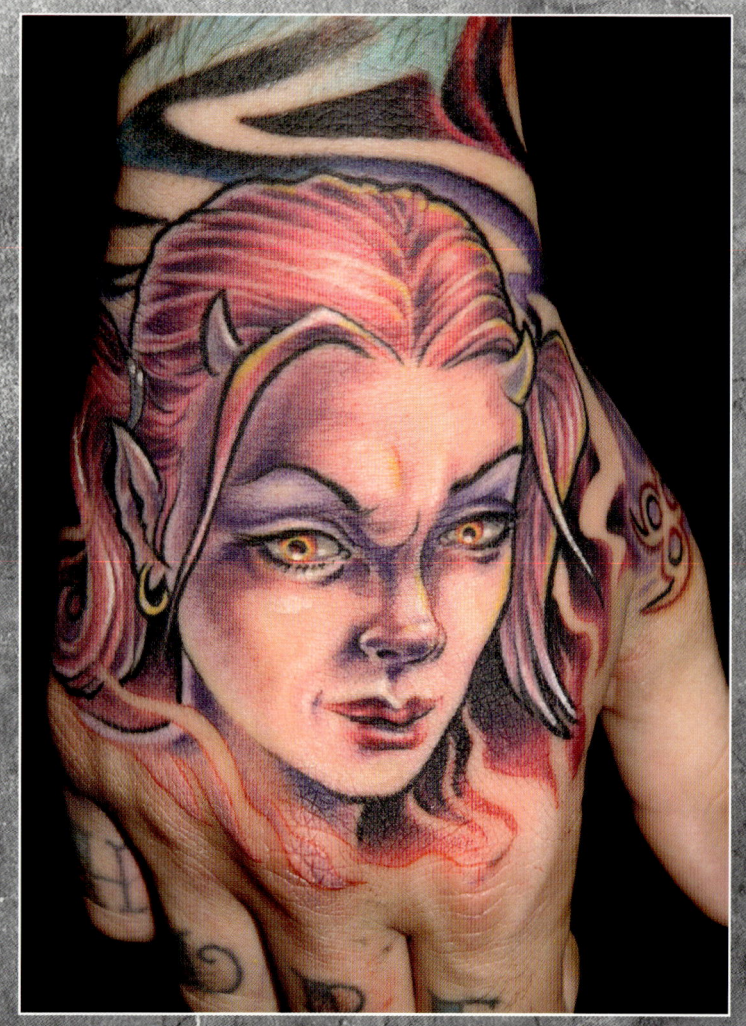

James Kern

Billy Weigler

Franco Vescovi

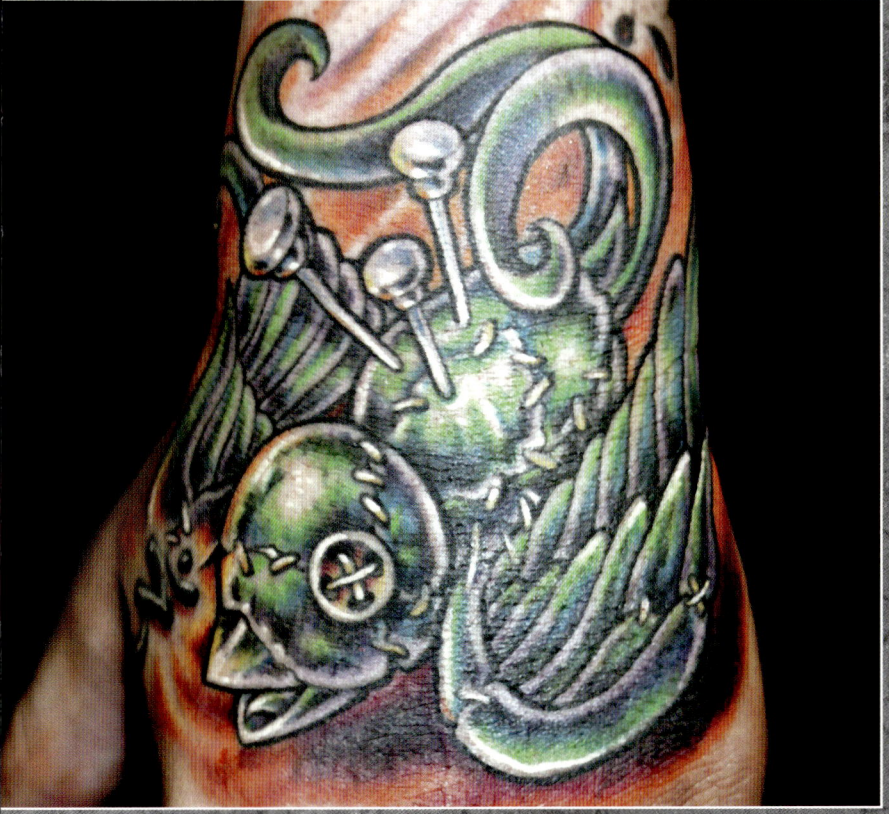

James Kern

Franco Vescovi

118

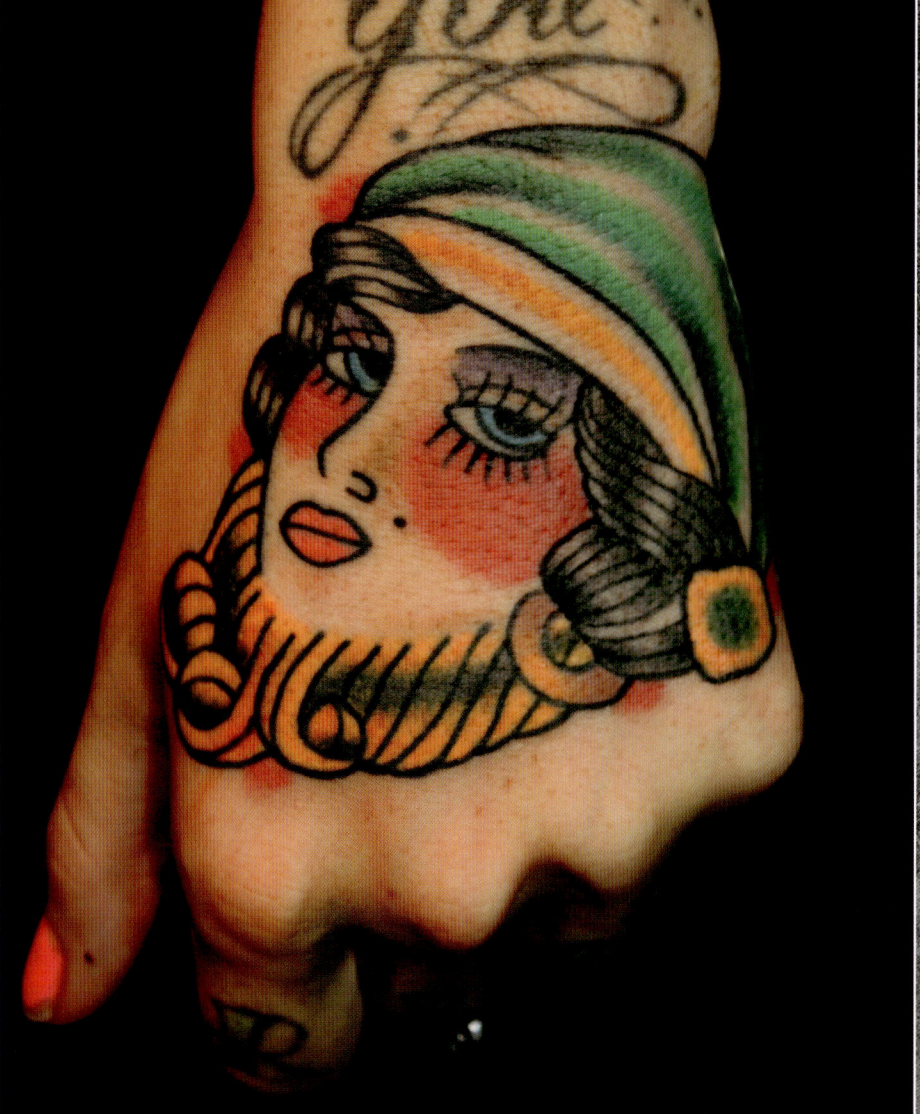

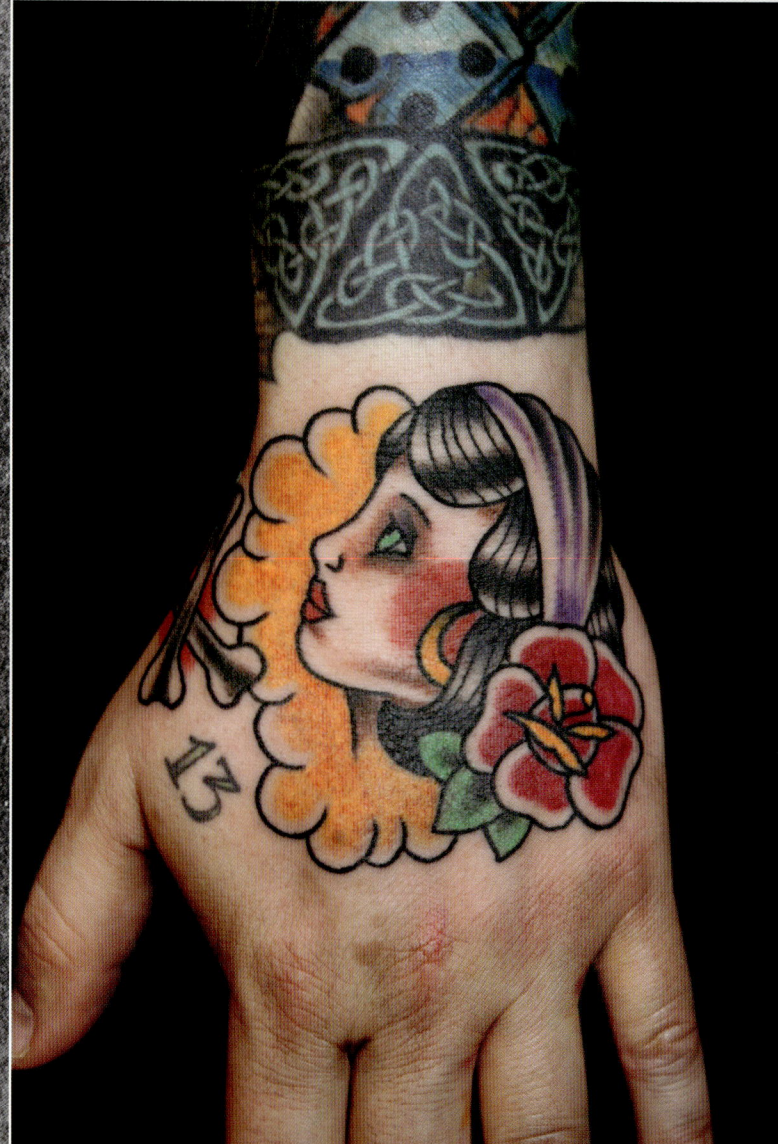

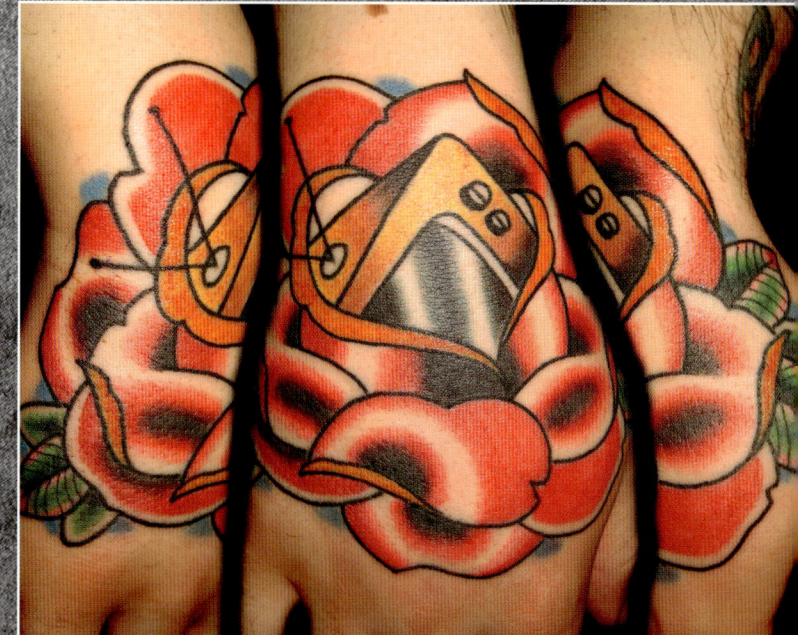

Myke Chambers

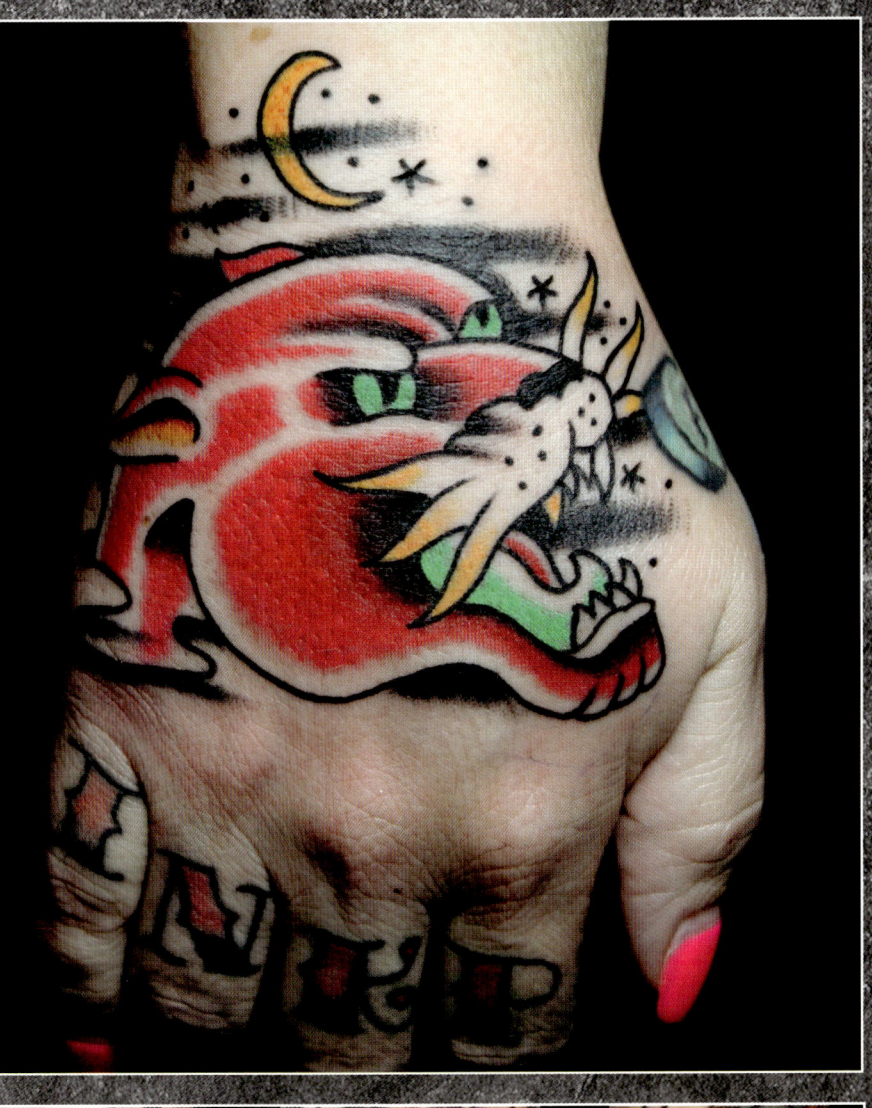

Myke Chambers

120

Sean Herman

Myke Chambers

Sean Herman

Sean Herman

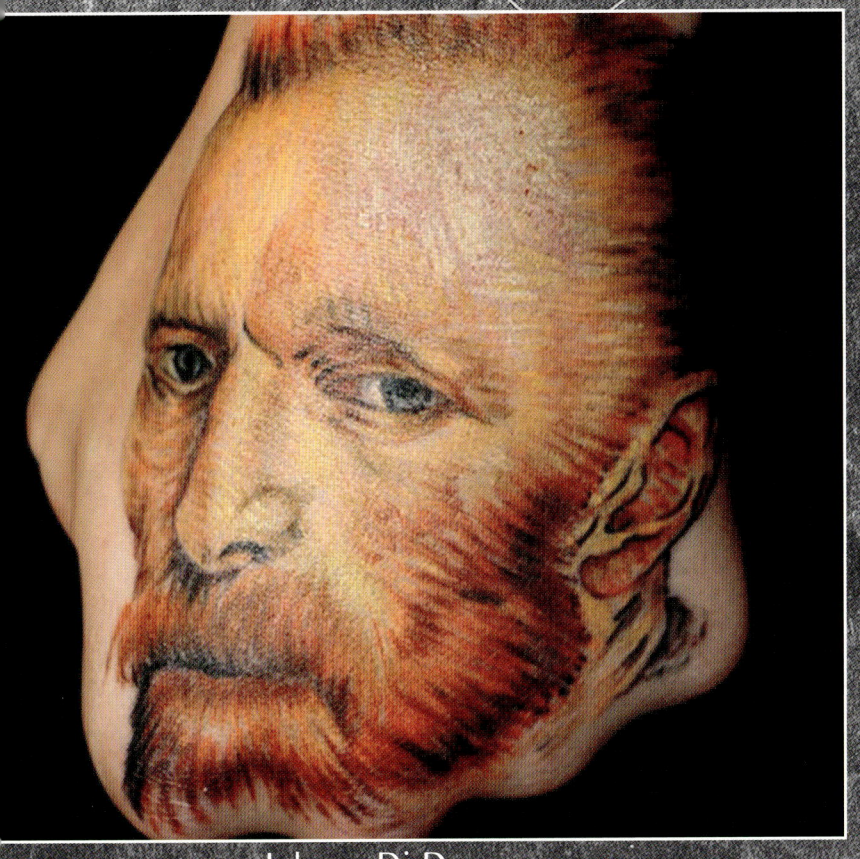

Johnny Di Donna

Jeff Croci

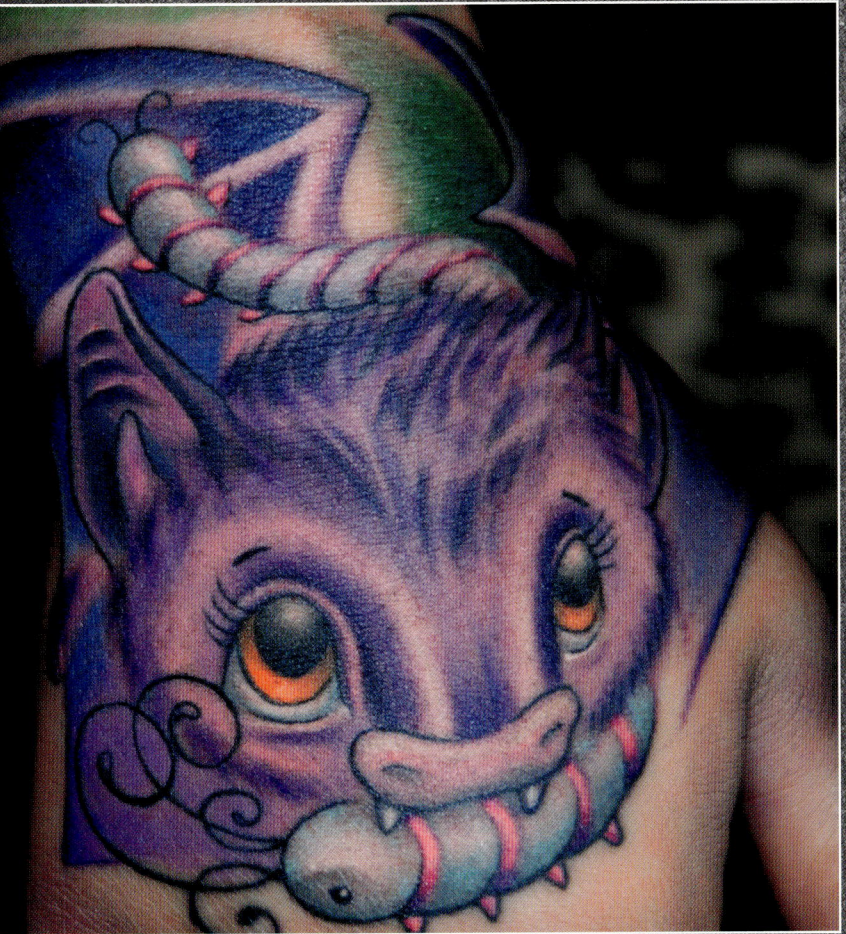

Sweet Laraine

Tutti Serra

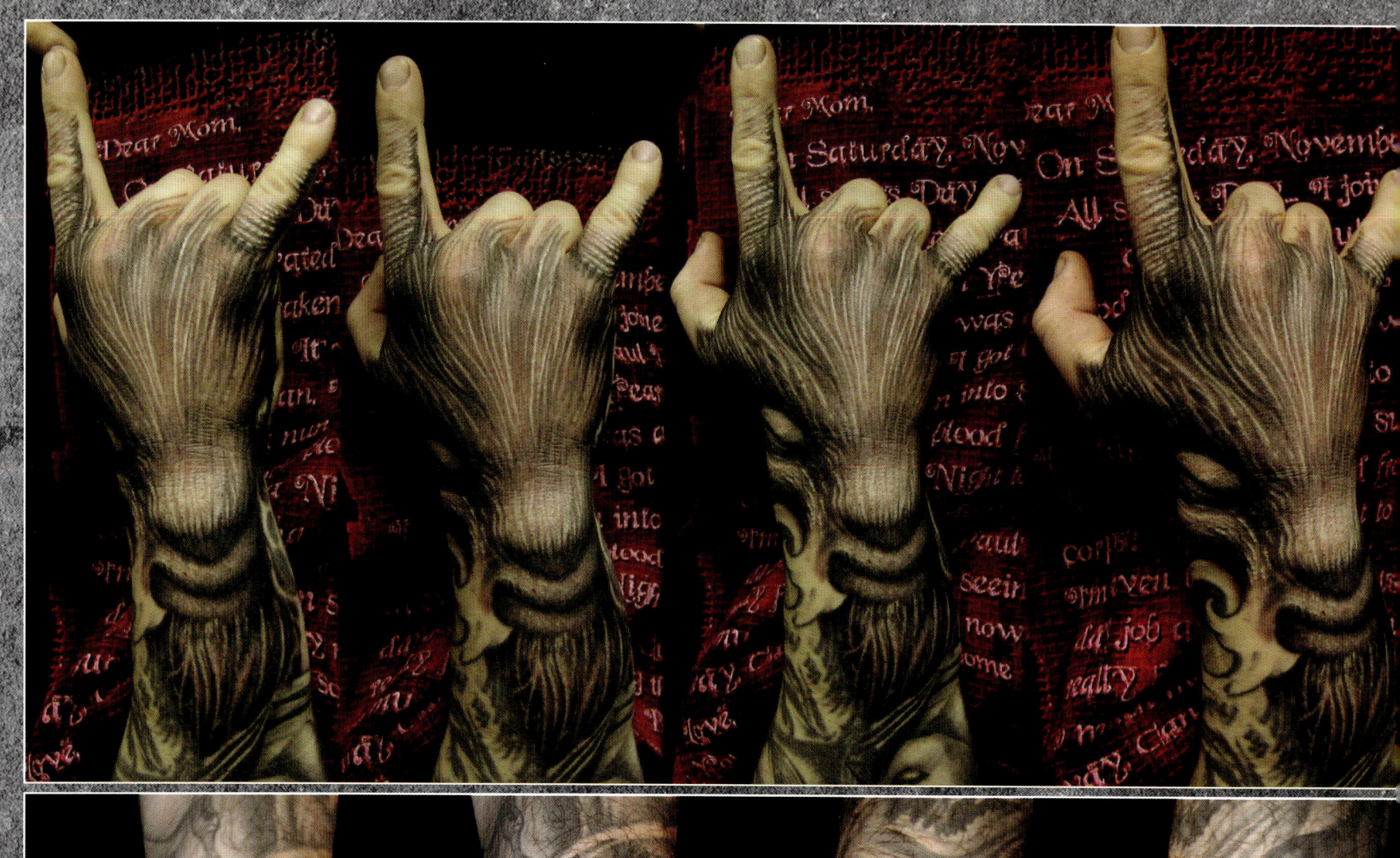

Paul Booth

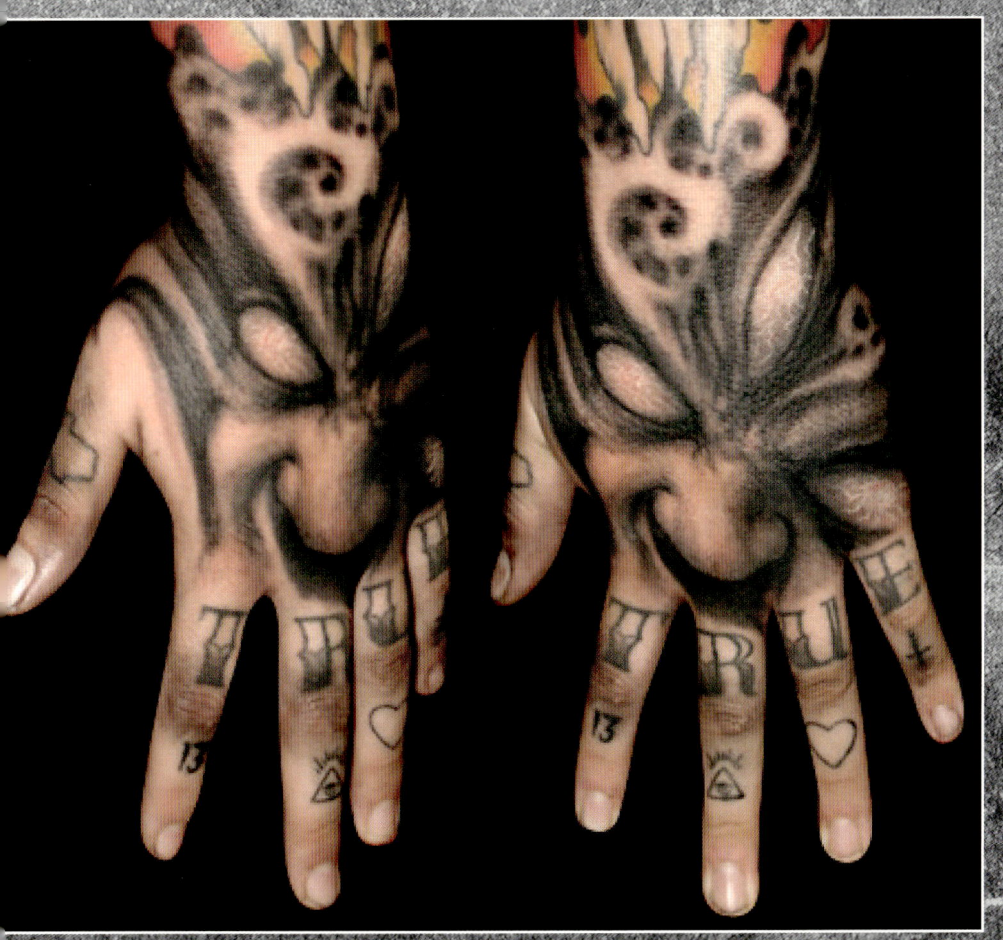

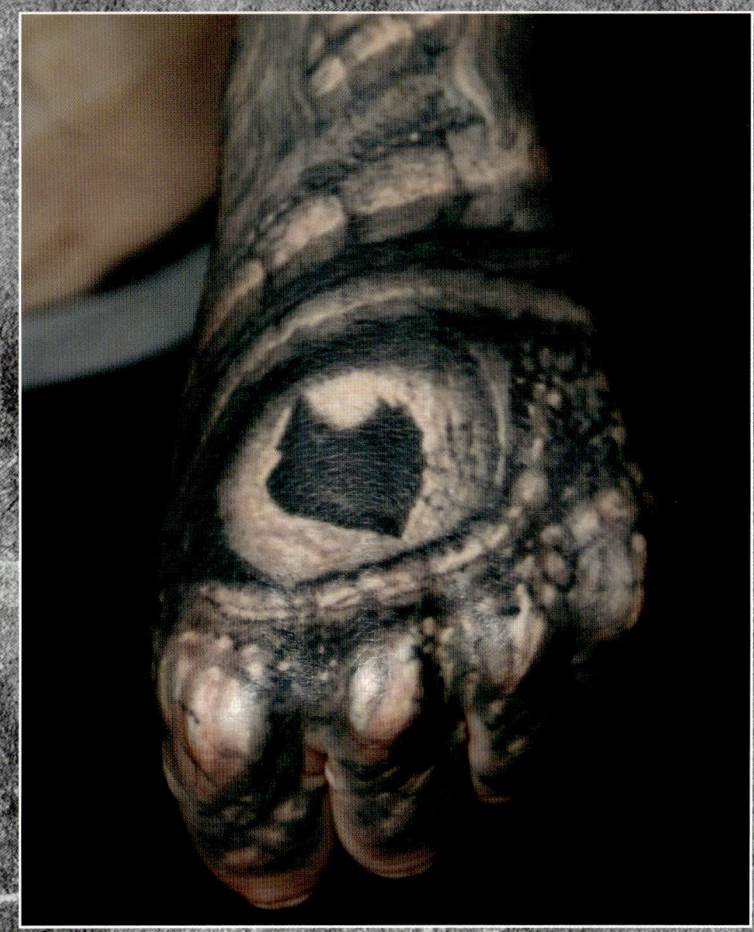

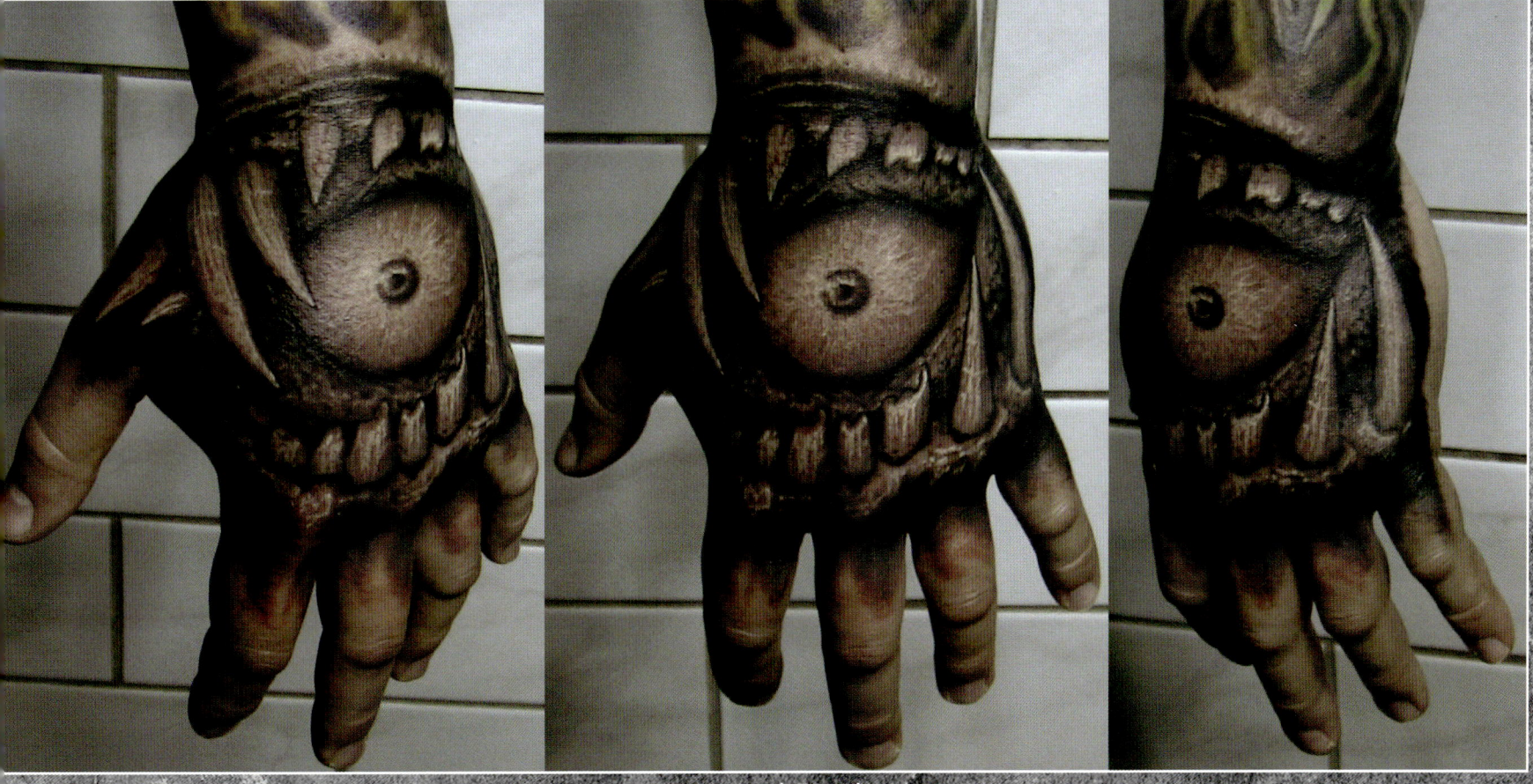

Paul Booth

Jeremy Miller

Mick Squires

Mick Squires

Kyle Cotterman

Kyle Cotterman

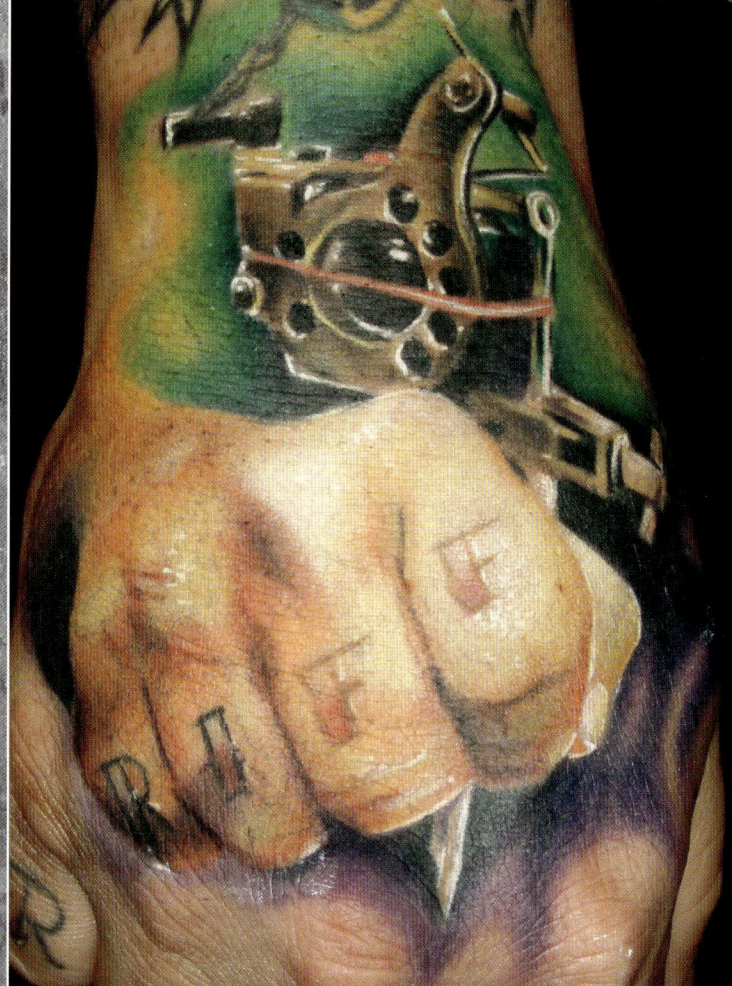

Kyle Cotterman

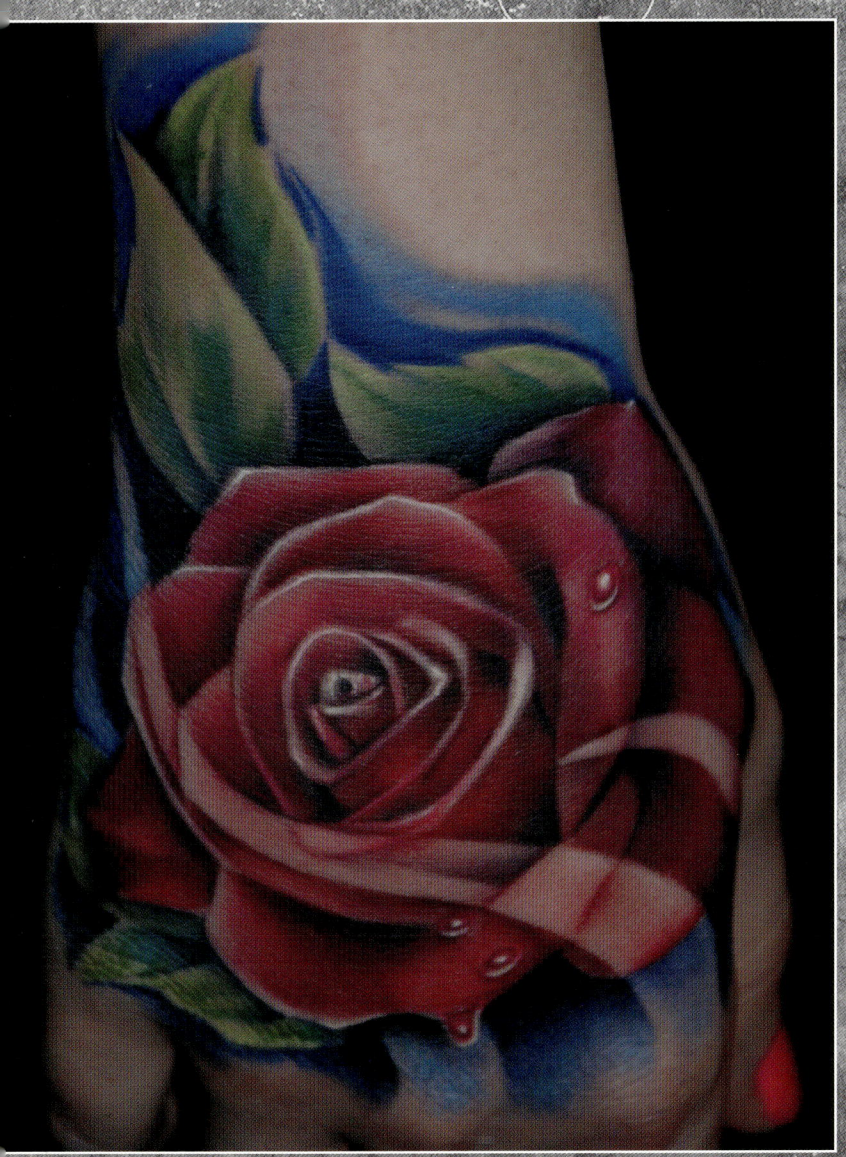

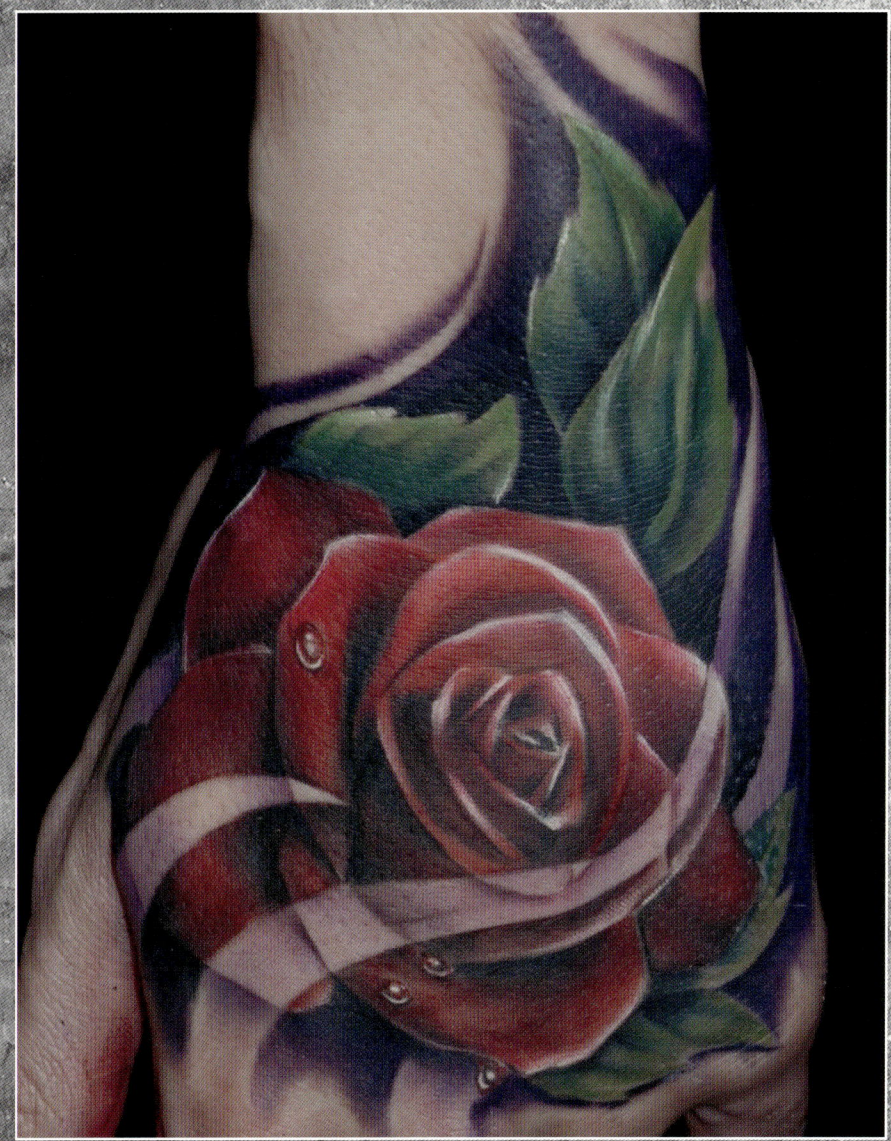

Kyle Cotterman

Stefano Alcantara

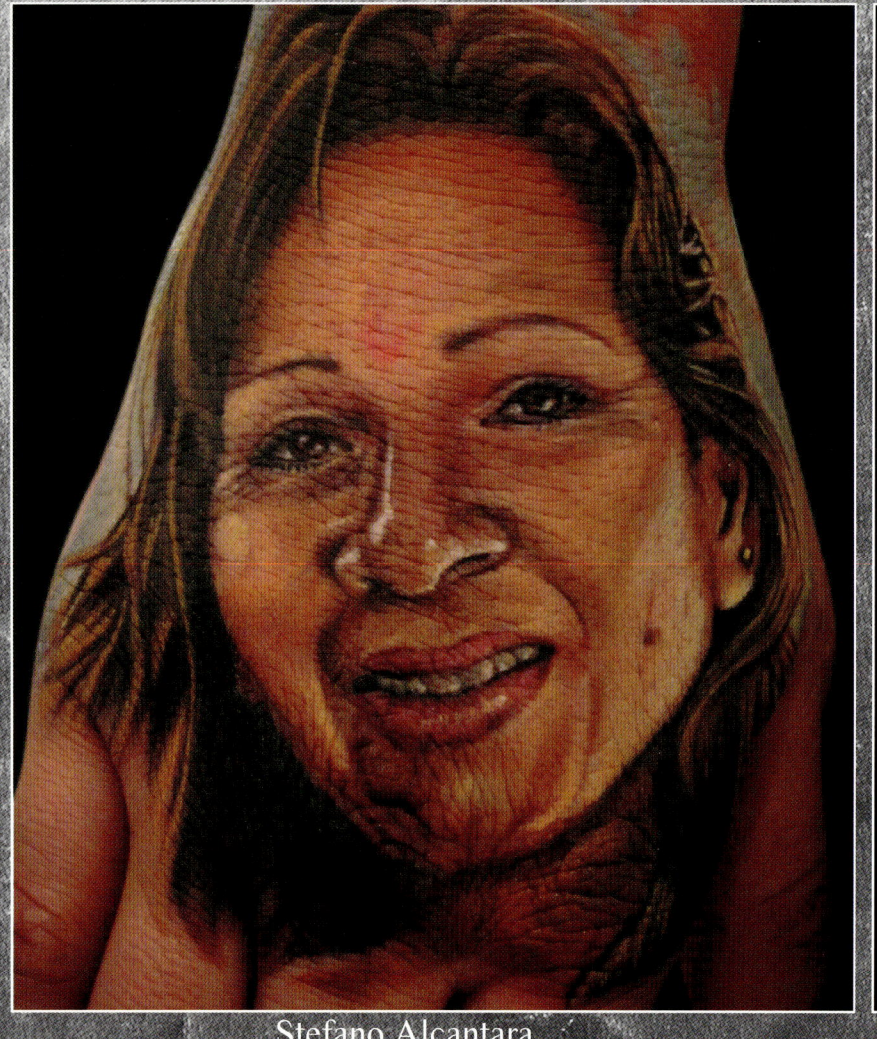

Stefano Alcantara

Stefano Alcantara

Krzysztof Wisniewski

Krzysztof Wisniewski

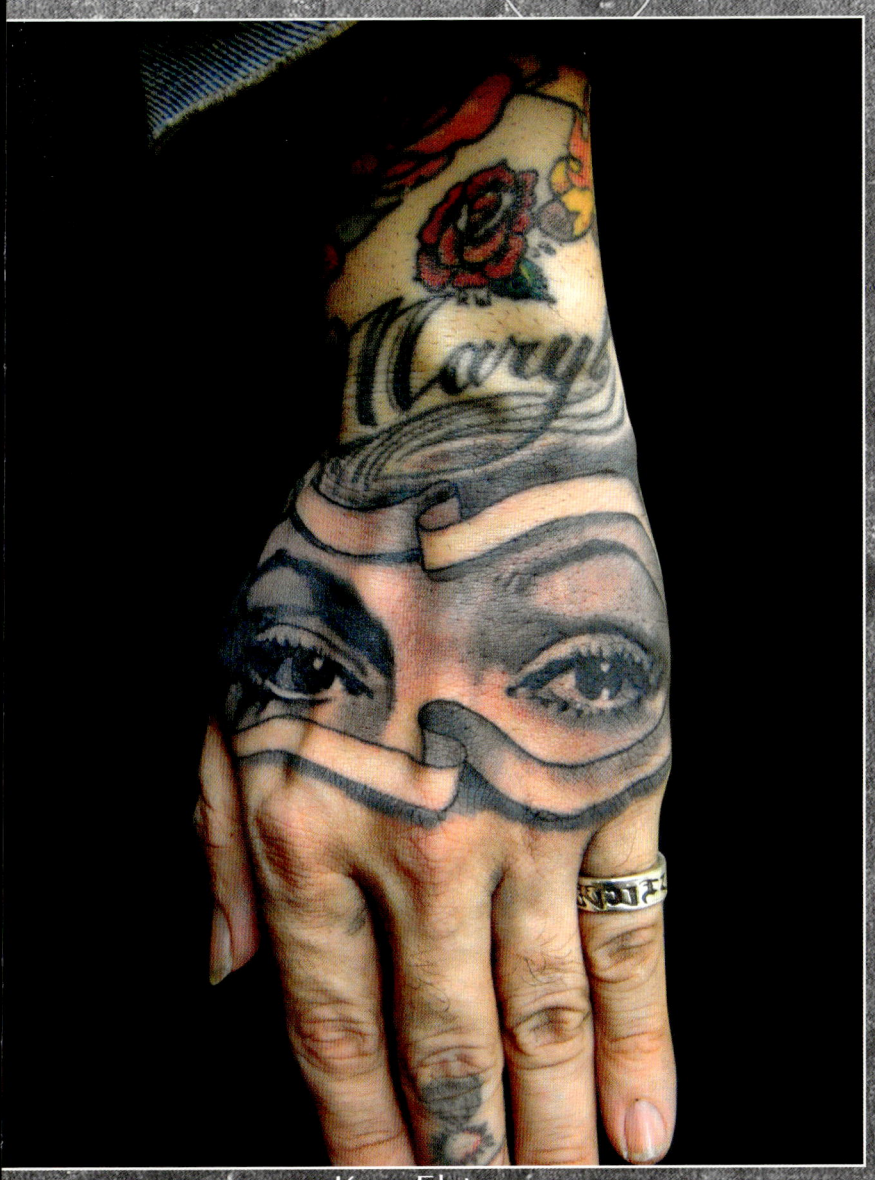

Kore Flatmo

Ben Stubbs

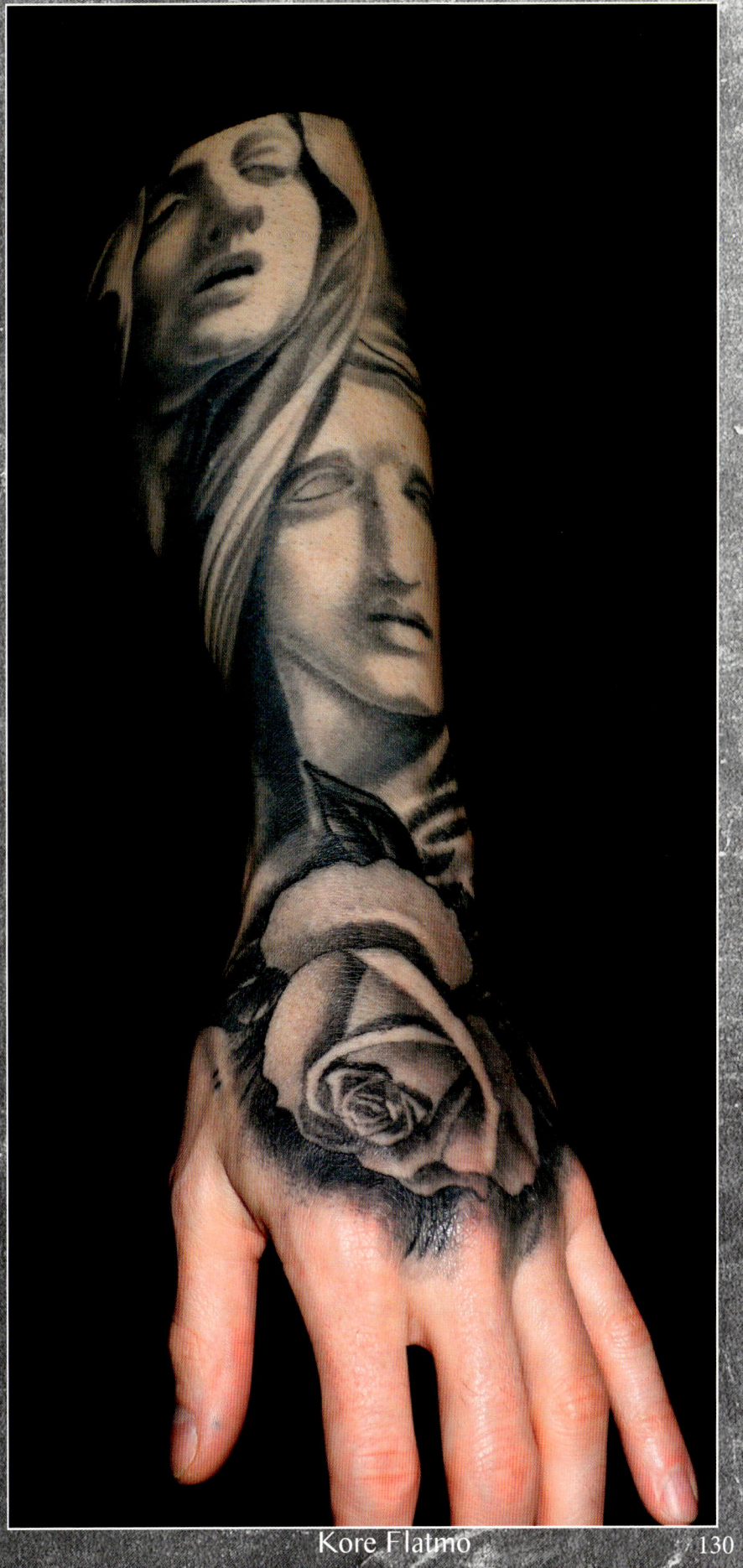

Kore Flatmo

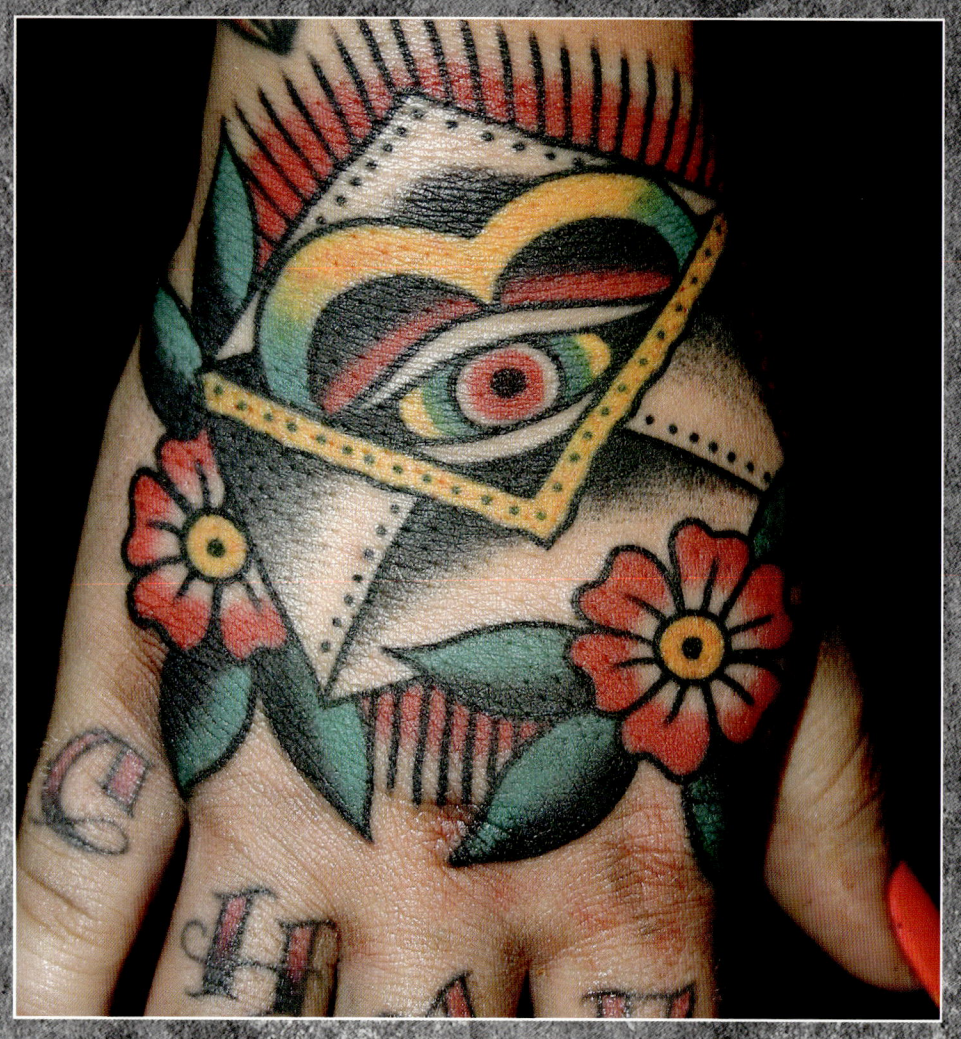

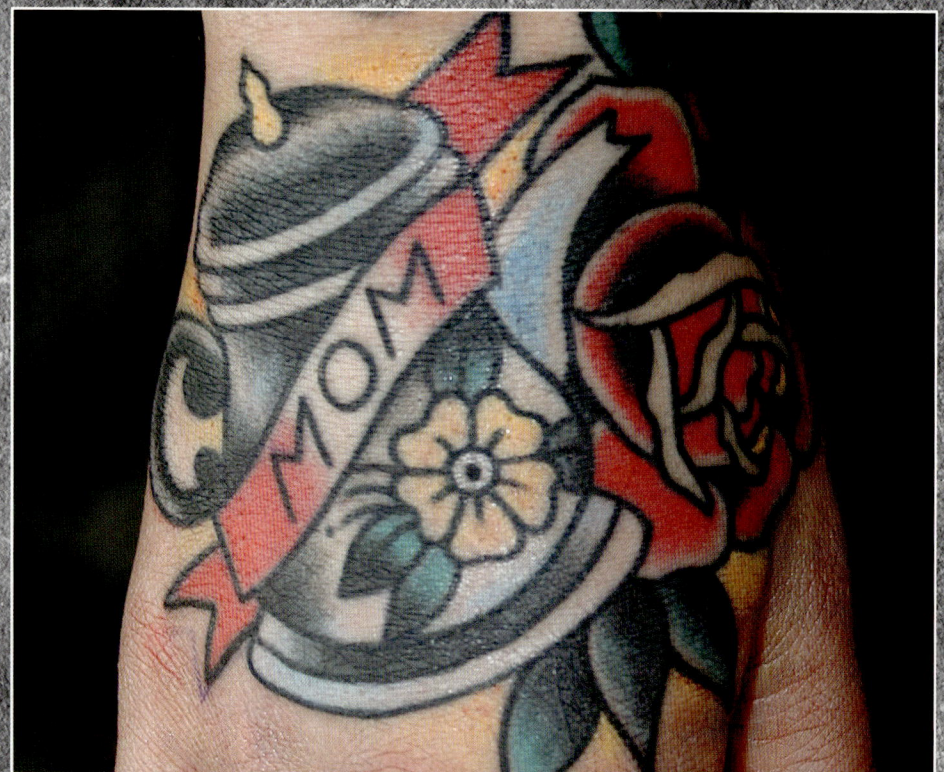

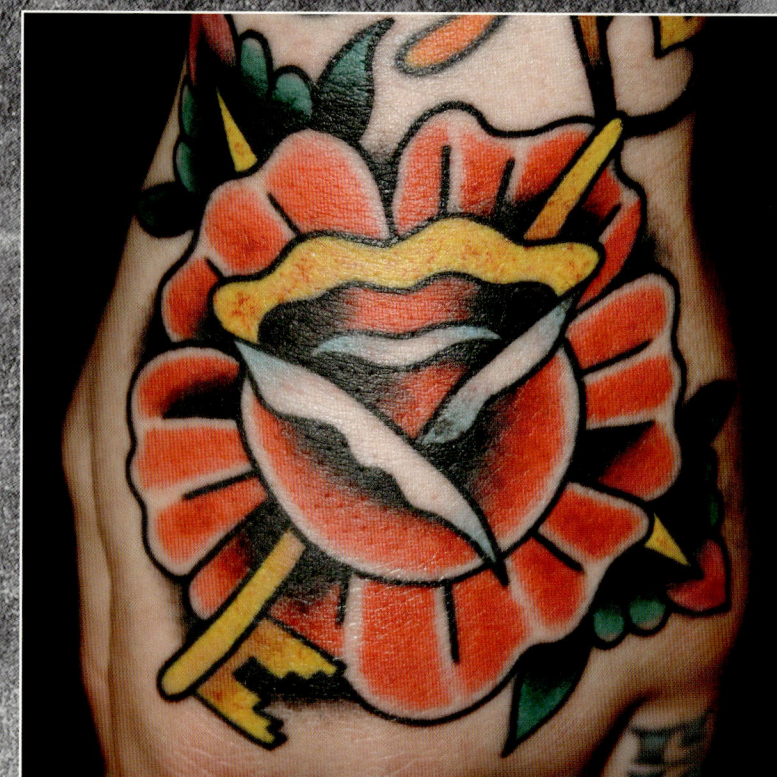

Steve Byrne

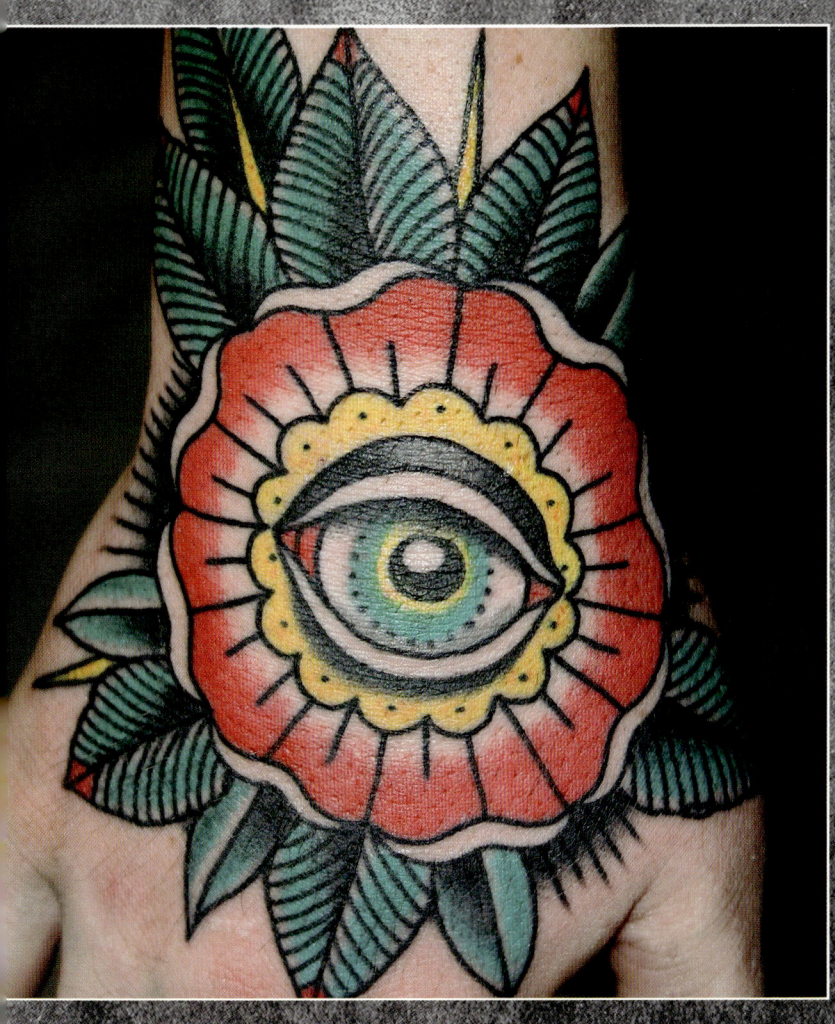

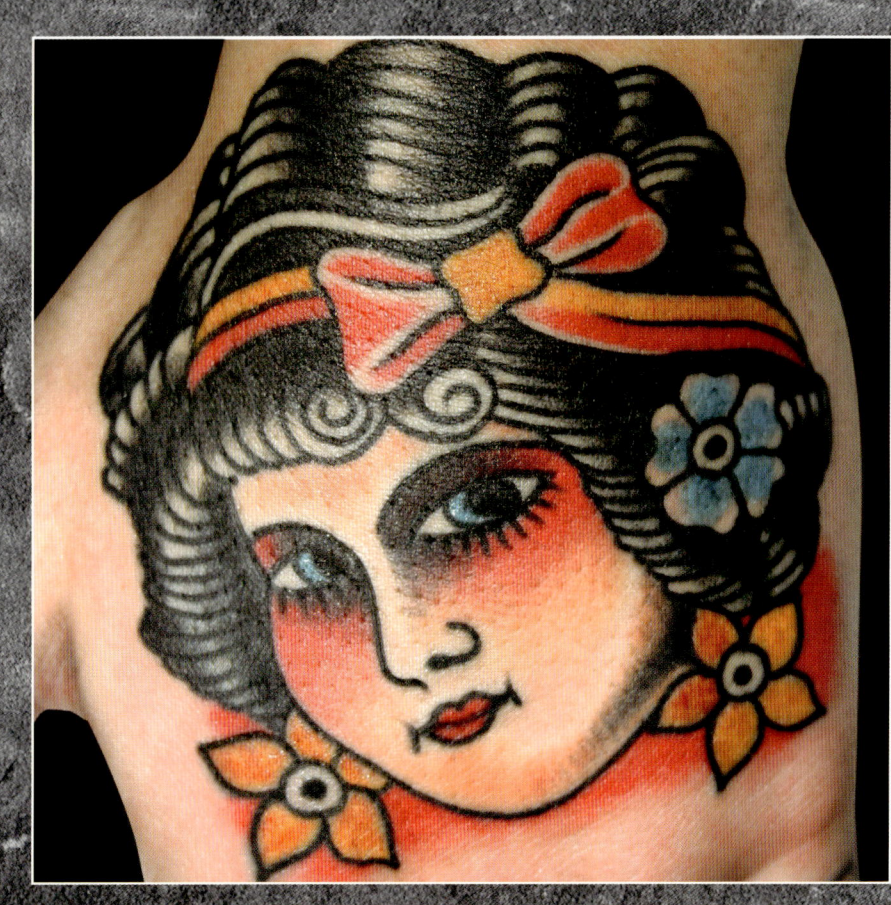

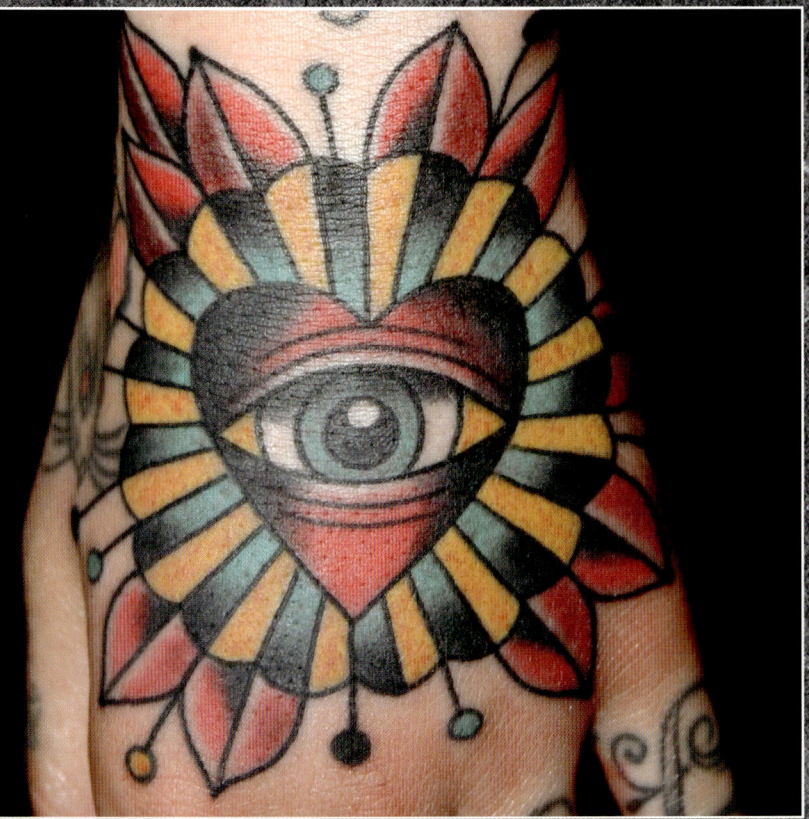

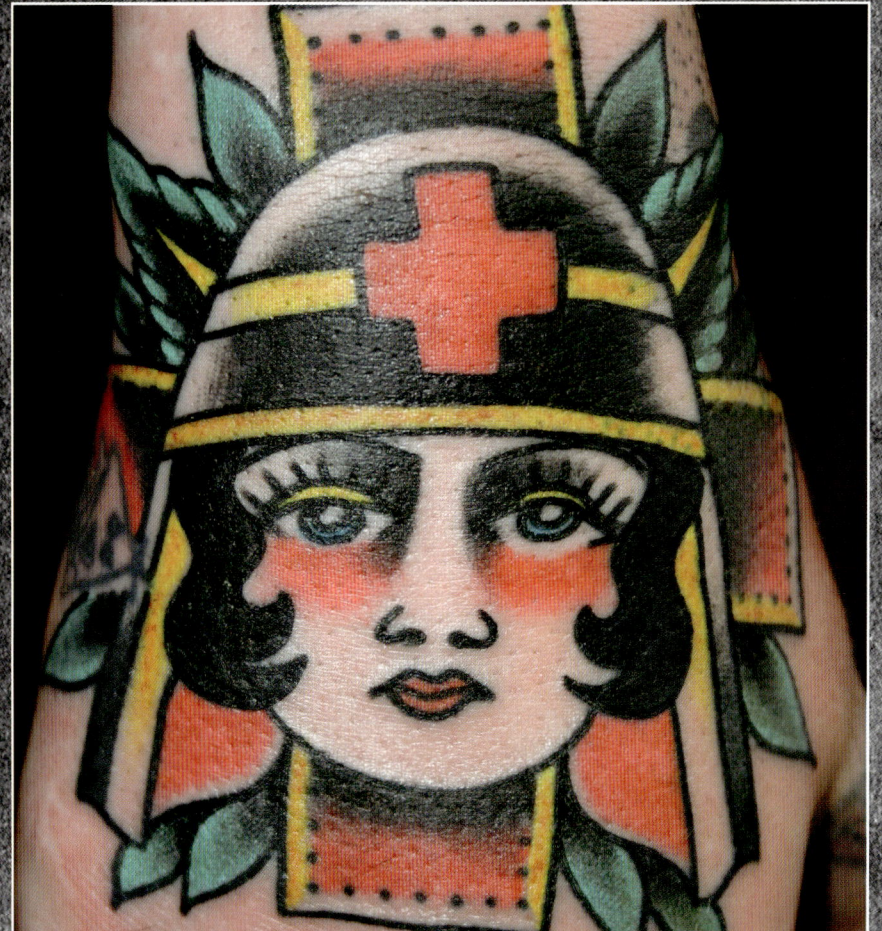

Steve Byrne

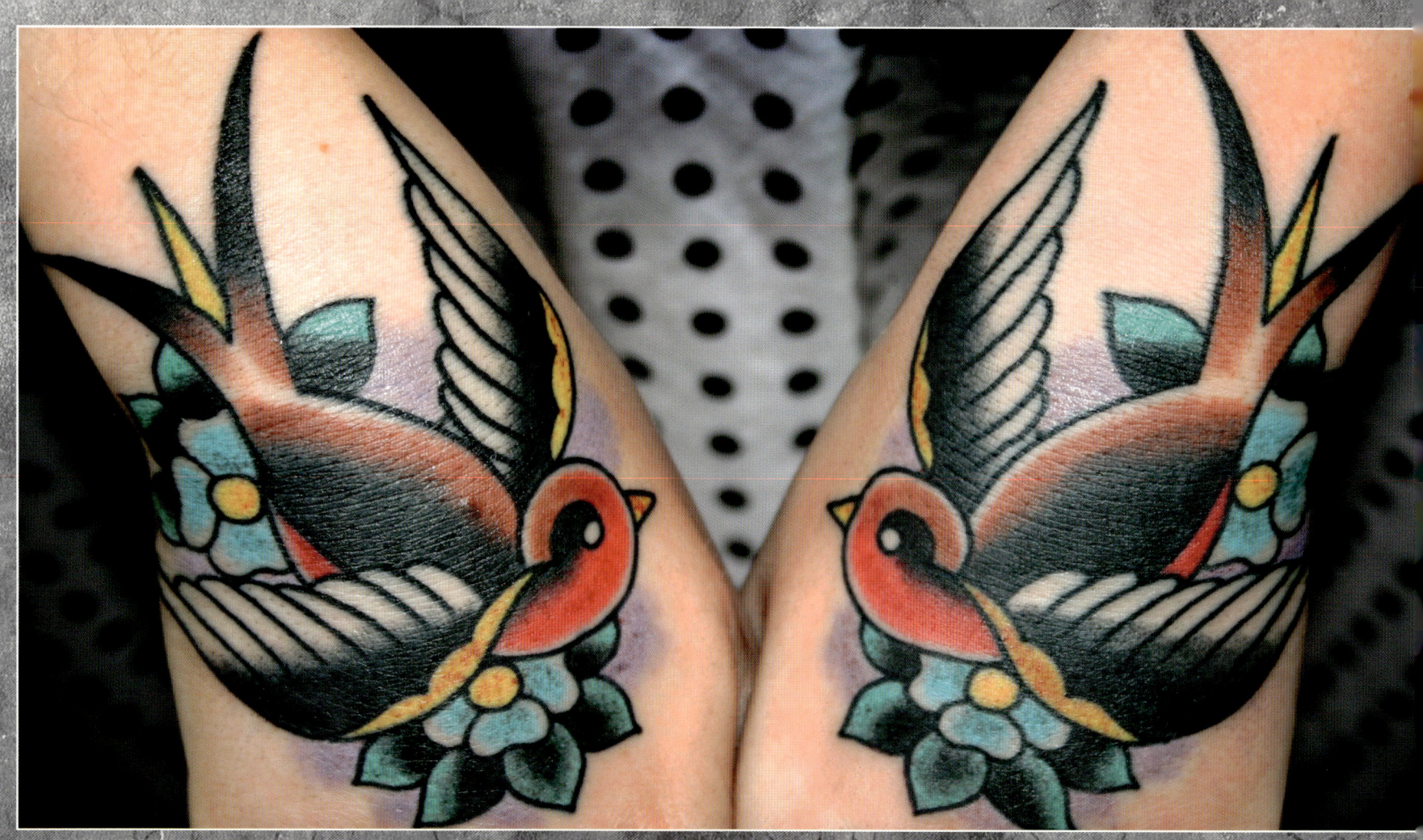

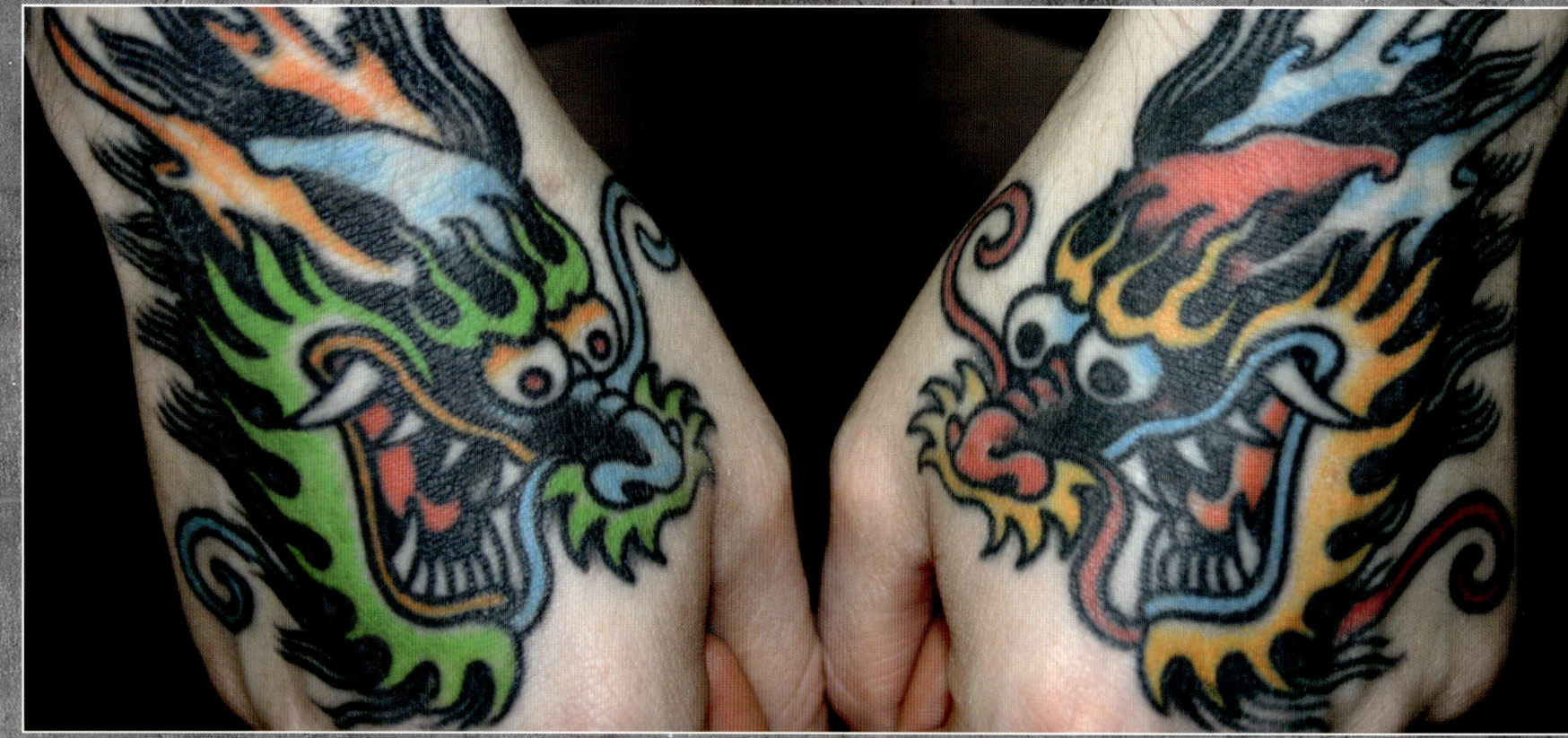

Steve Byrne

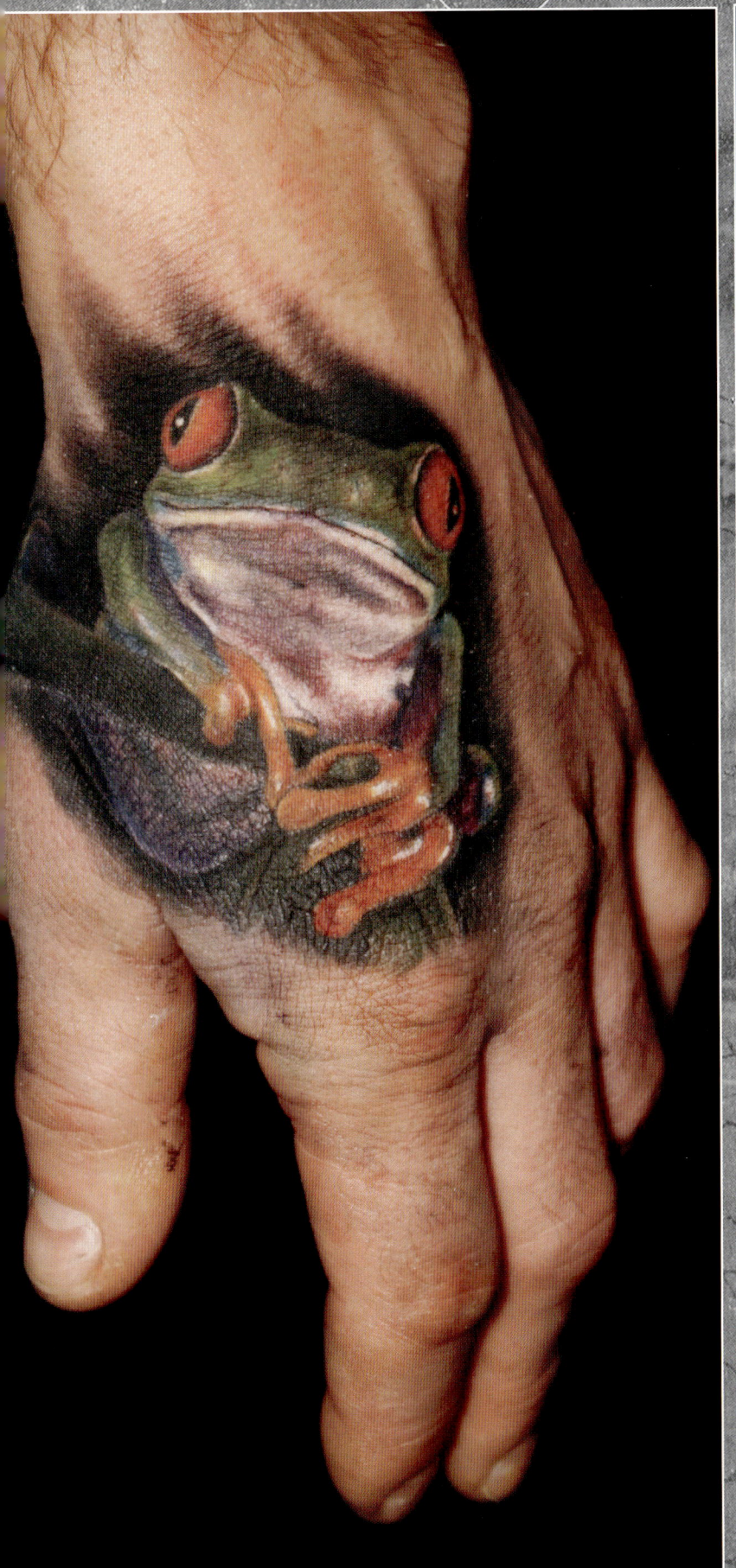

Petri Syrjälä

Steve Byrne

Steve Byrne

134

Steve Byrne

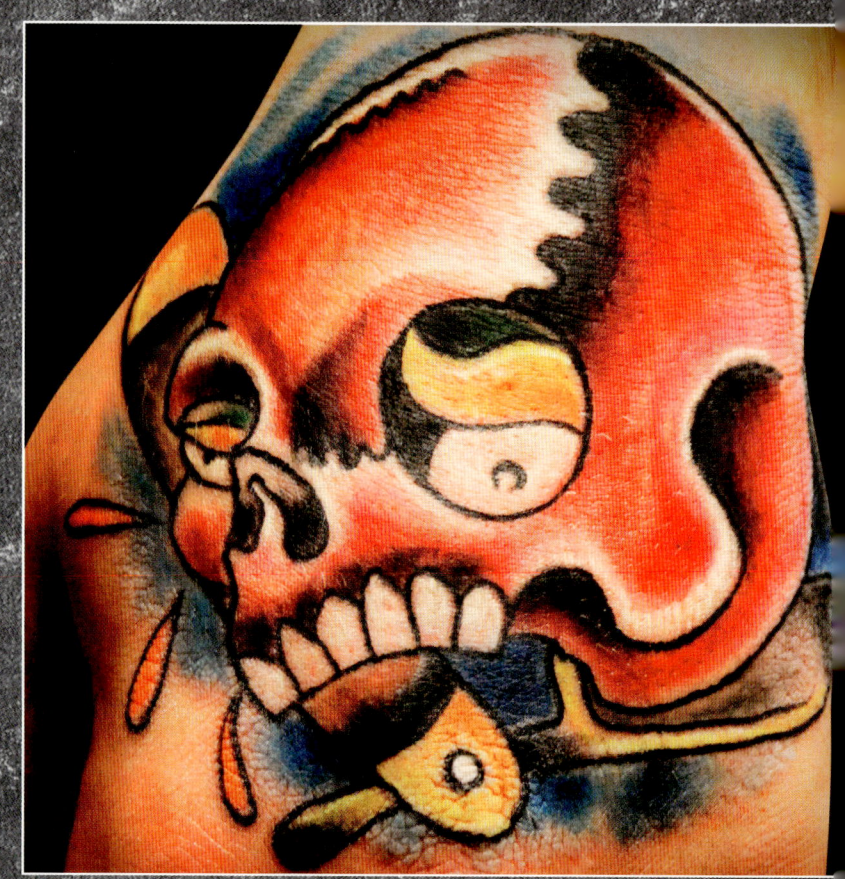

Johnny Jackson

Johnny Jackson

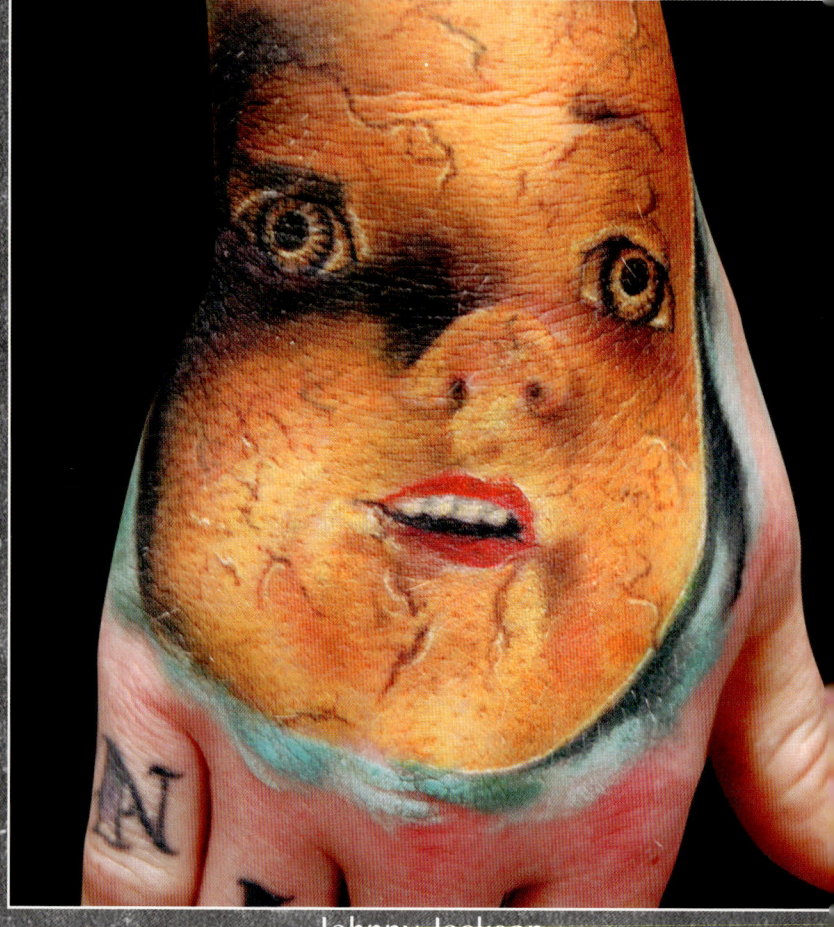

Johnny Jackson

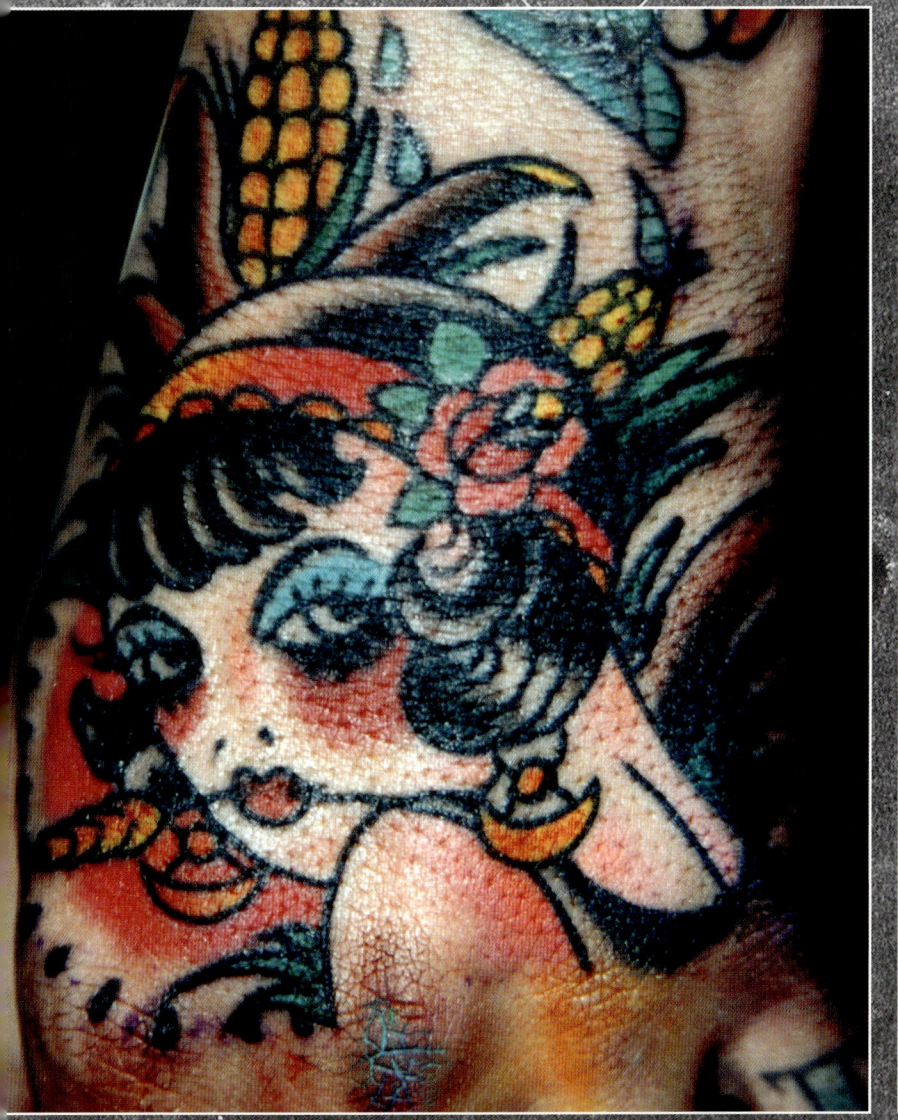

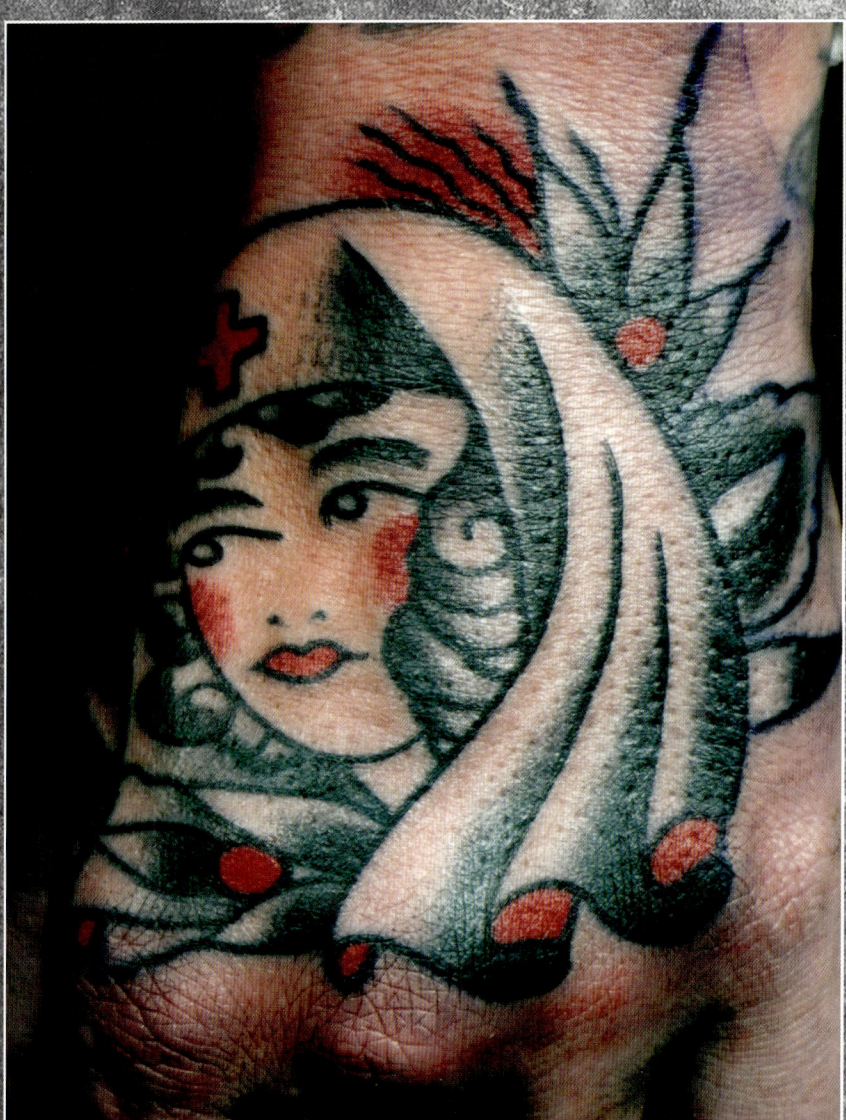

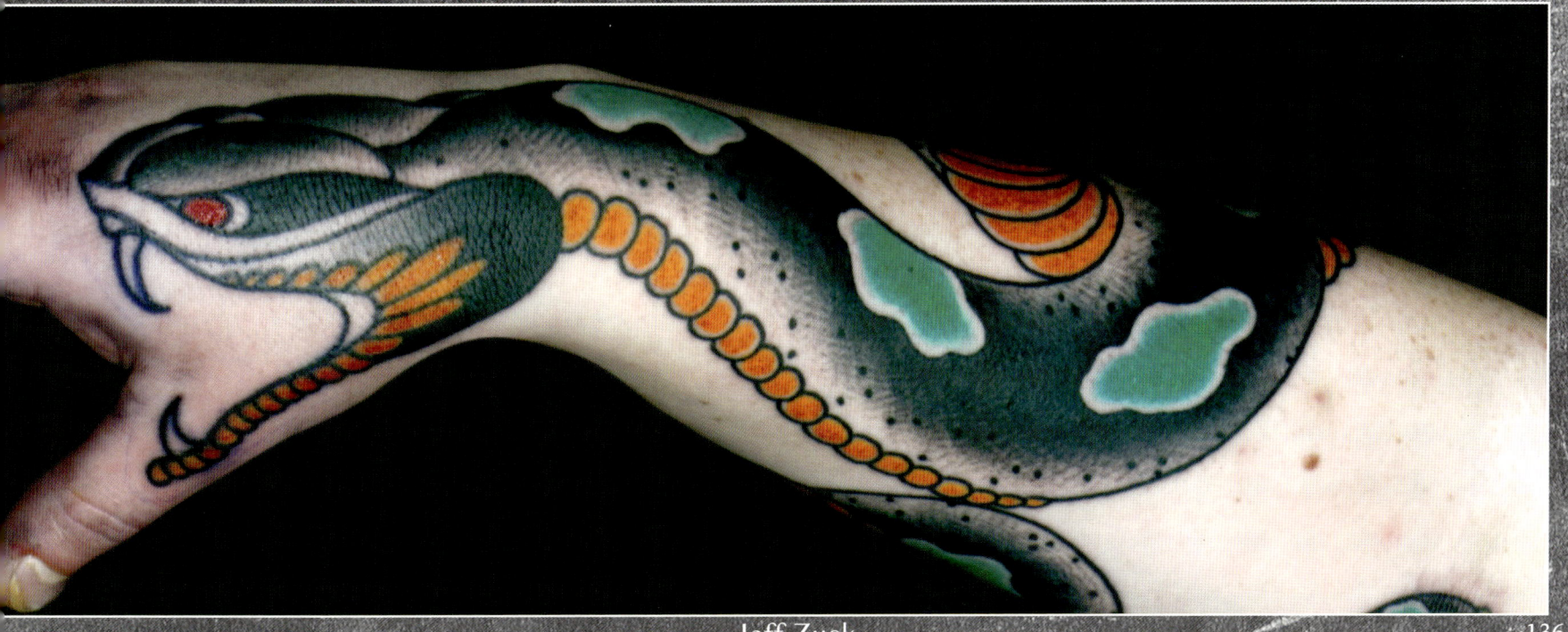

Jeff Zuck

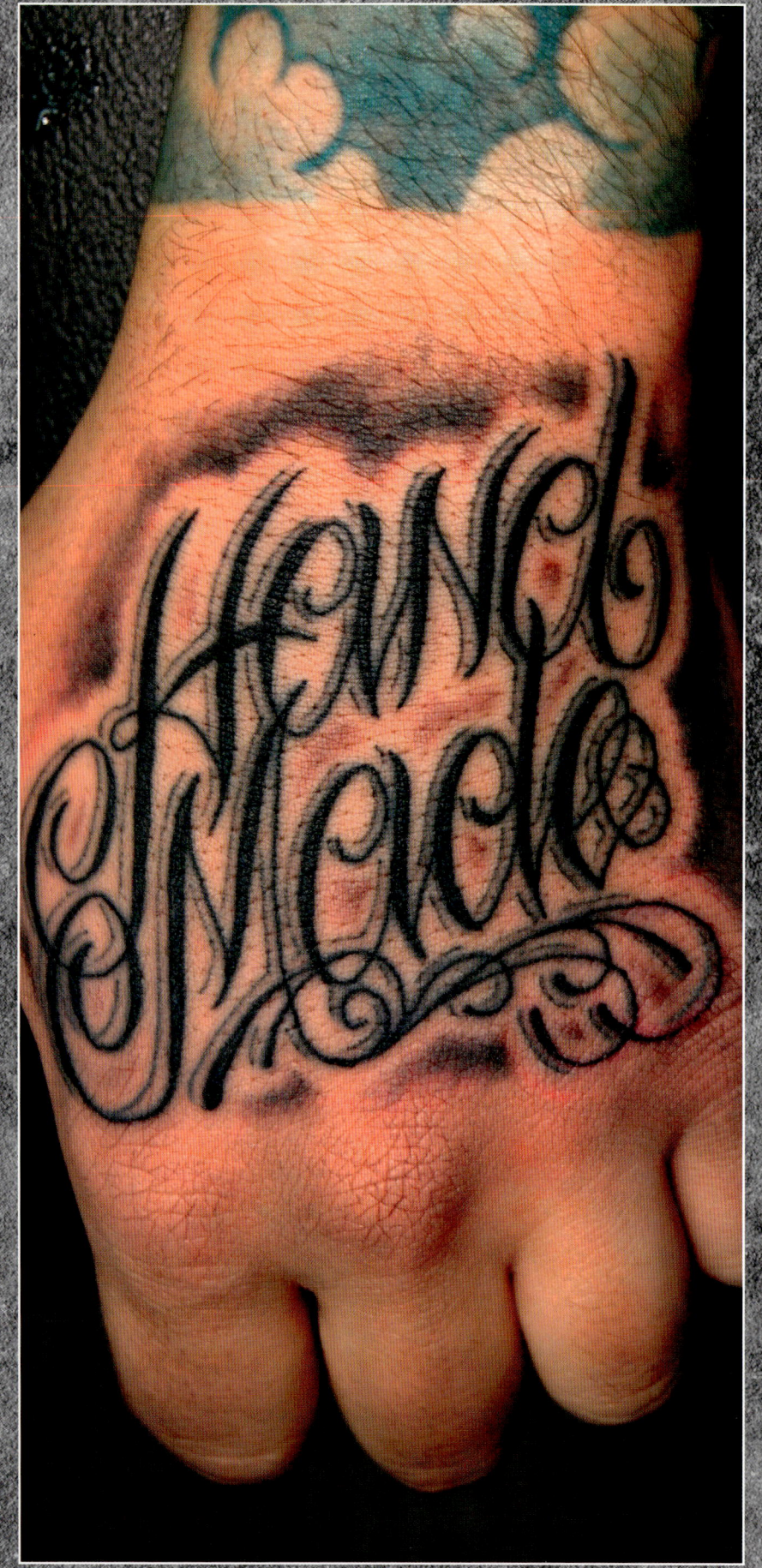

Jim Sylvia

Jim Sylvia

Kirk Alley

Jeff Zuck

Jeff Zuck

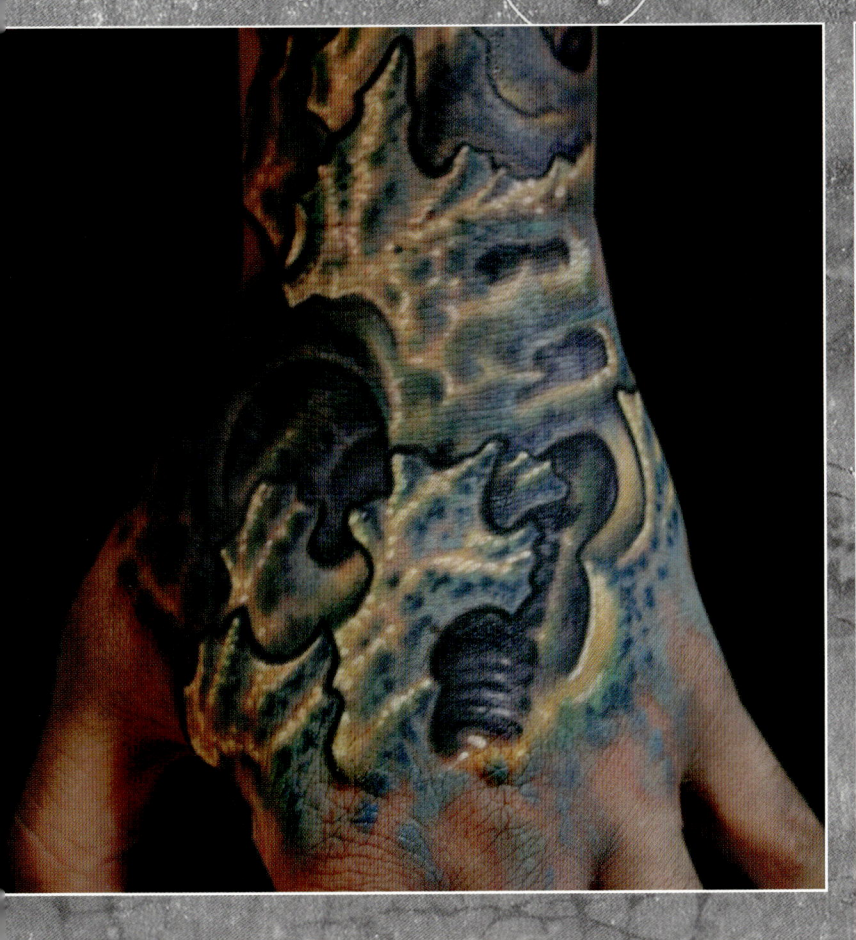

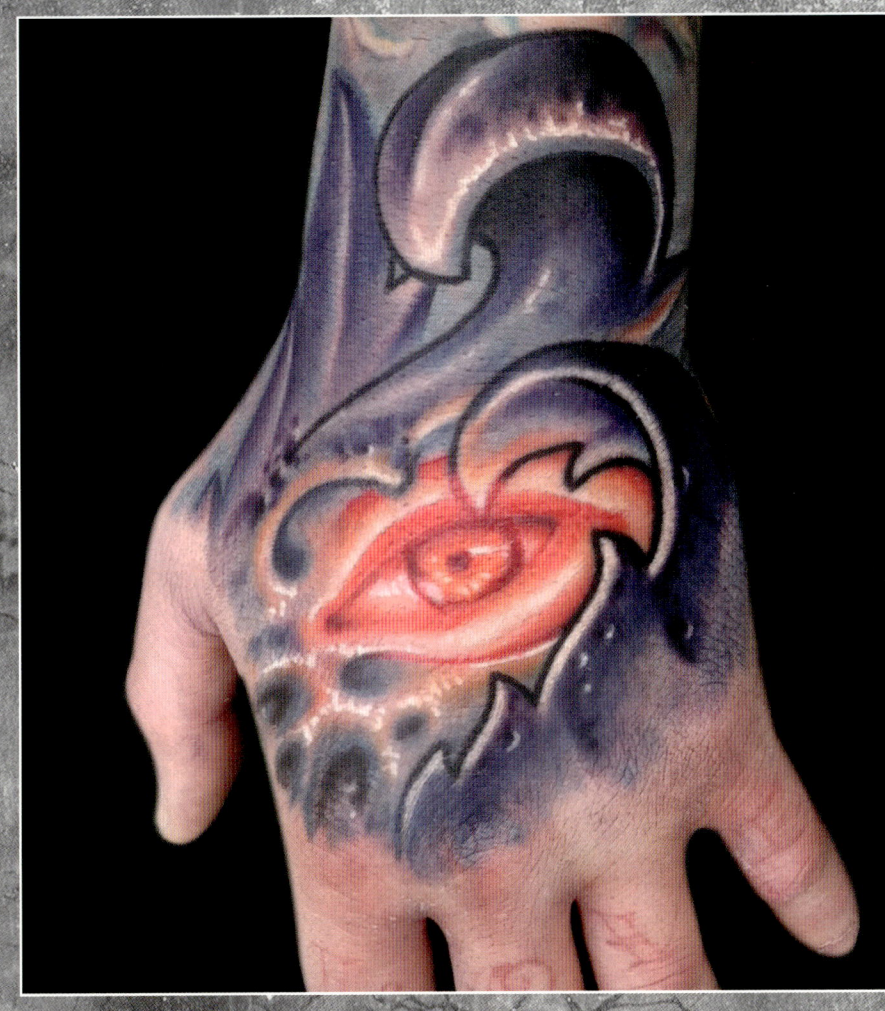

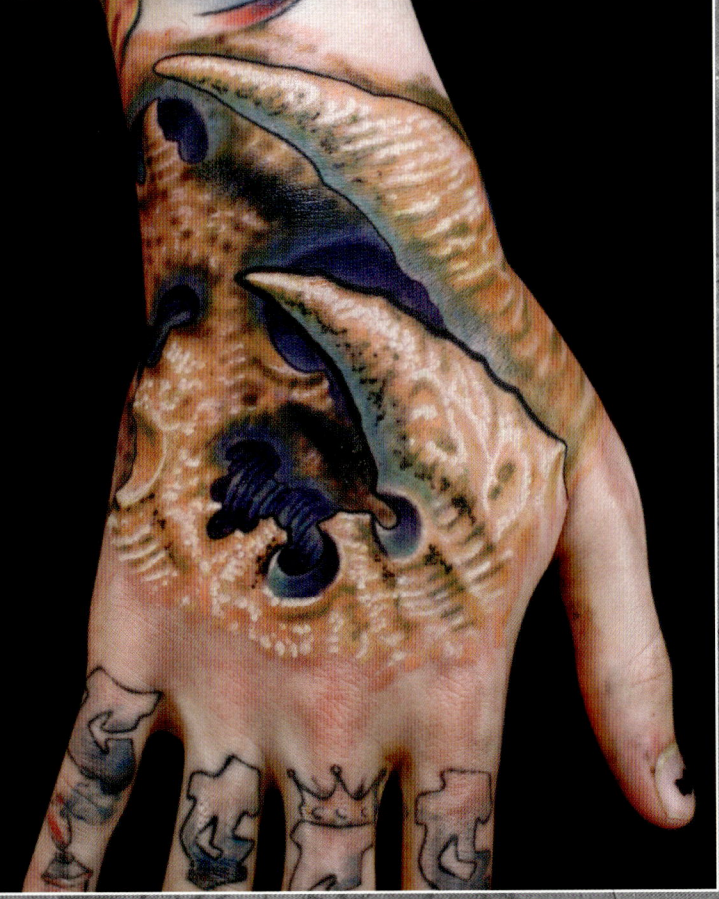

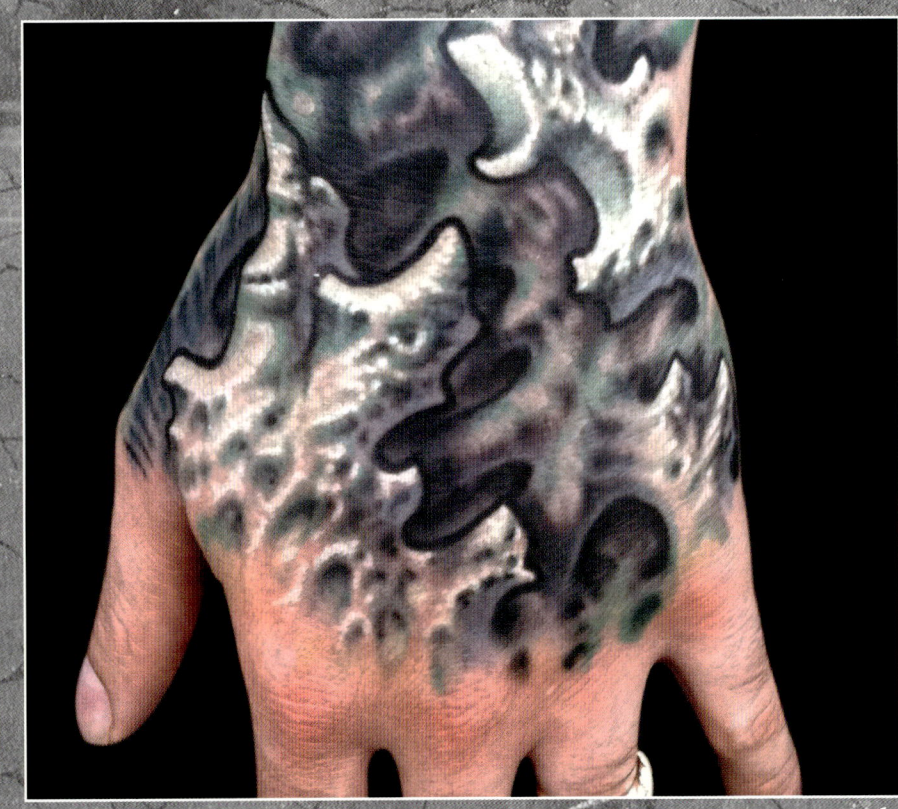

Ty McEwen

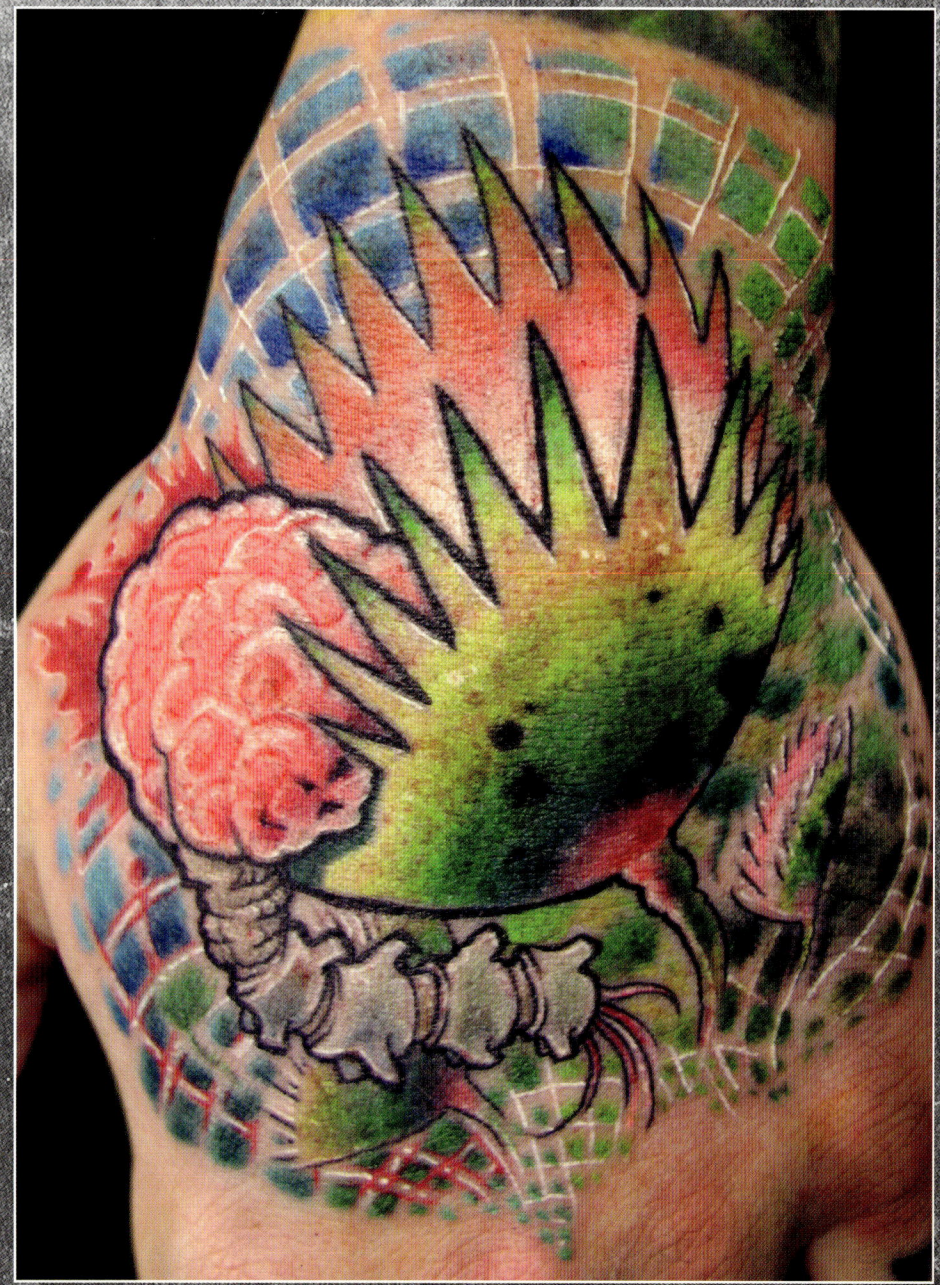

Tim Pangburn

Timothy Boor

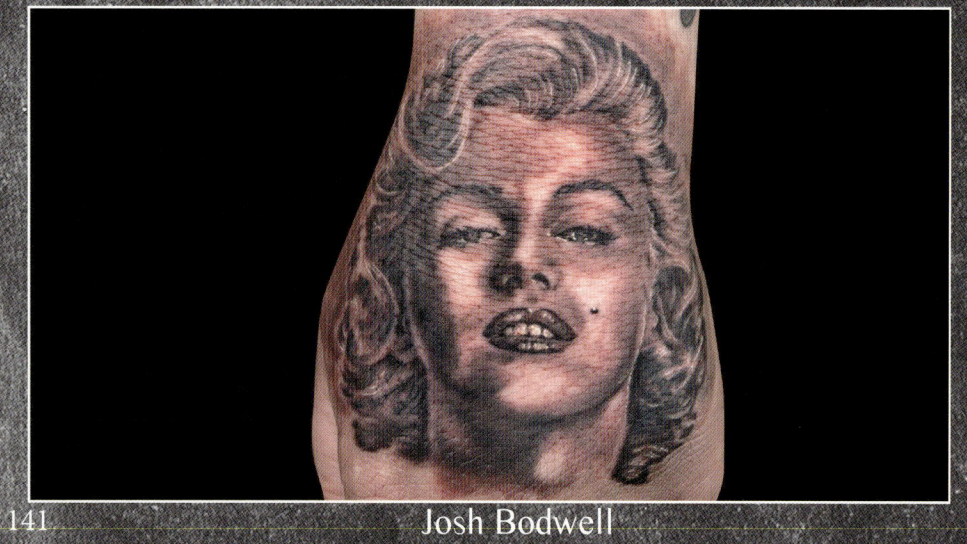

Josh Bodwell

Ty McEwen

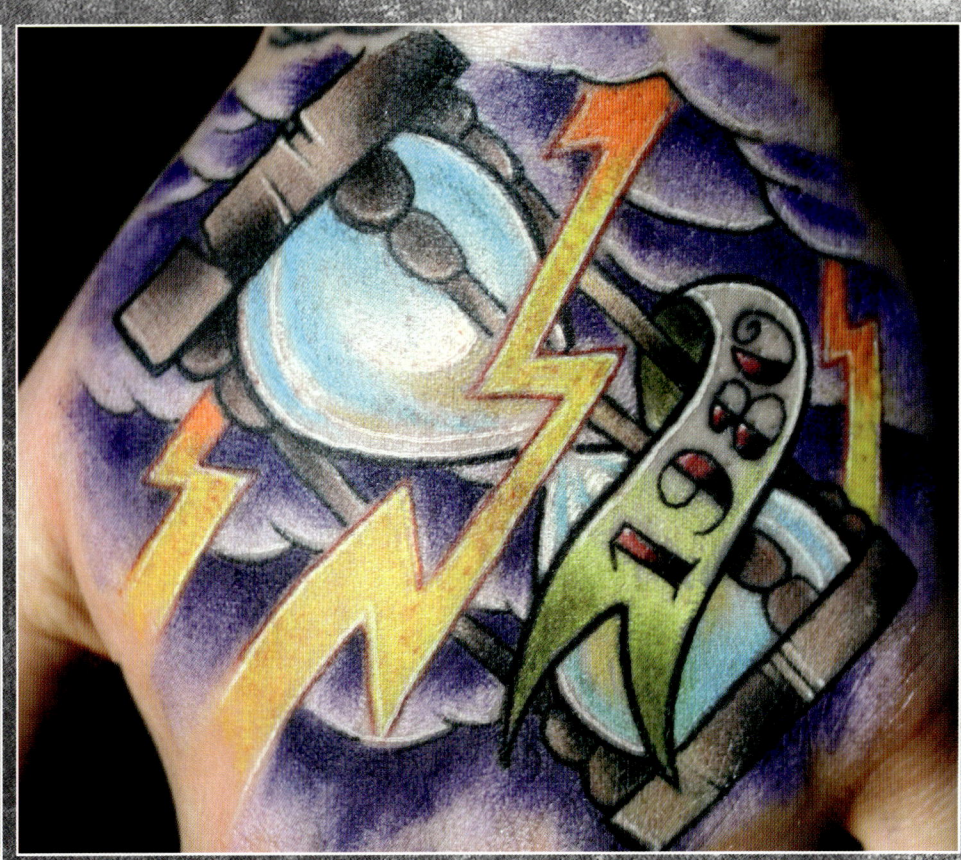

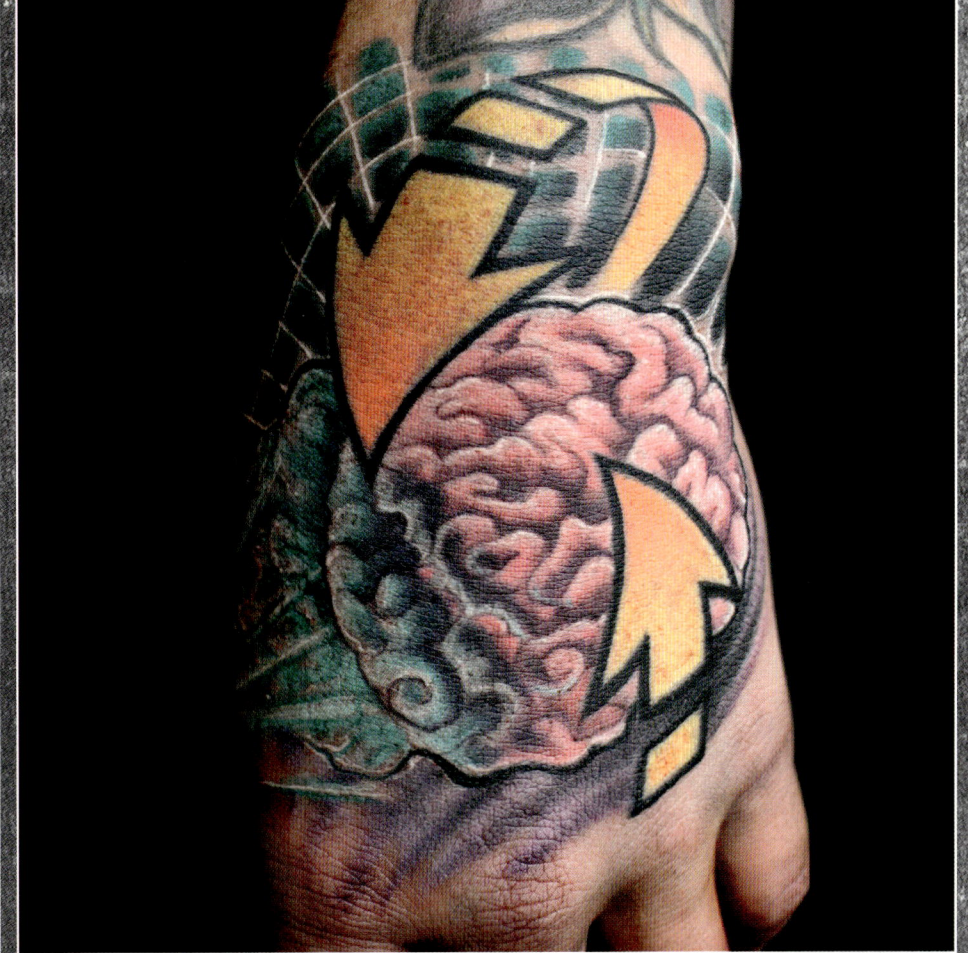

Tim Pangburn

Jess Yen

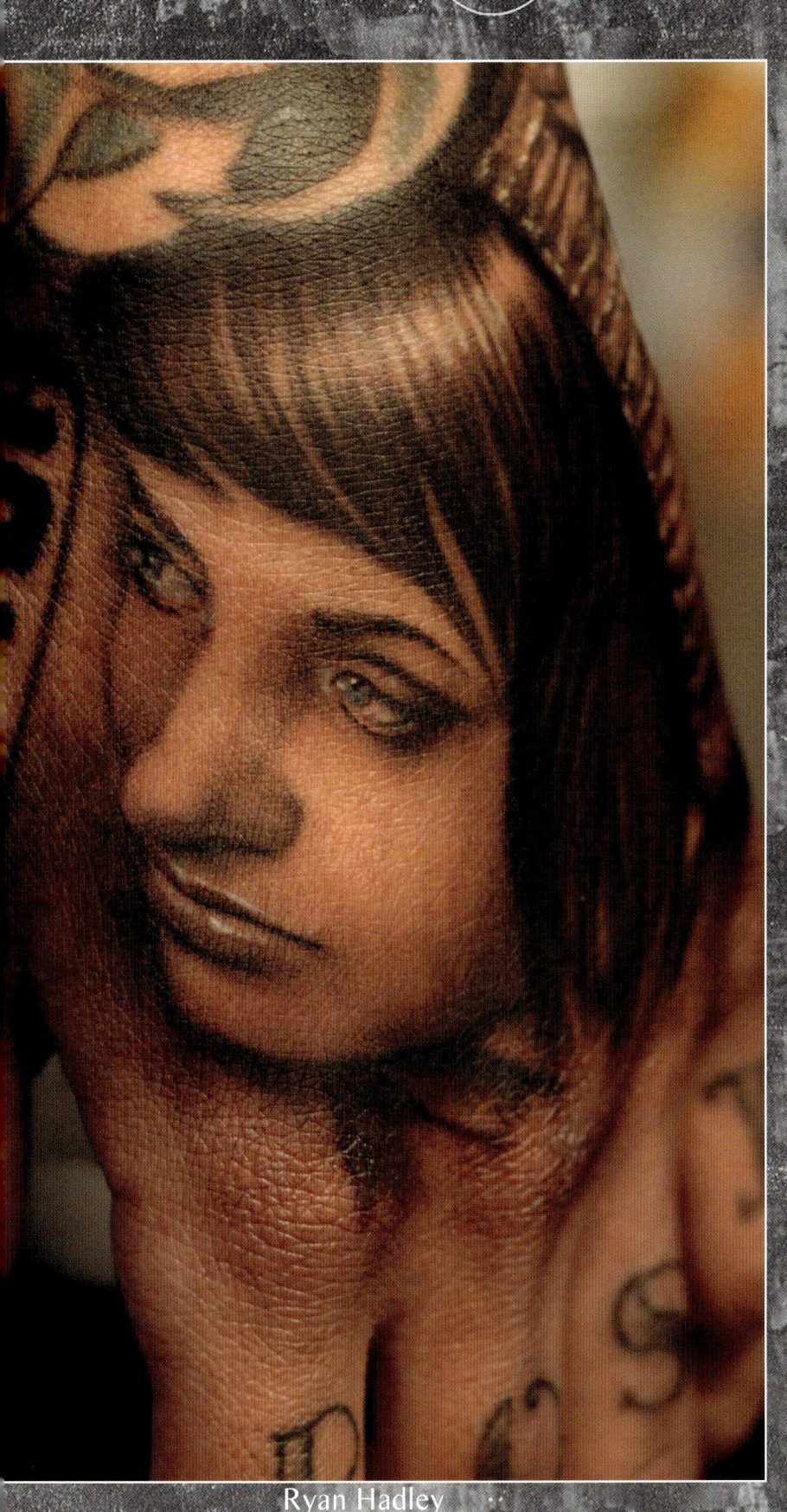

Ryan Hadley

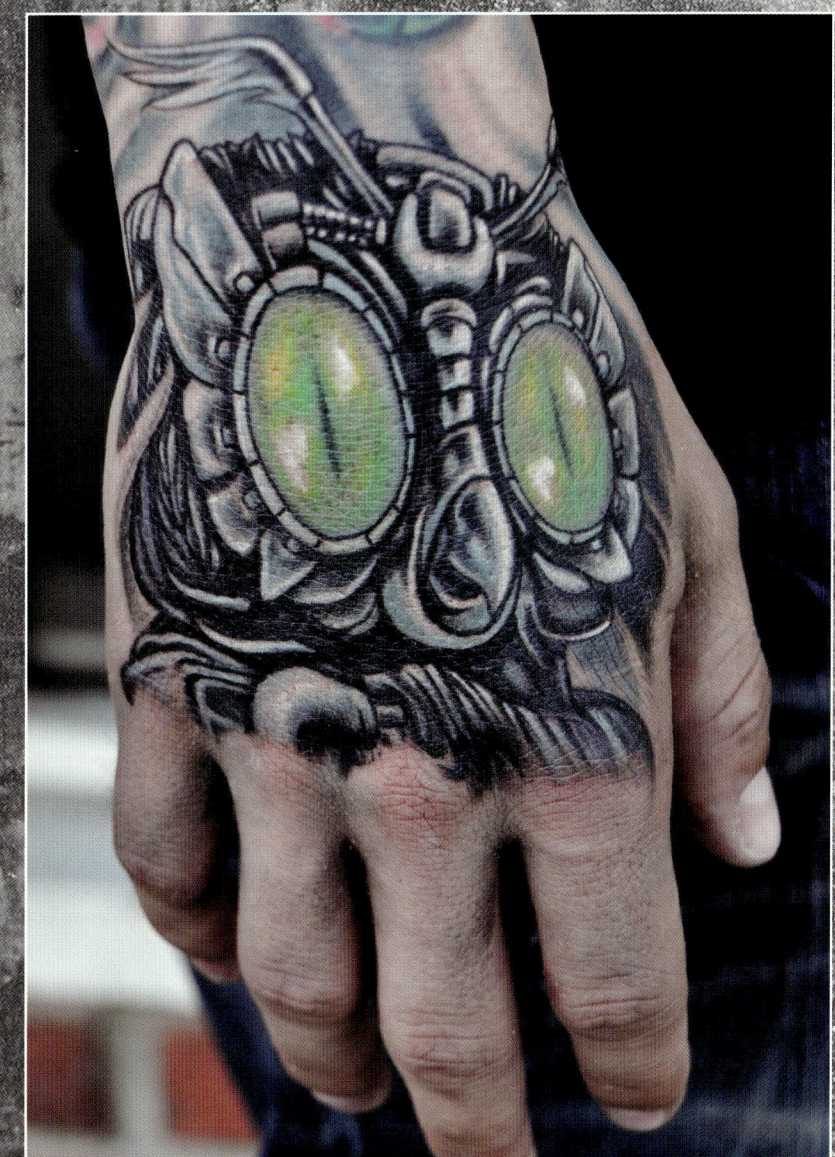

Mike DeVries

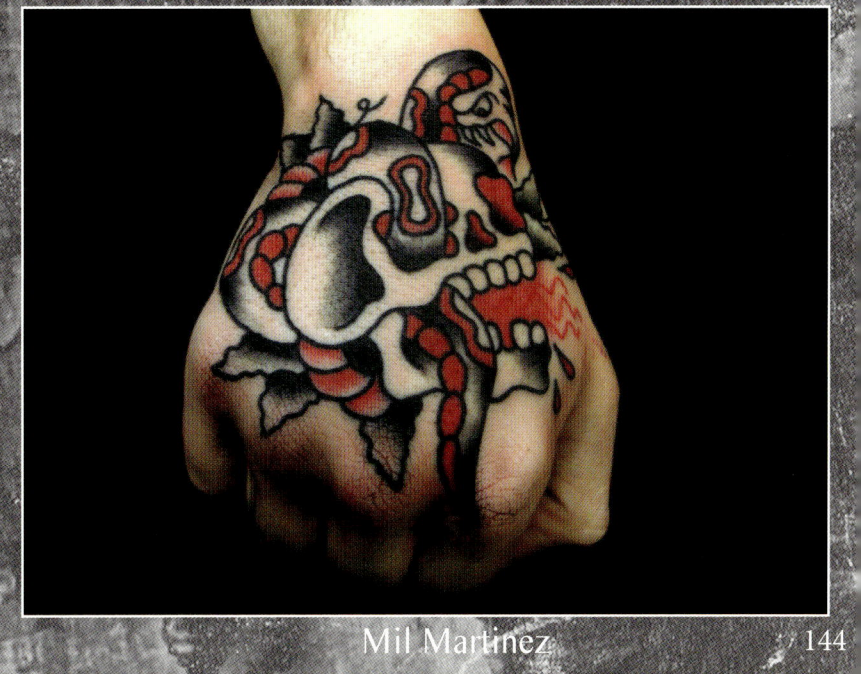

Mil Martinez

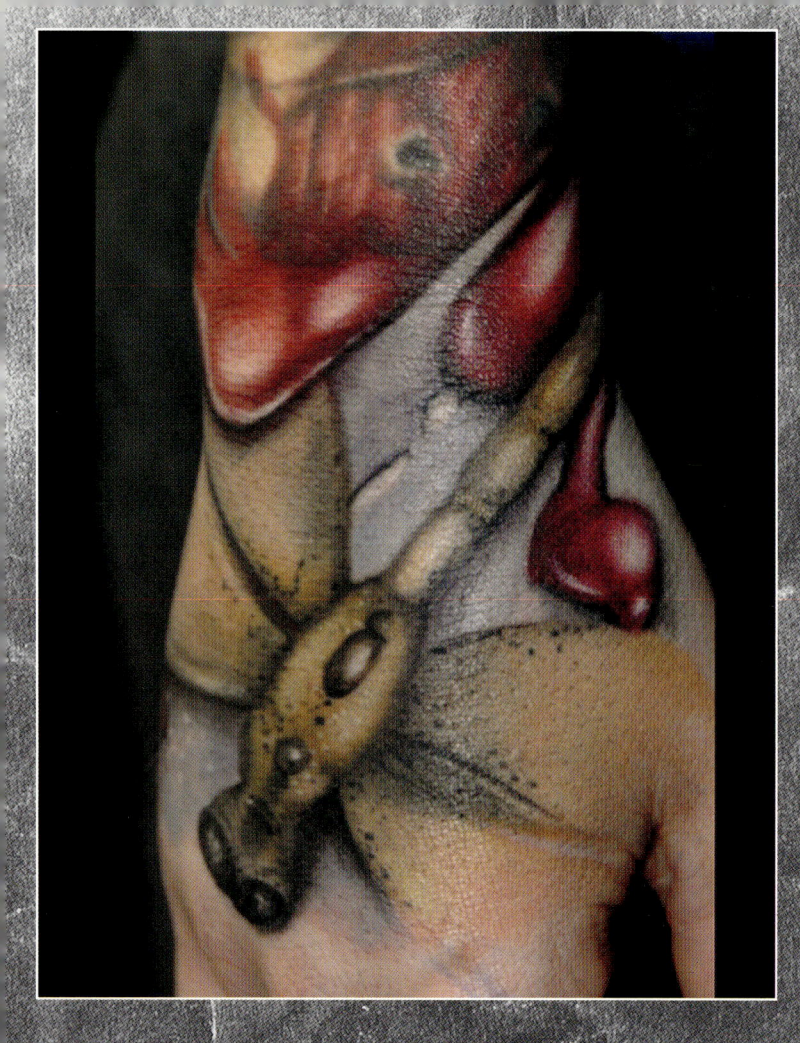

Ryan Hadley

Robert Hendrickson

Phil Watkins

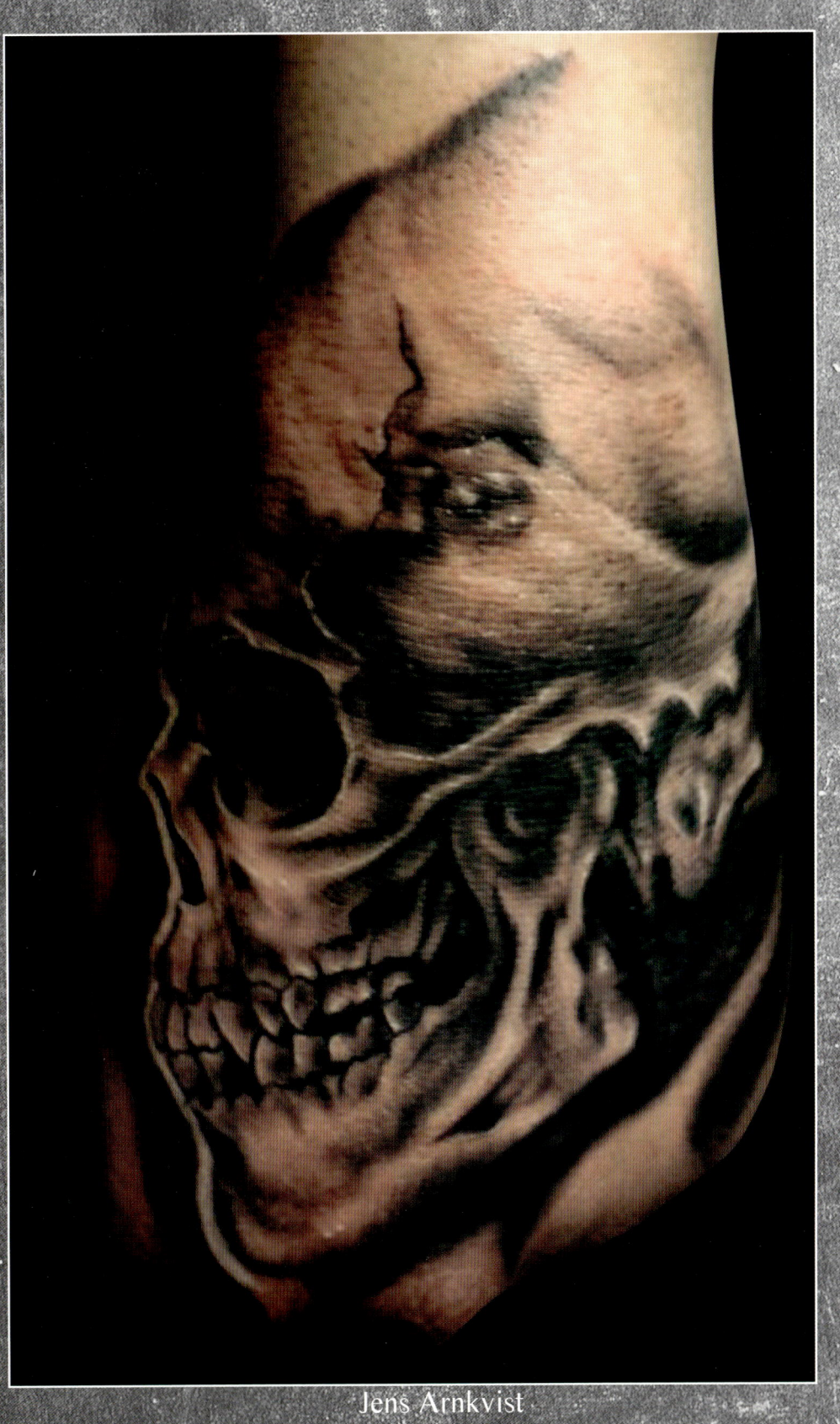

Jens Arnkvist

146

Miro Tomas

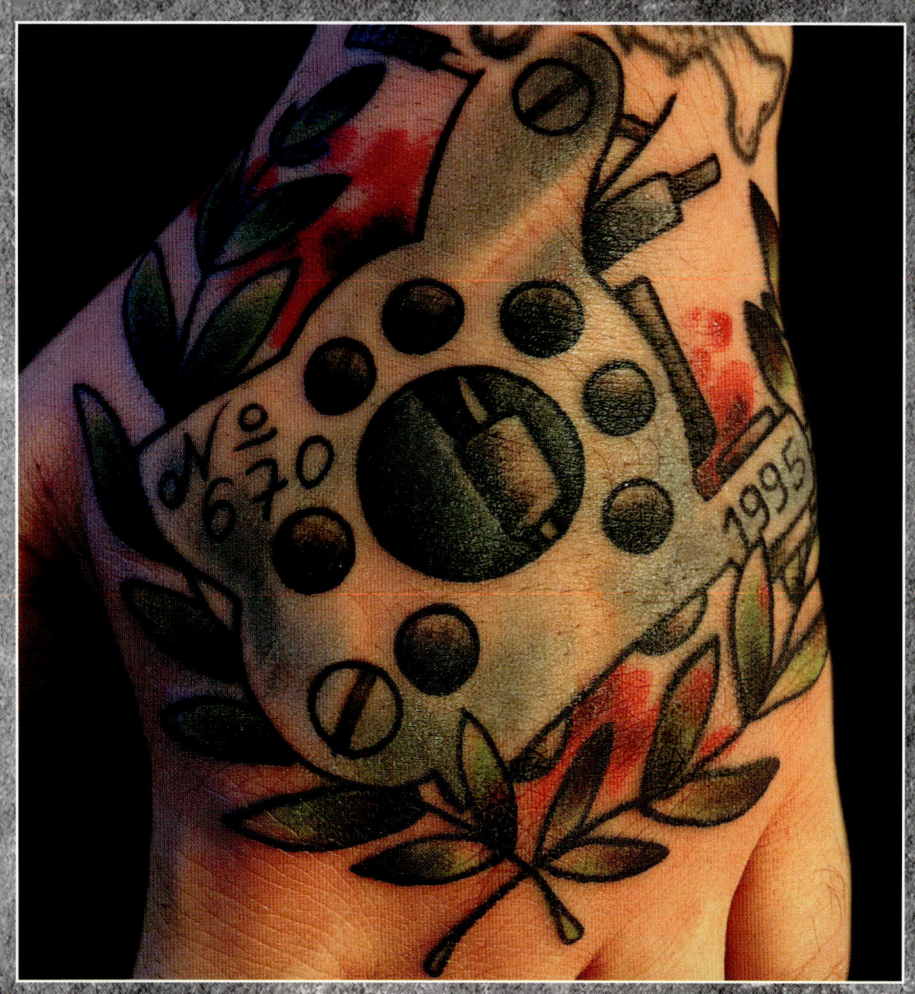

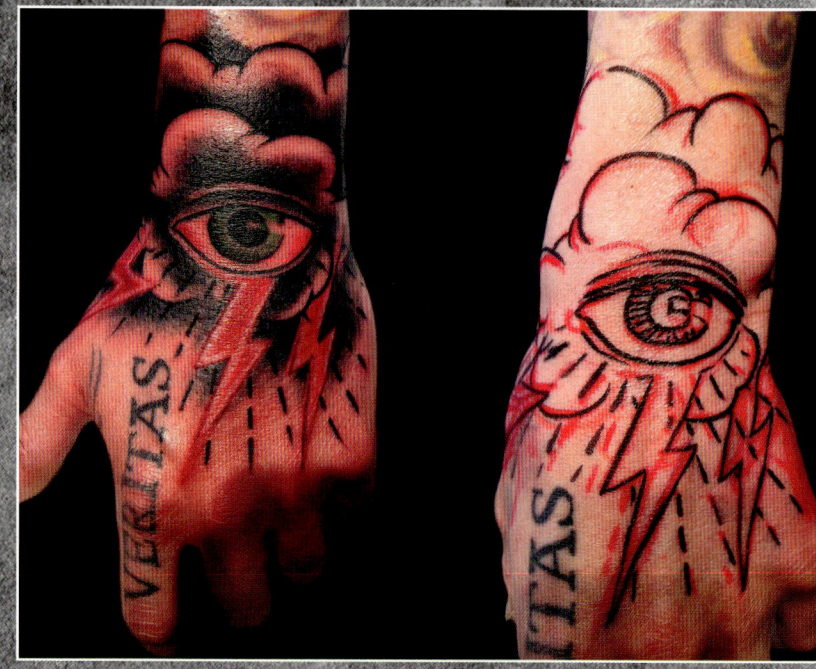

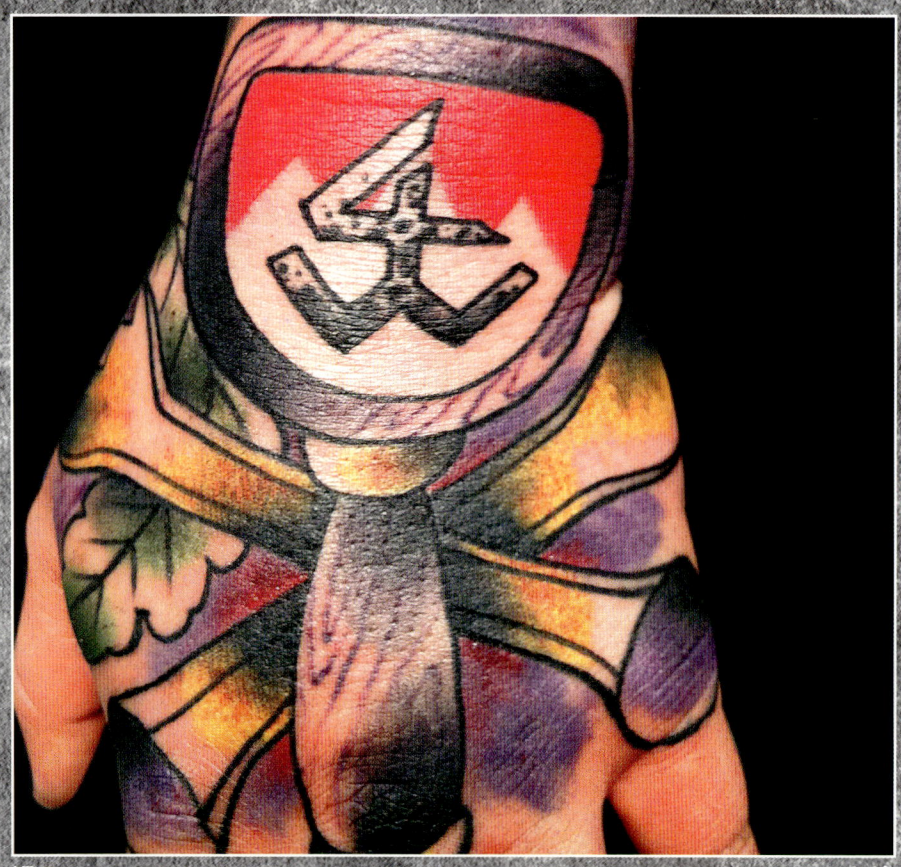

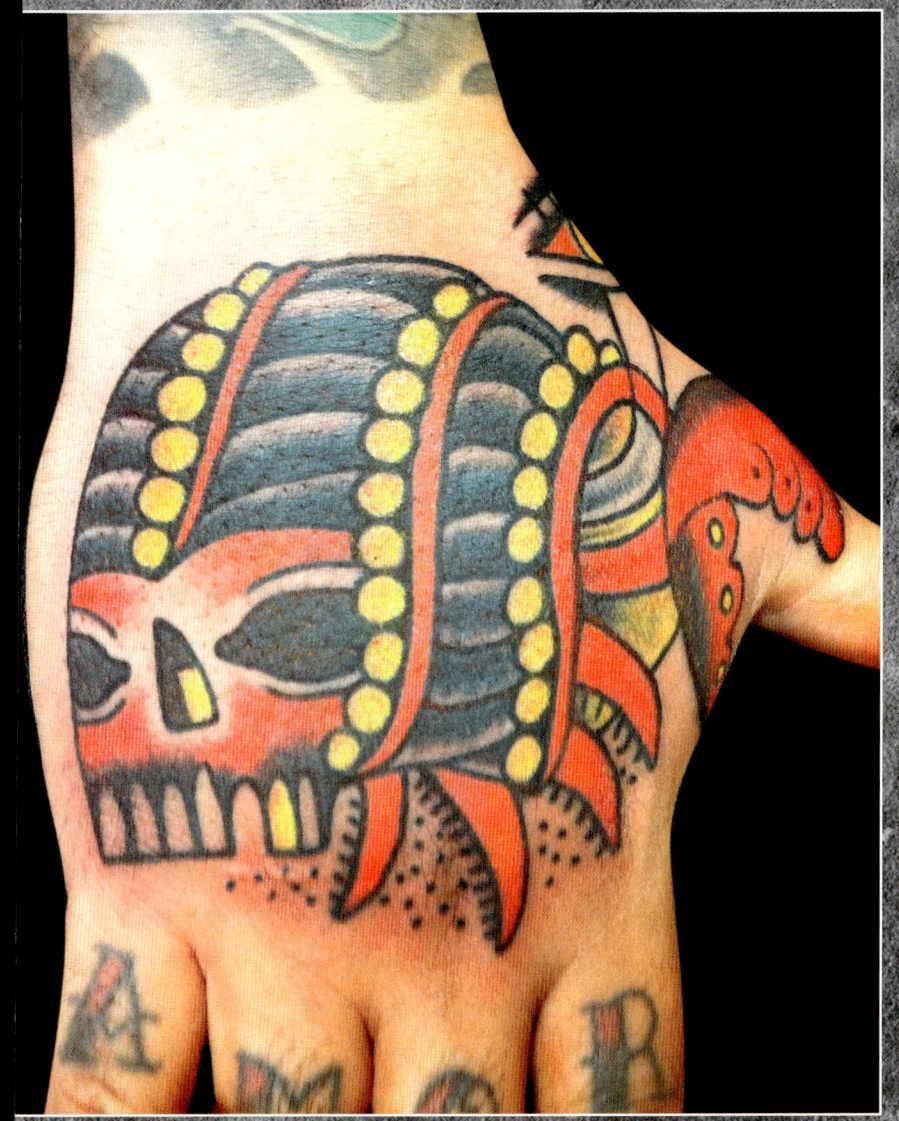

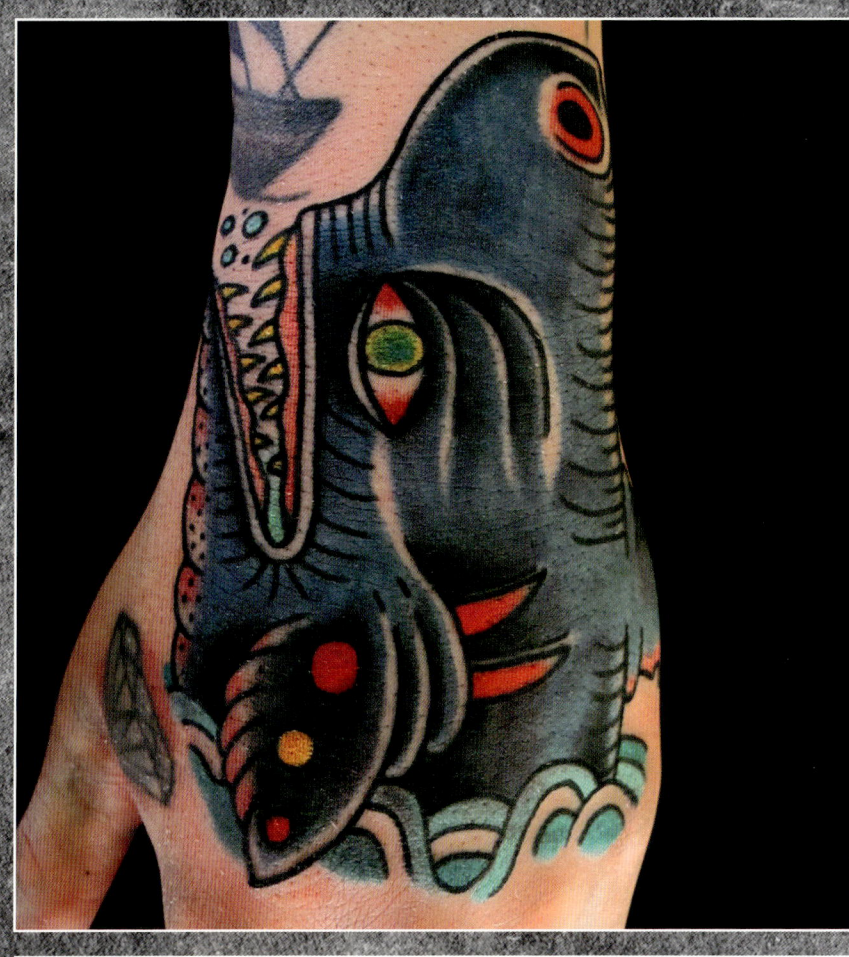

Deno

Demon

David Bruehl

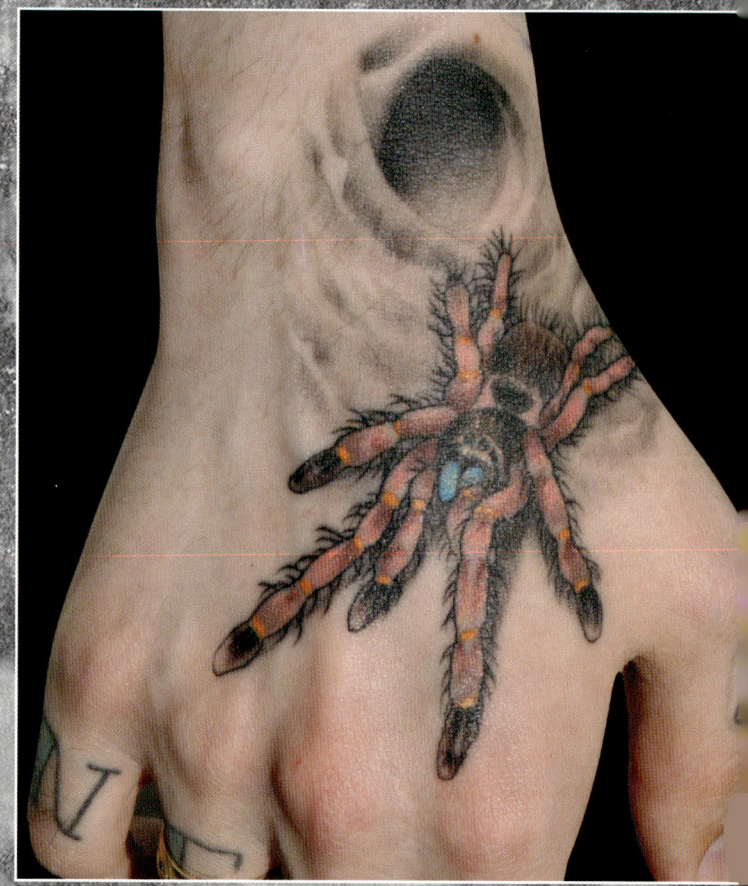

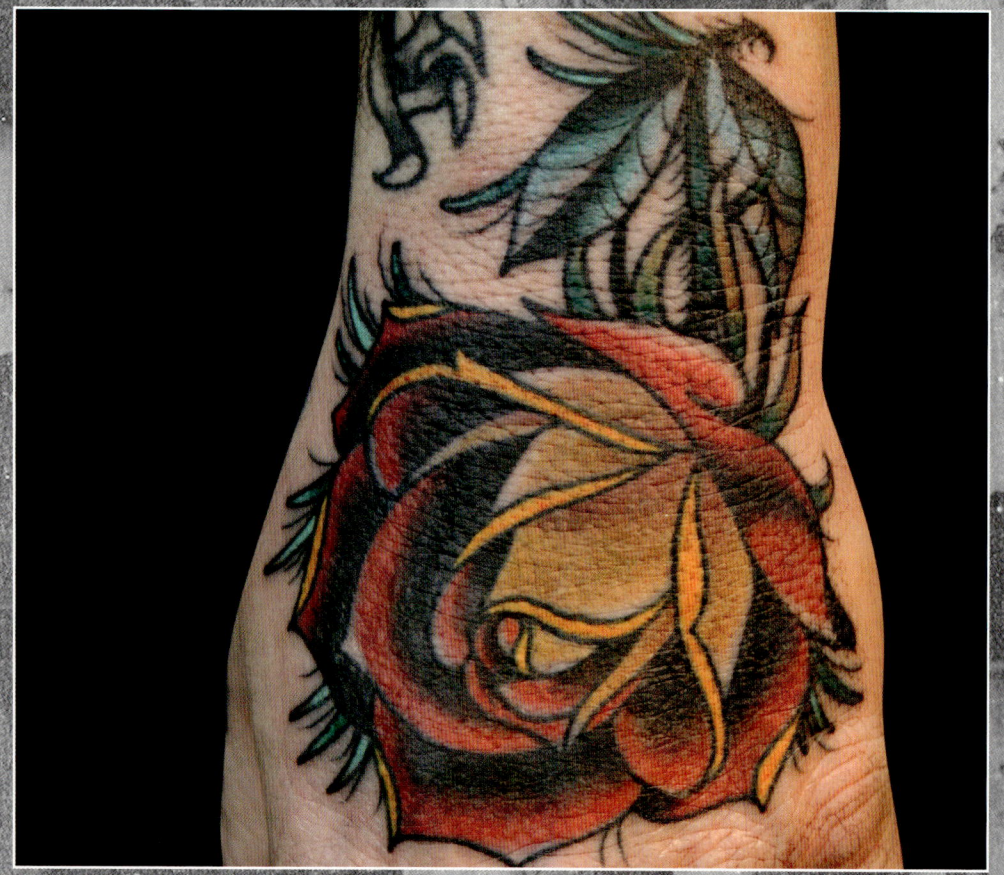

David Bruehl

David Bruehl

David Bruehl

David Bruehl

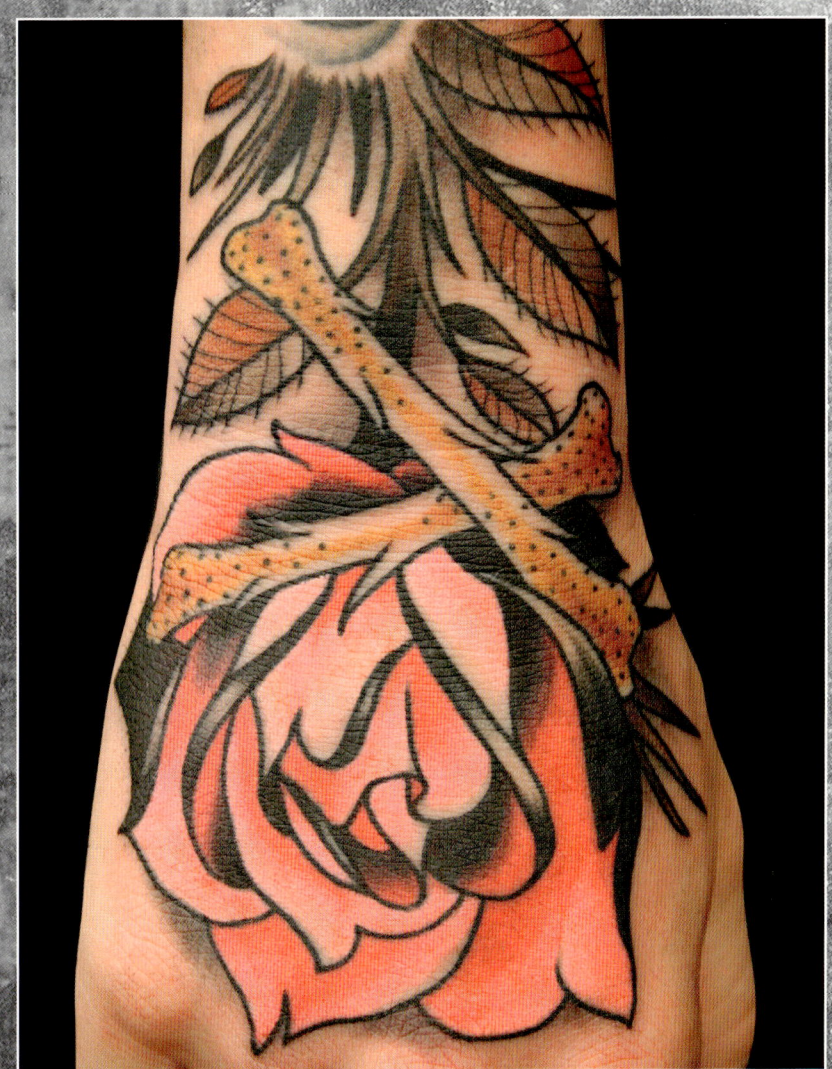

Phillip Spearman

Richard Holt

Matt Griffith

Phillip Spearman

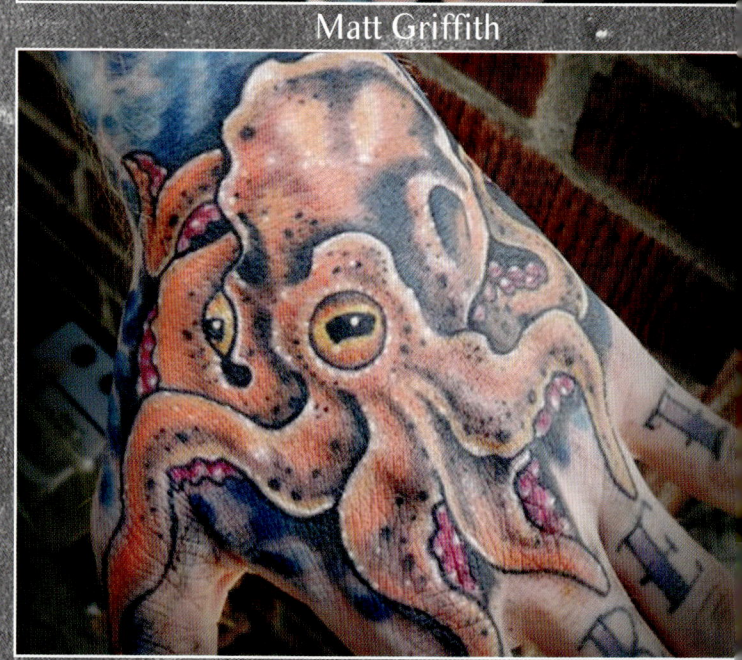

Matt Griffith

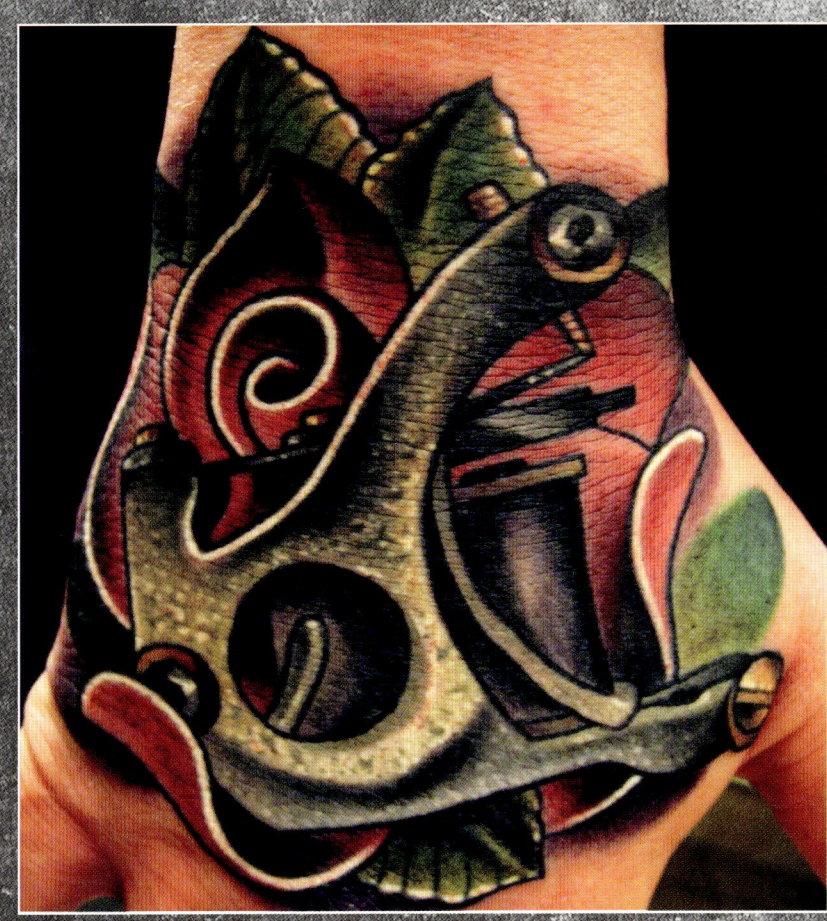

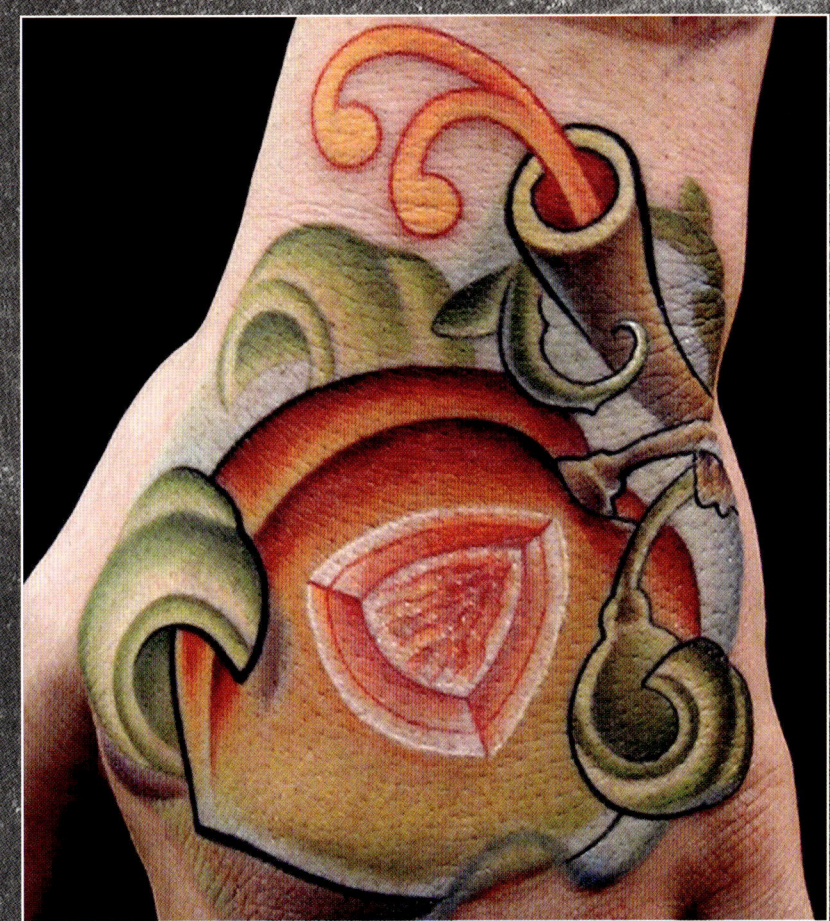

Jonathan Montalvo

Leigh Oldcorn

Jonathan Montalvo

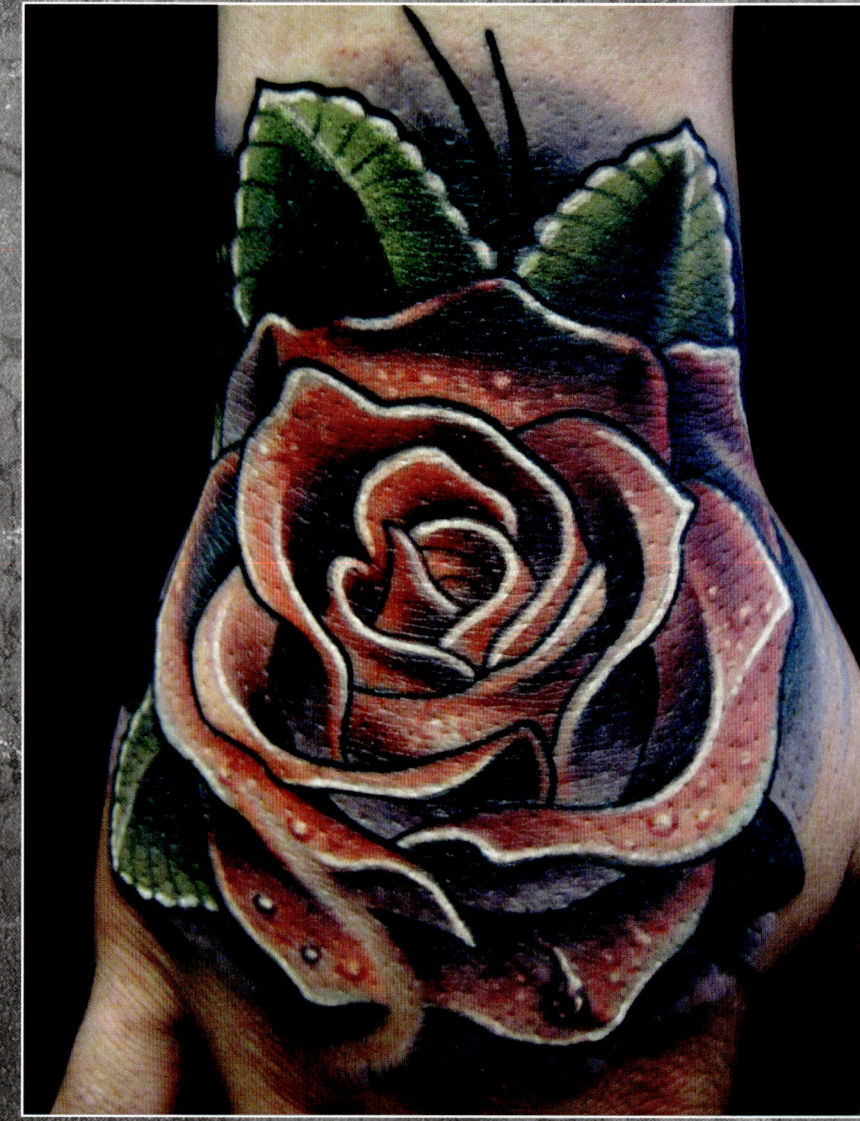

Jonathan Montalvo

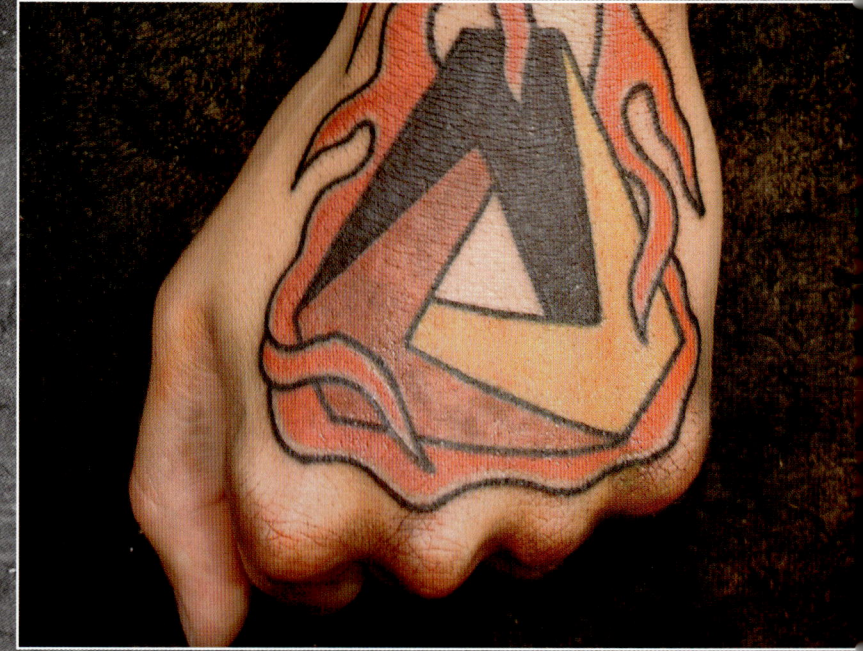

DestroyTroy

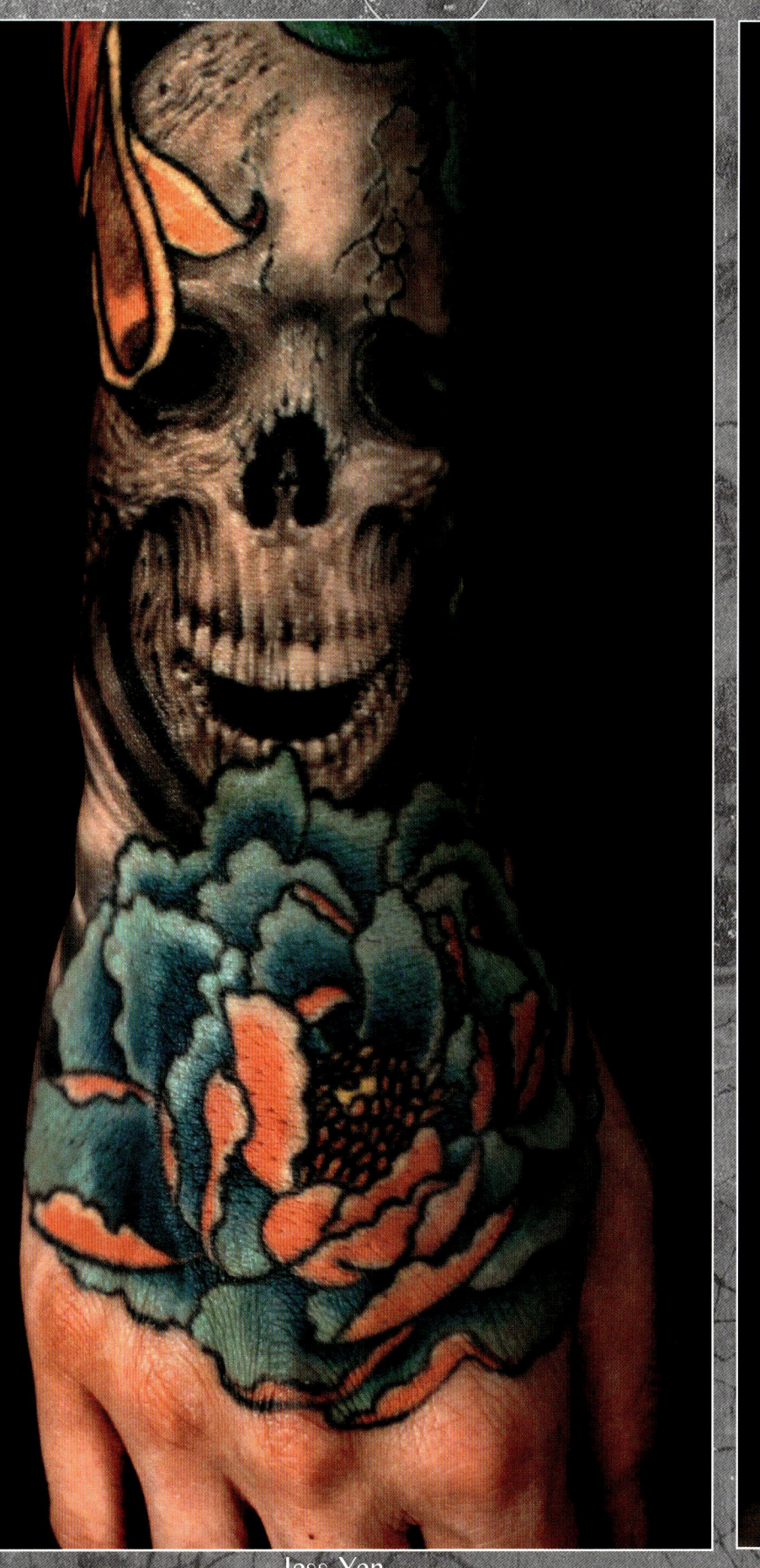

Jess Yen

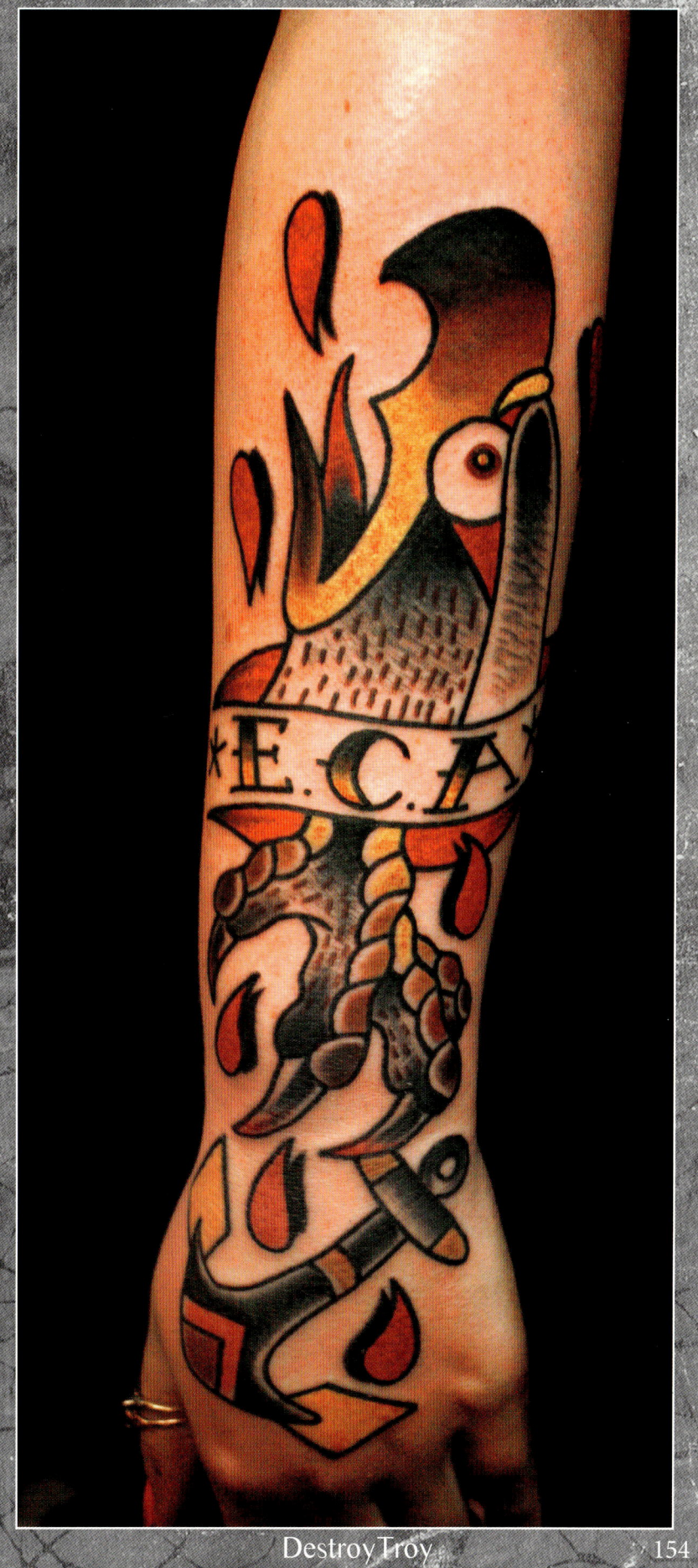

Destroy Troy

Jo Harrison

Jo Harrison

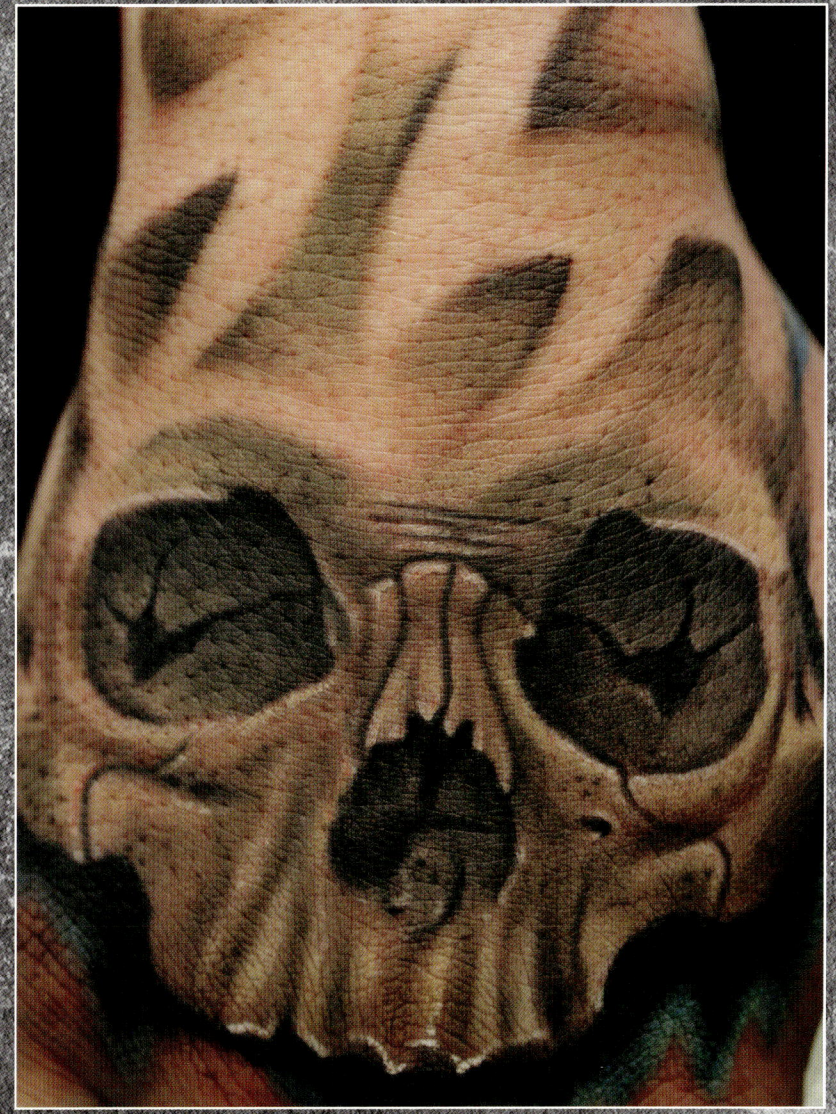

Jason Morrow

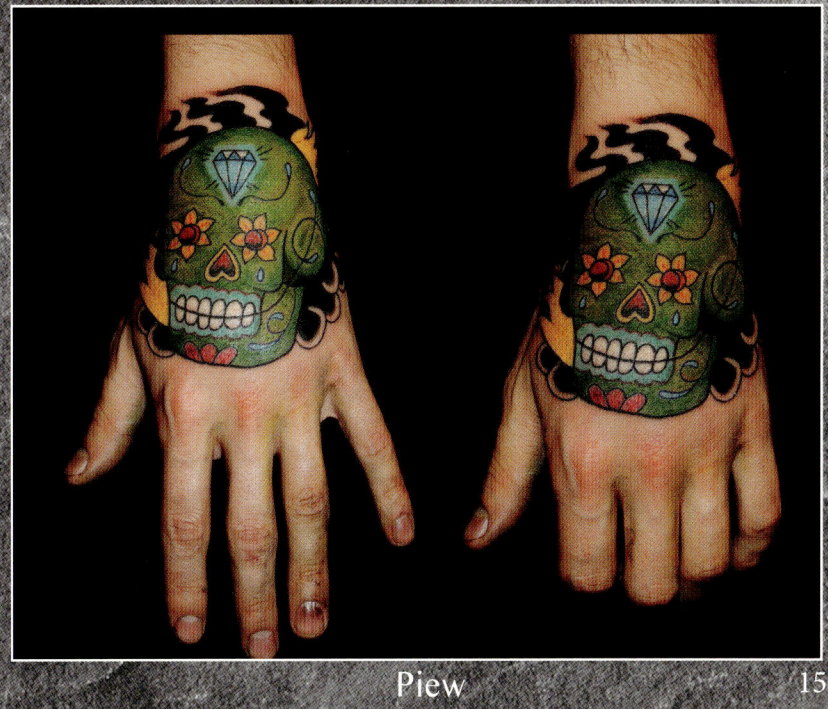

Piew

Piew

156

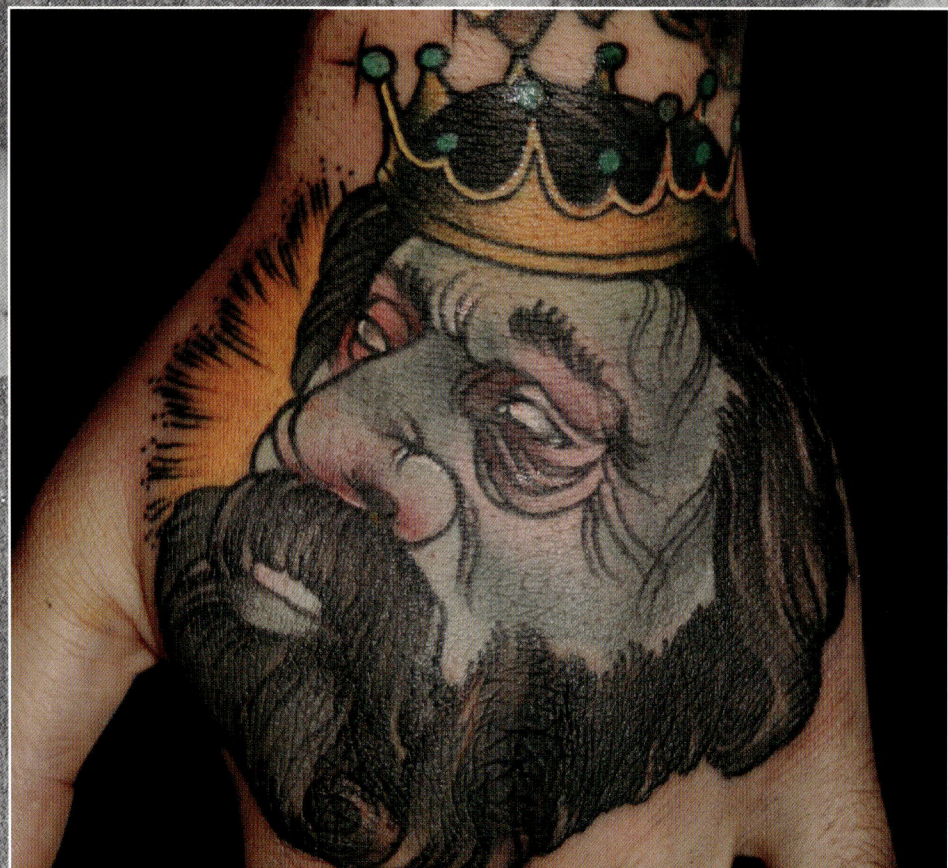

Jasmin Austin

Daniel Chashoudian

Michele Turco

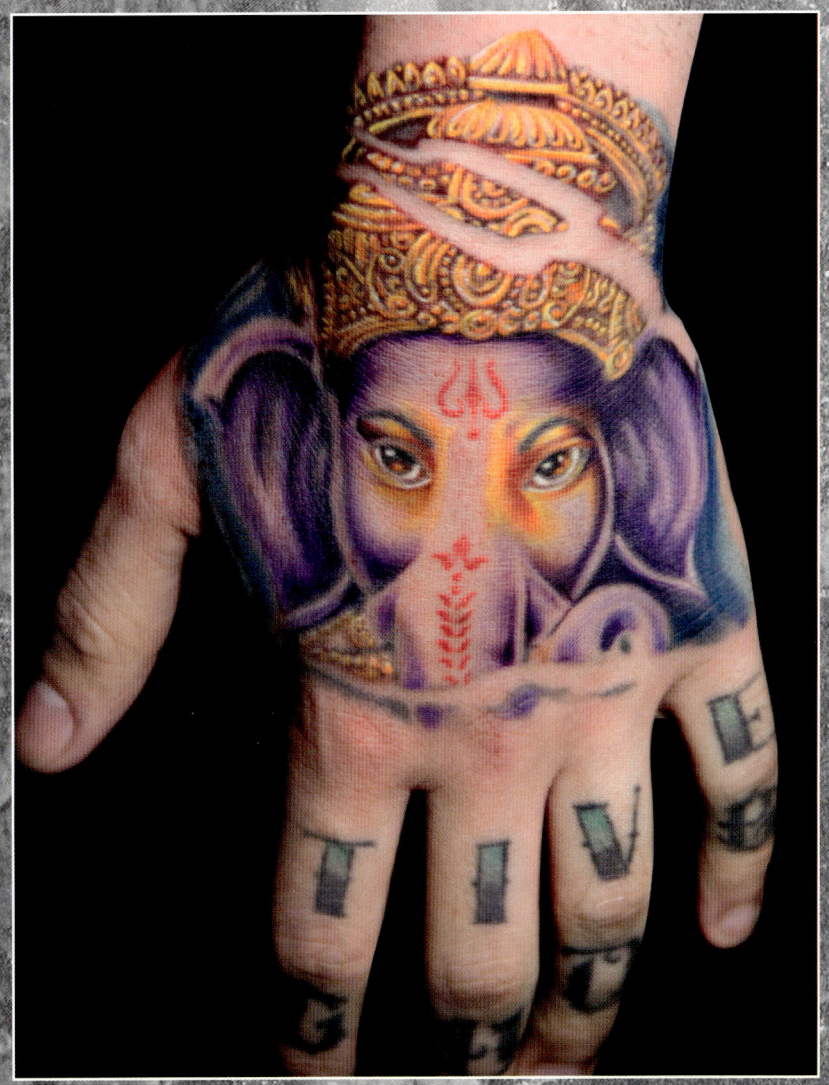

André Tenório

Buddah Quinones

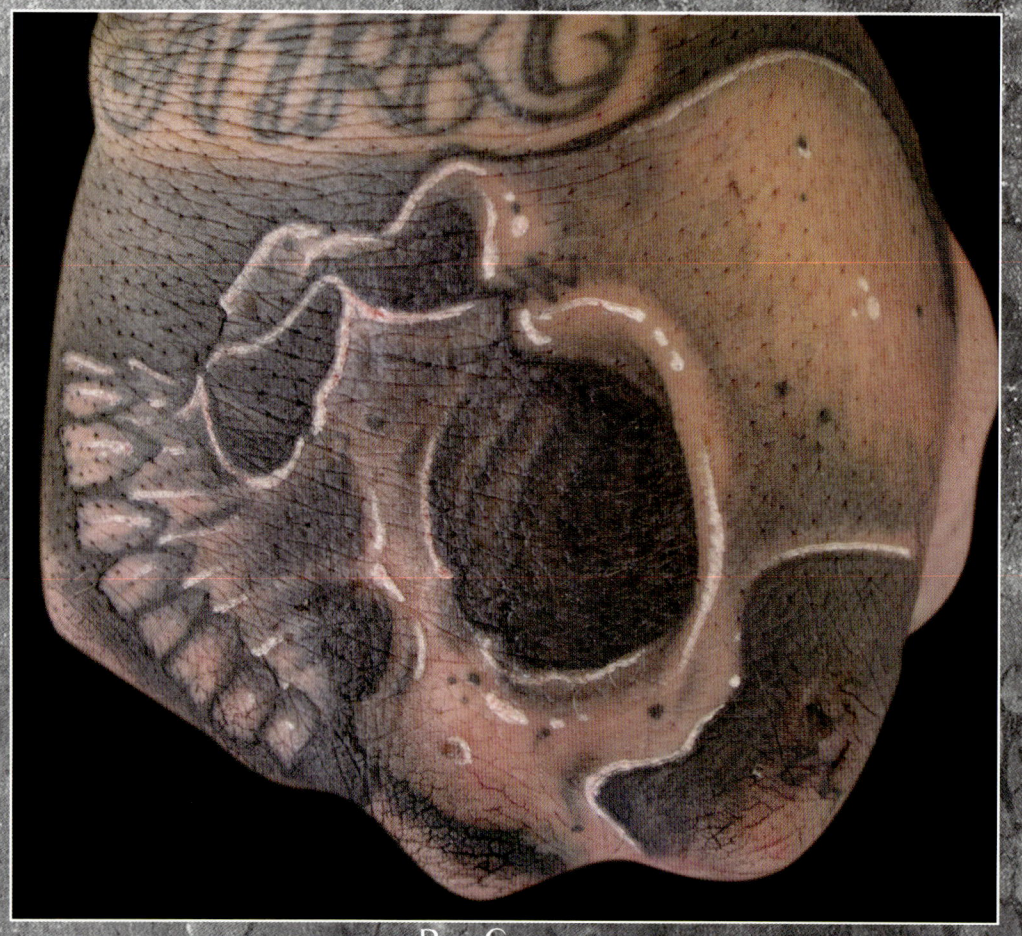

Ben Corn

Ben Corn

Ben Corn

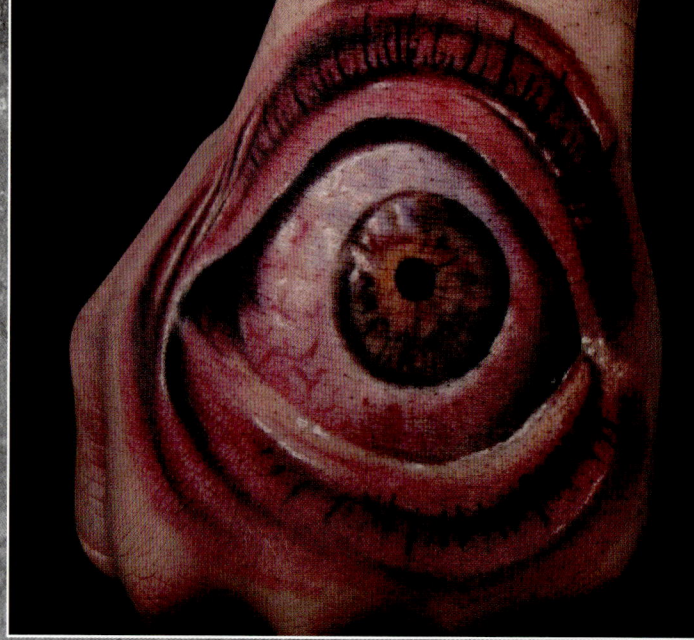

Daniel Chashoudian

Leigh Oldcorn

Don Mcdonald

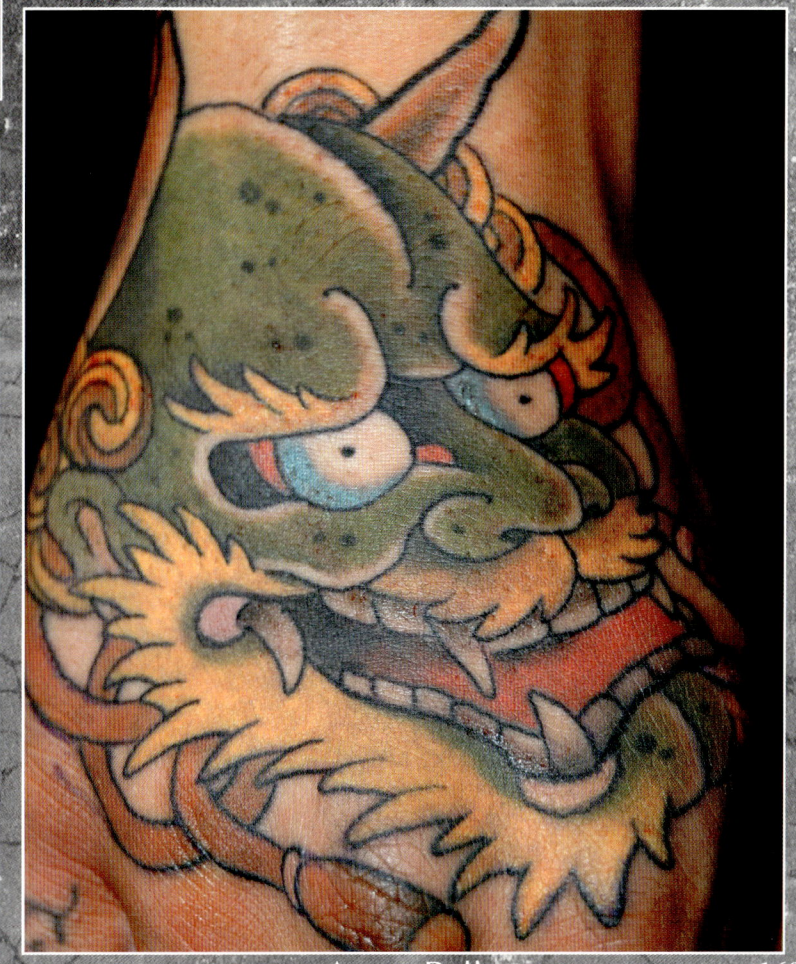

Aaron Bell

Aaron Bell

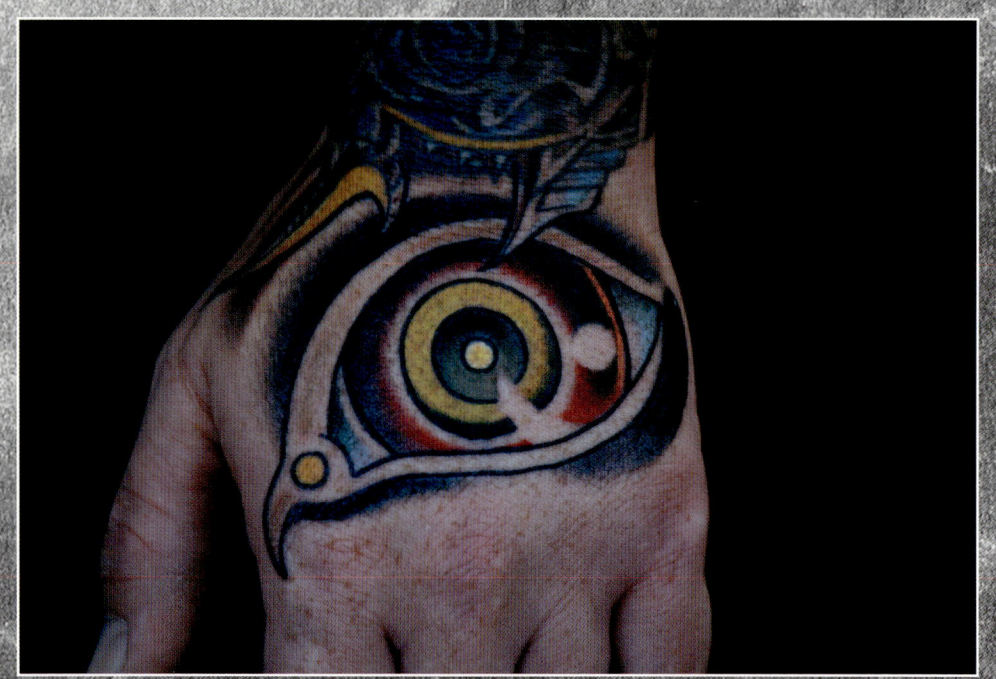

Zack Spurlock

Zack Spurlock

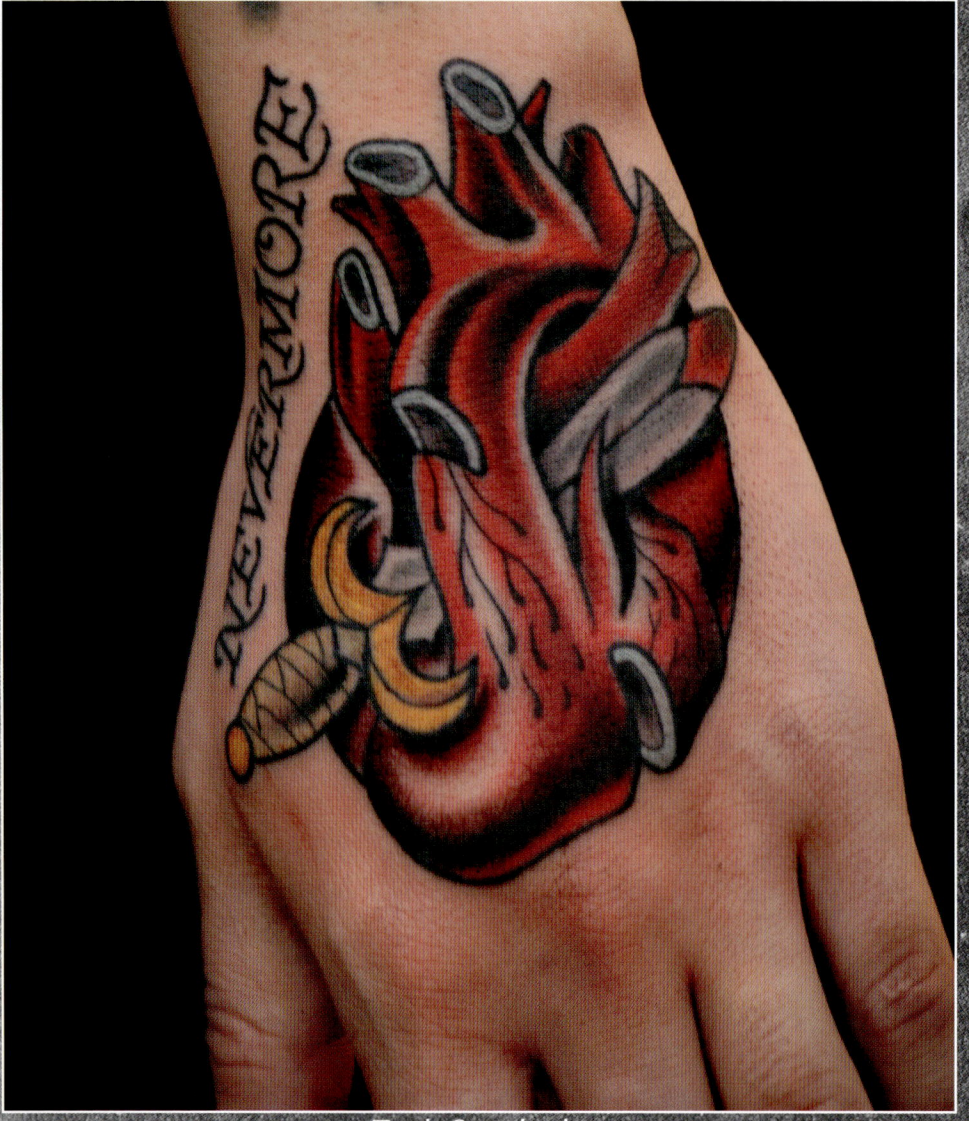

Zack Spurlock

Timothy Boor

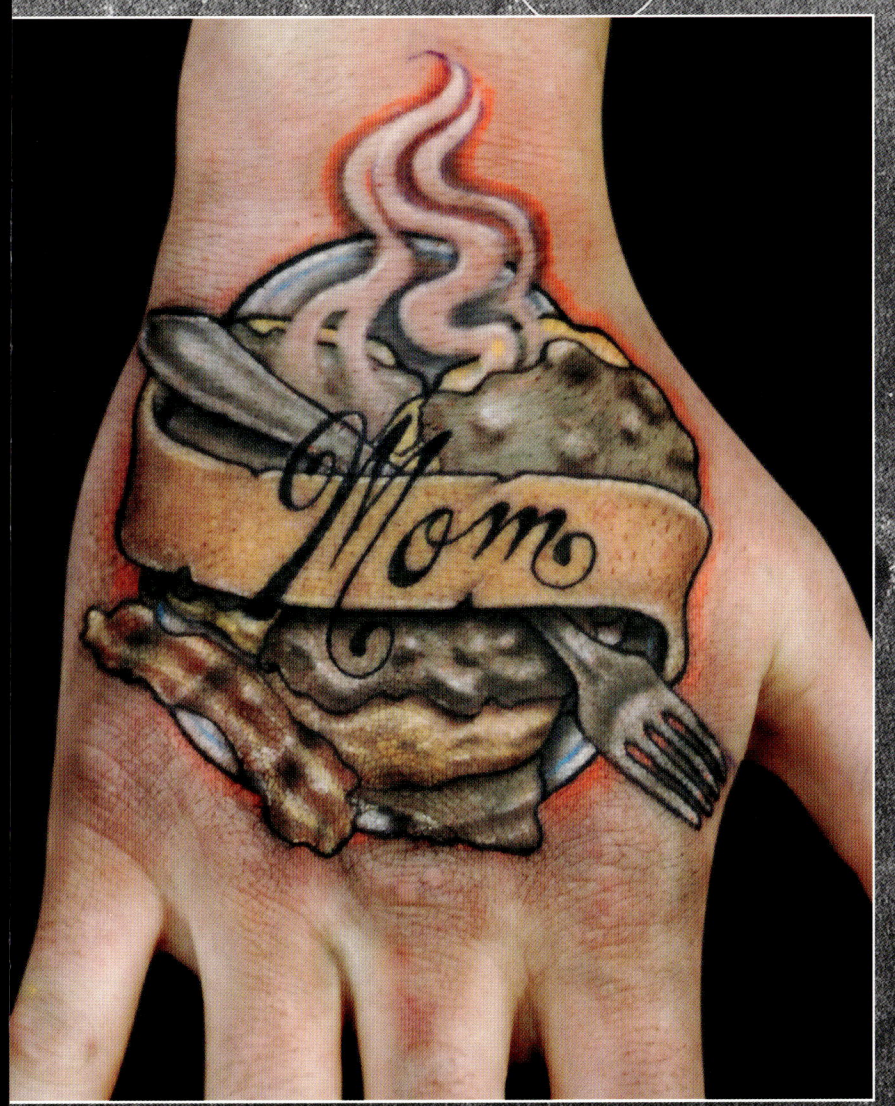

Timothy Boor

Juan Salgado

Juan Salgado

Juan Salgado

Juan Salgado

Jeremiah Barba

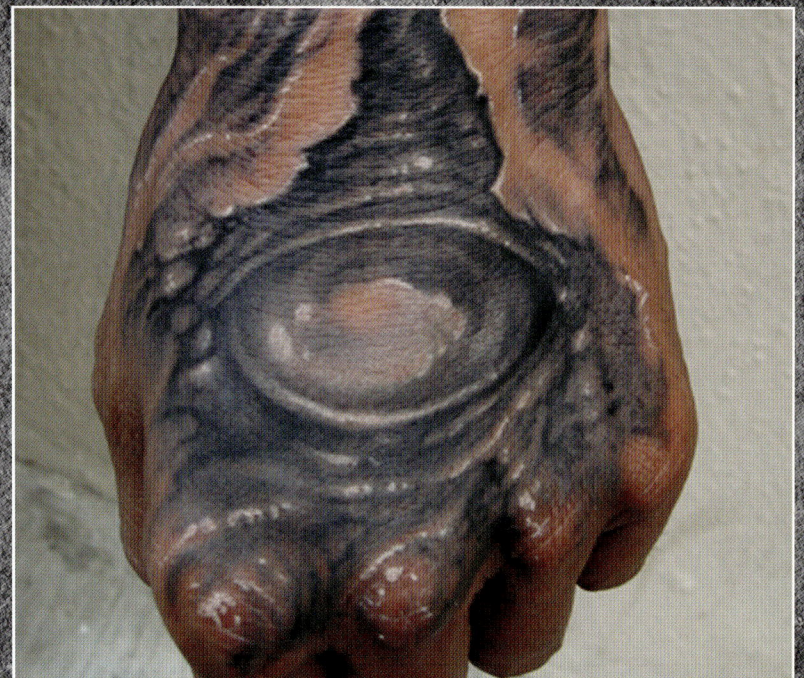

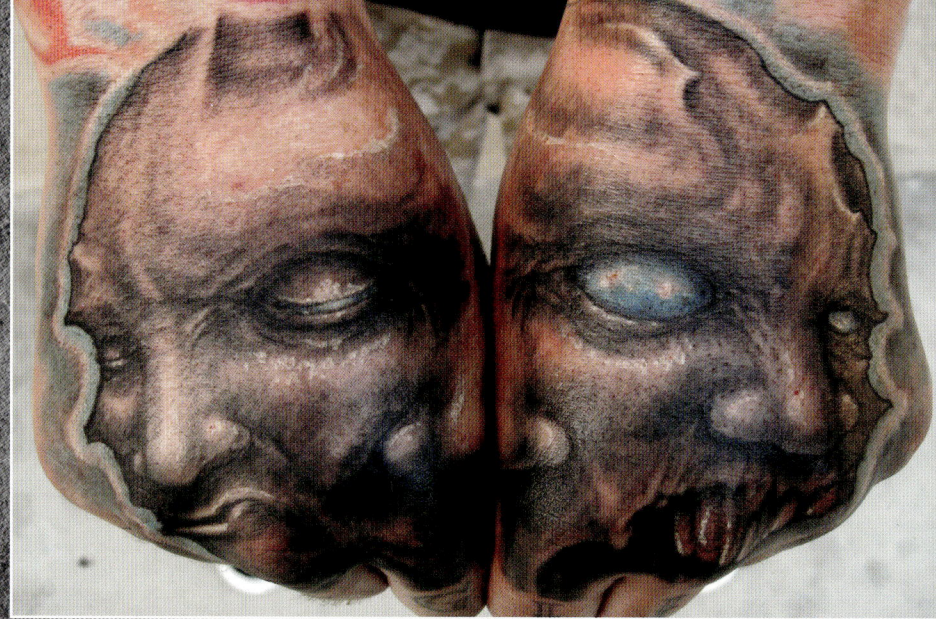

Jeremiah Barba

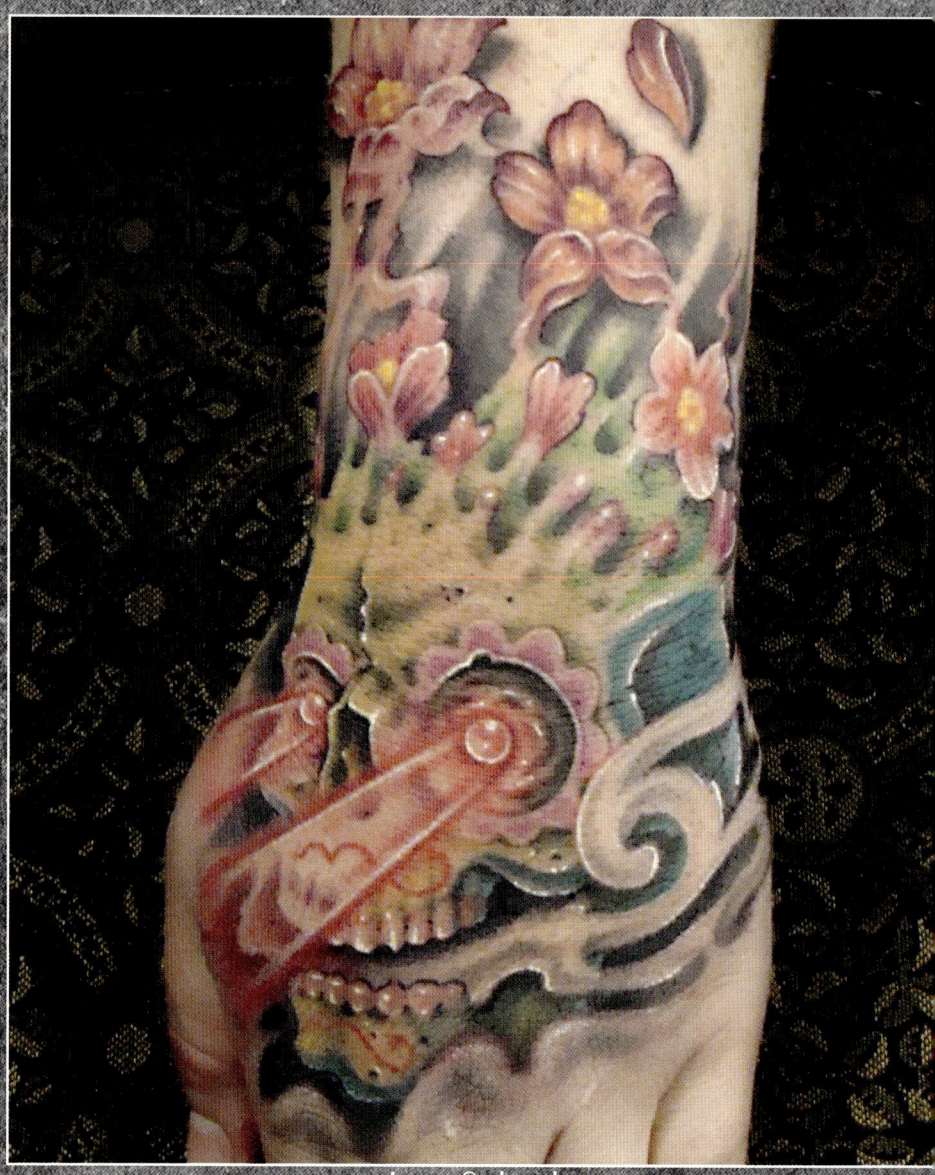

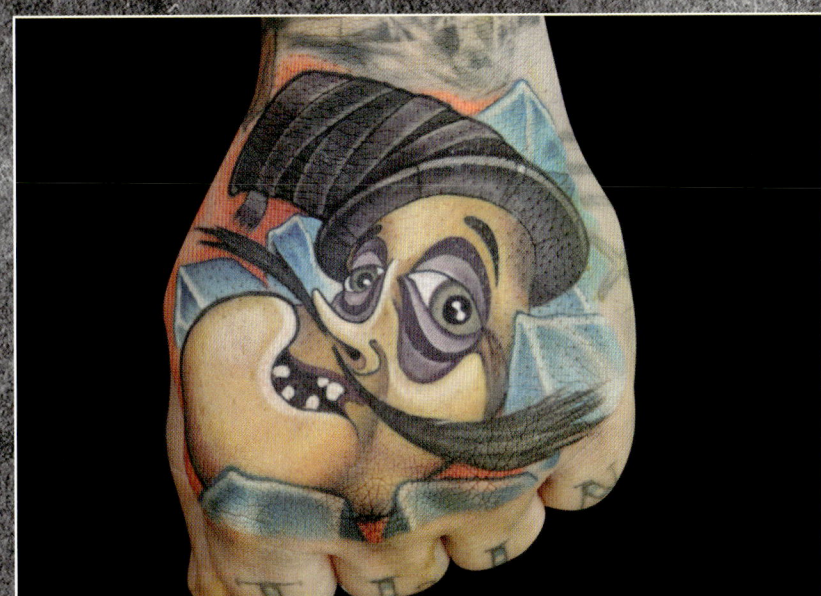

Jason Stephan

Gunnar

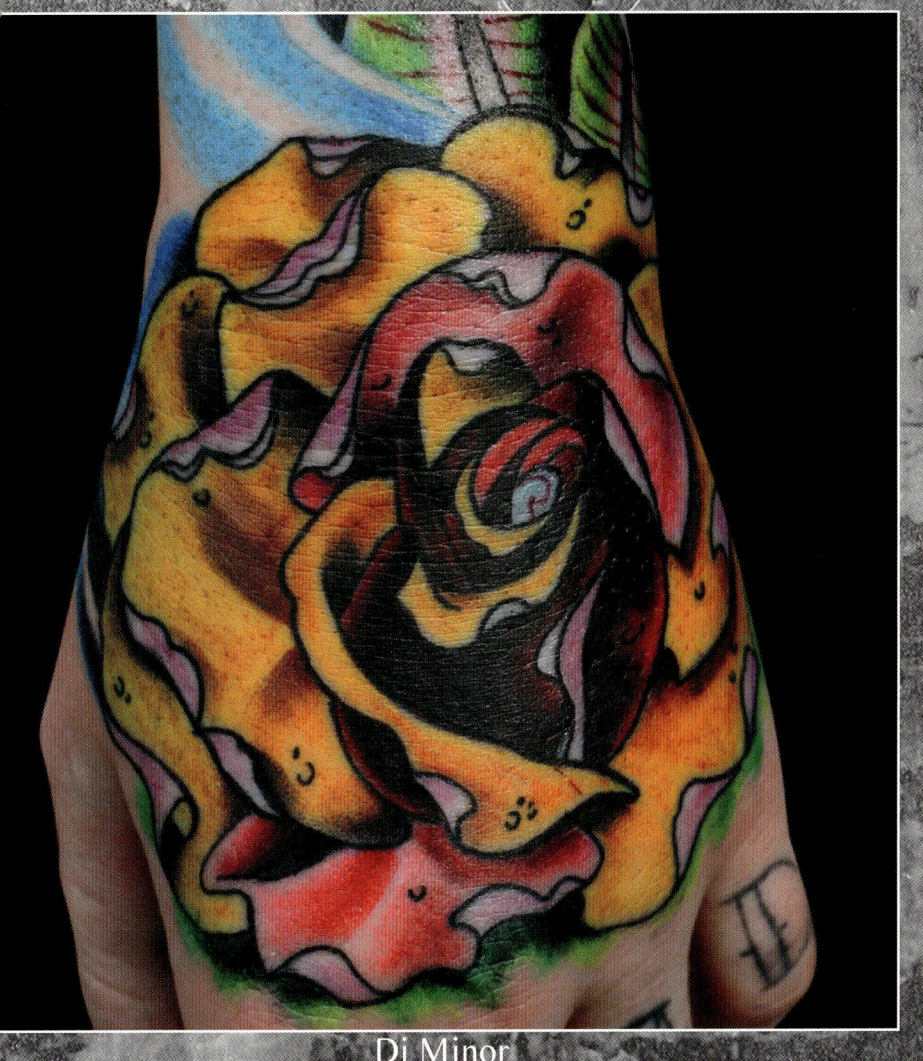

Dj Minor

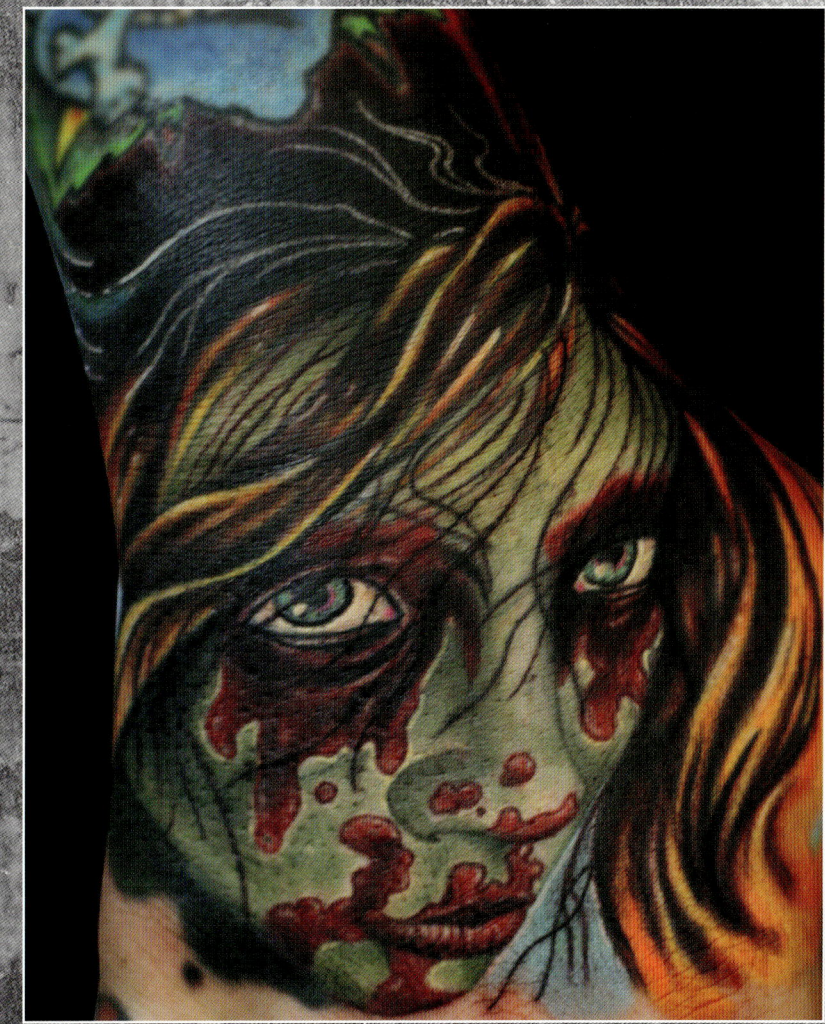

Dj Minor

Gunnar

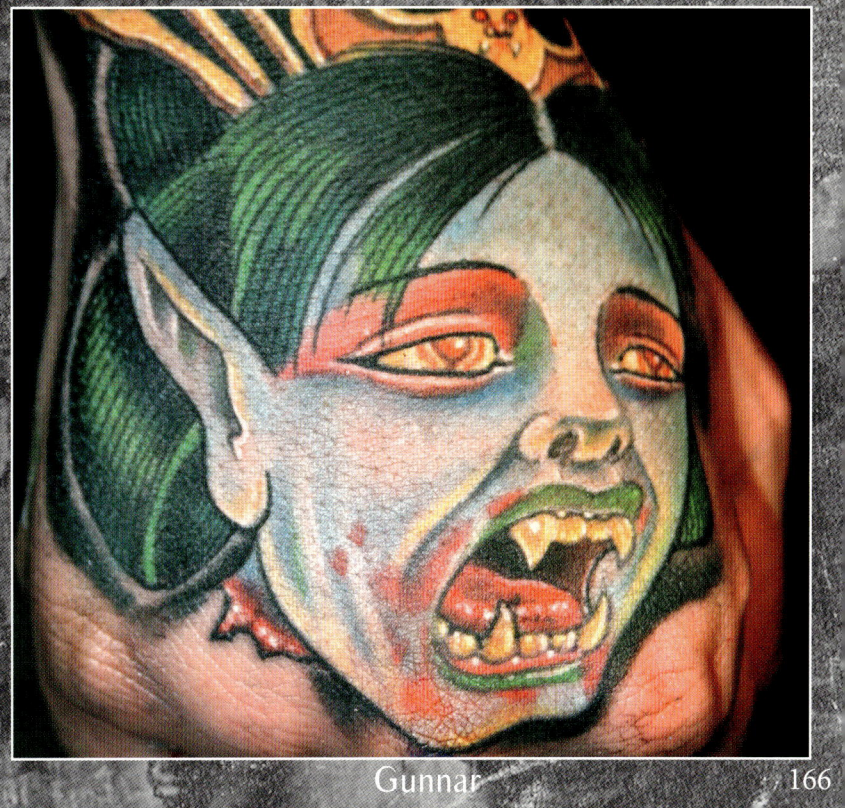

Gunnar

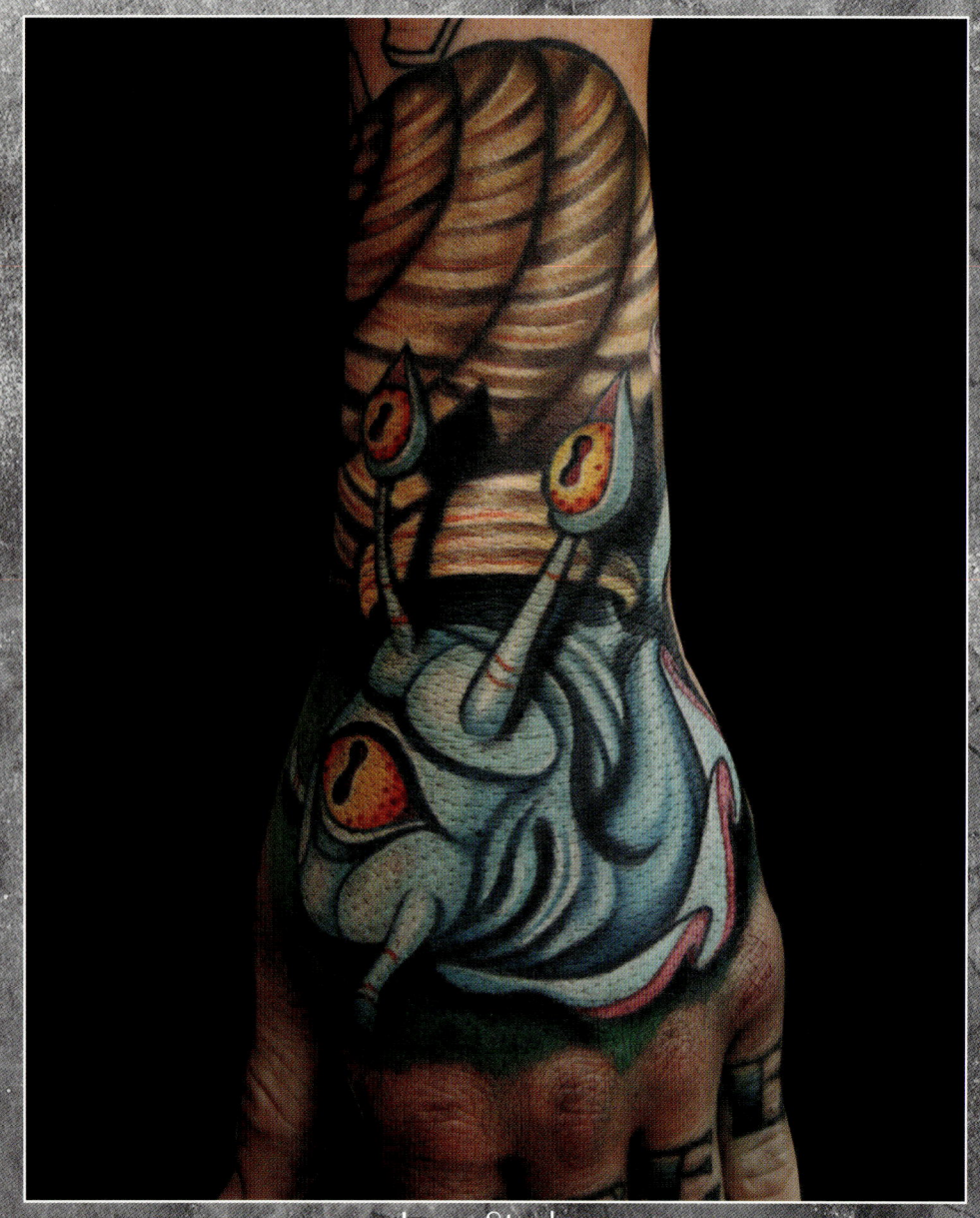

Jason Stephan

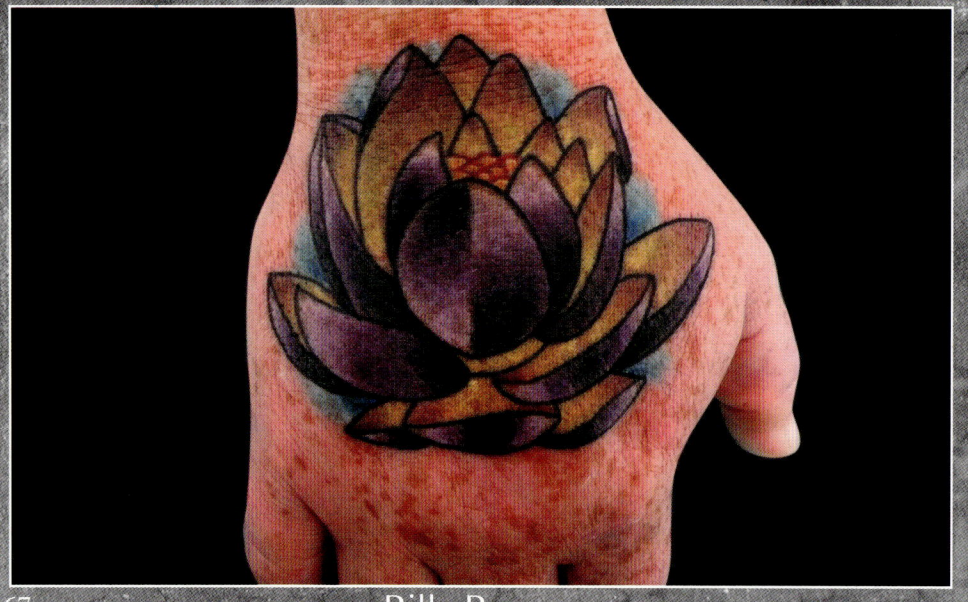

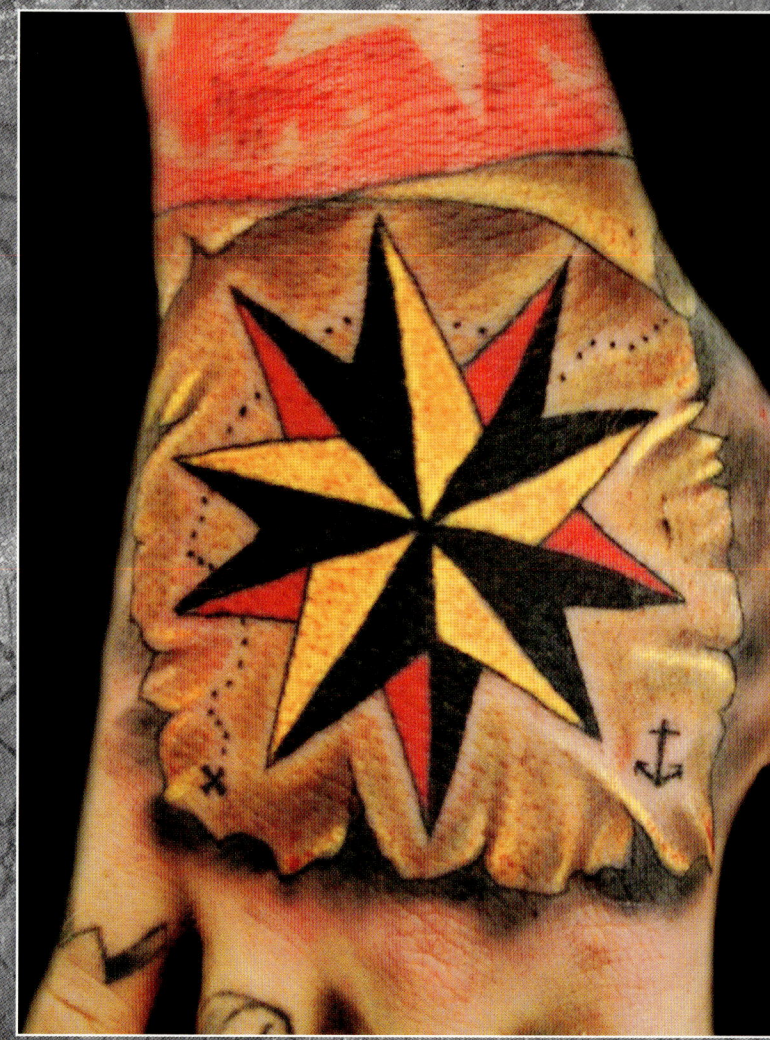

Brandon Bond

Billy Beans

Billy Beans

Sean Herman

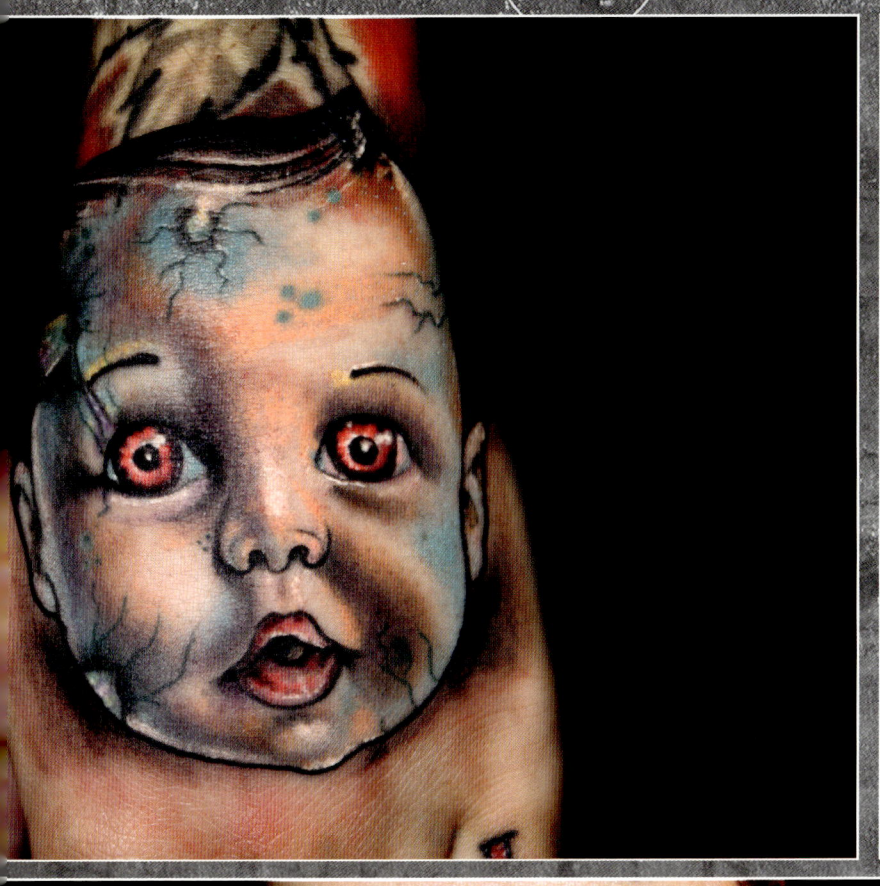

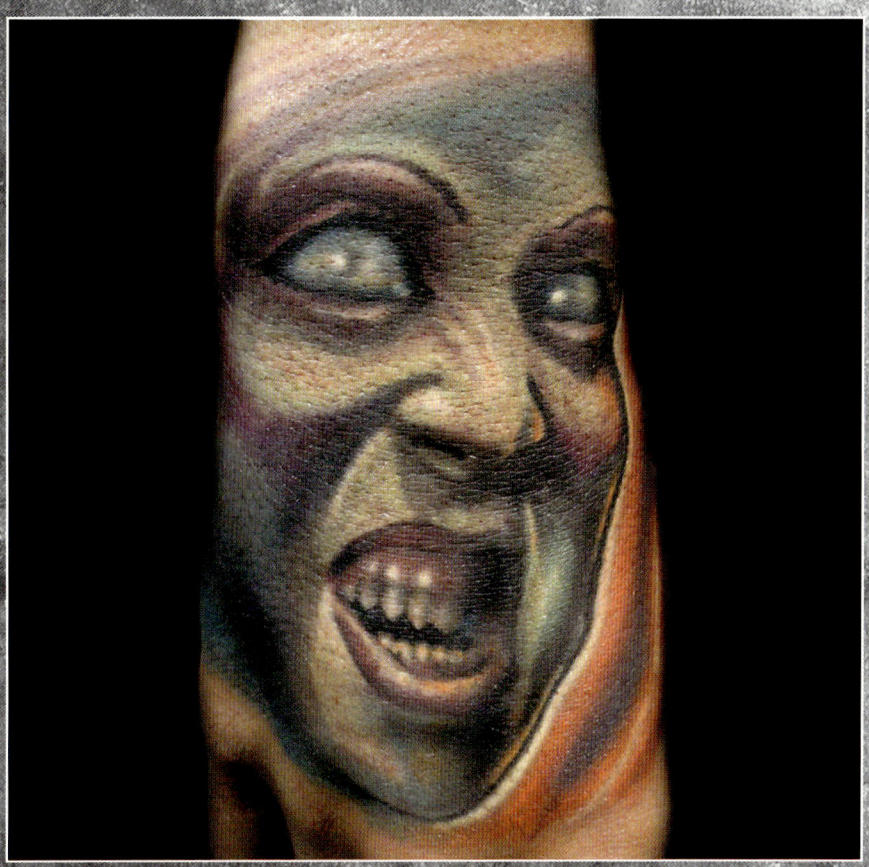

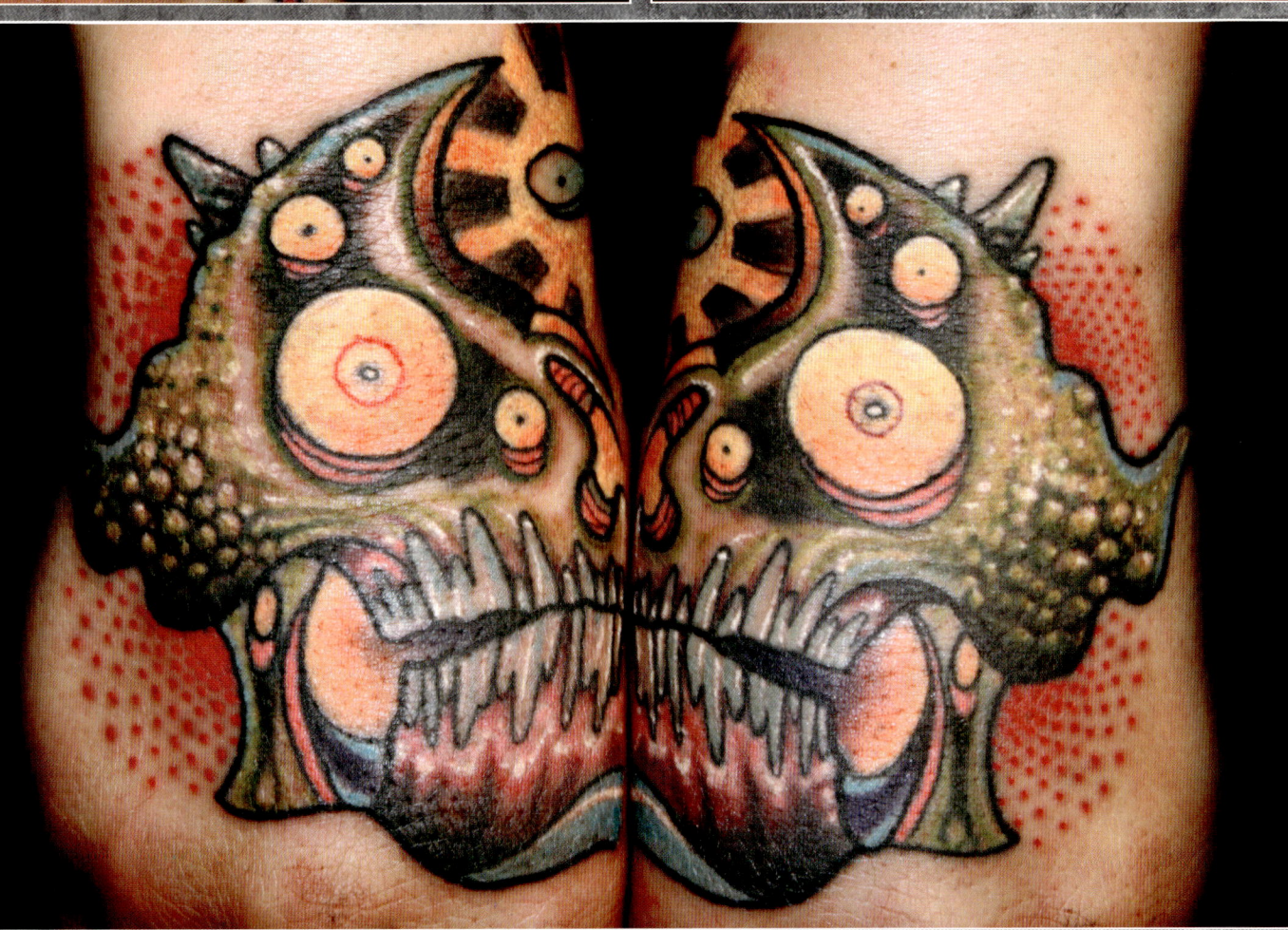

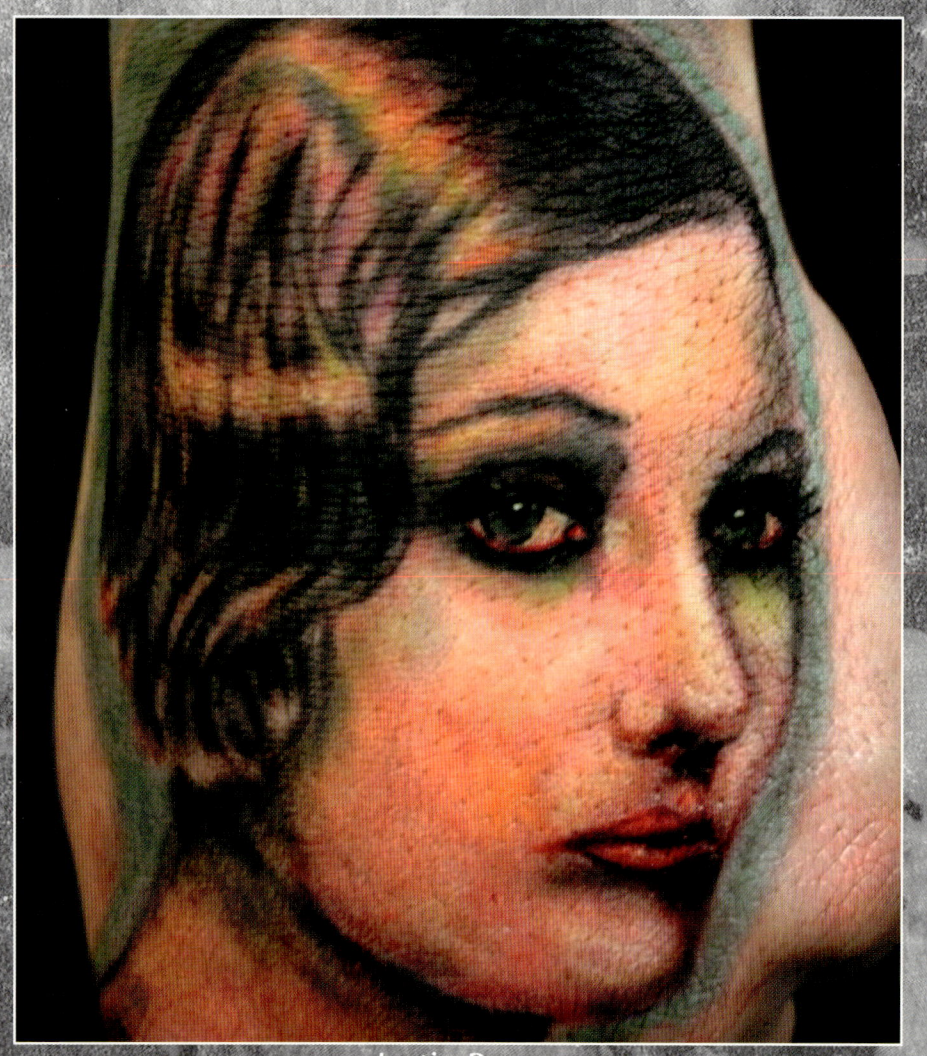

Justin Page

Justin Page

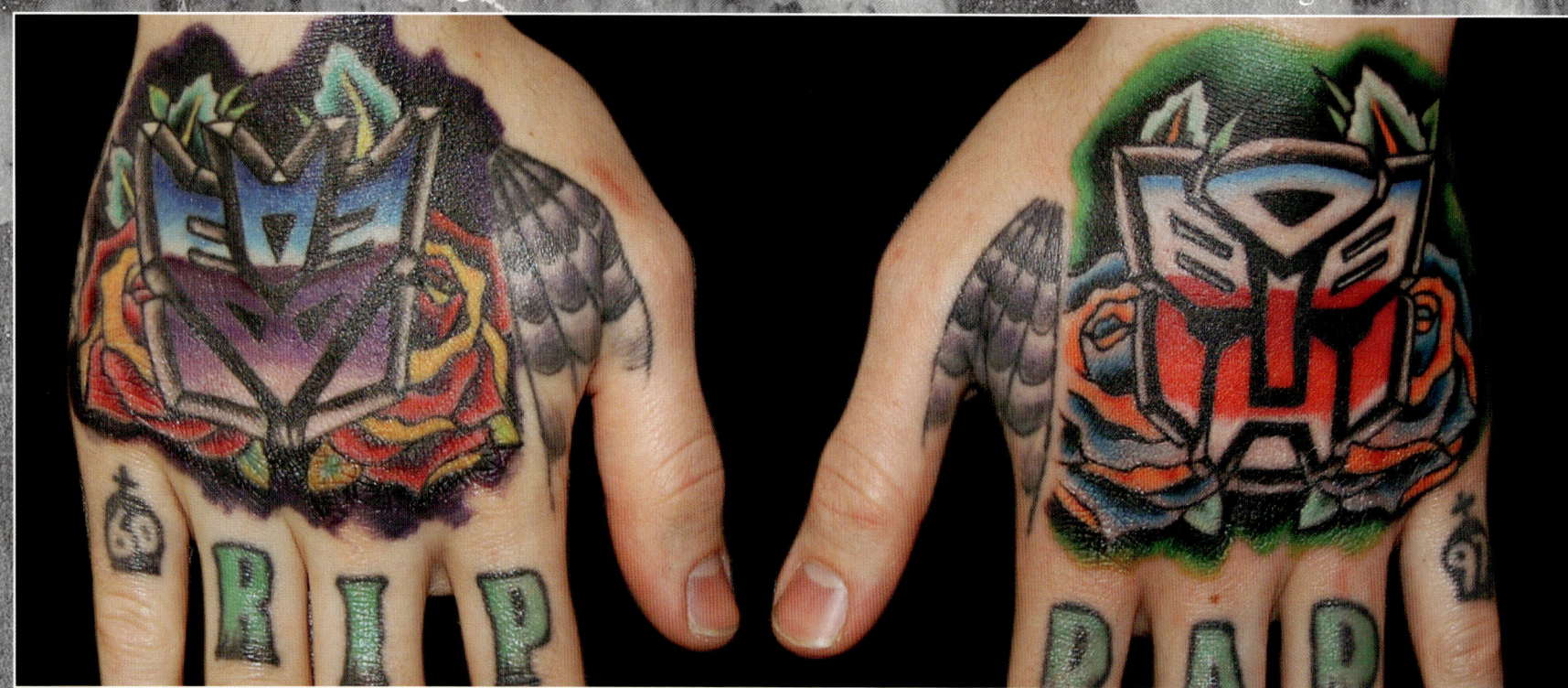

Pablo Jara

Andy Engel

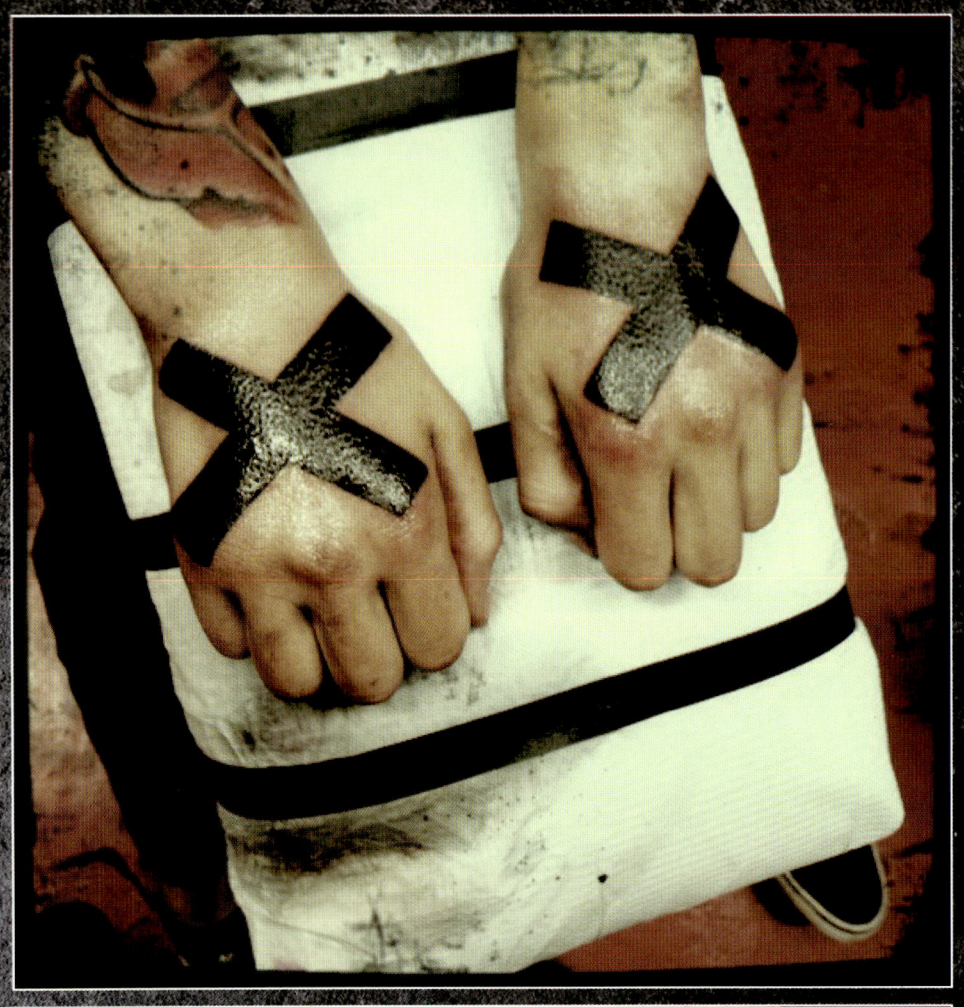

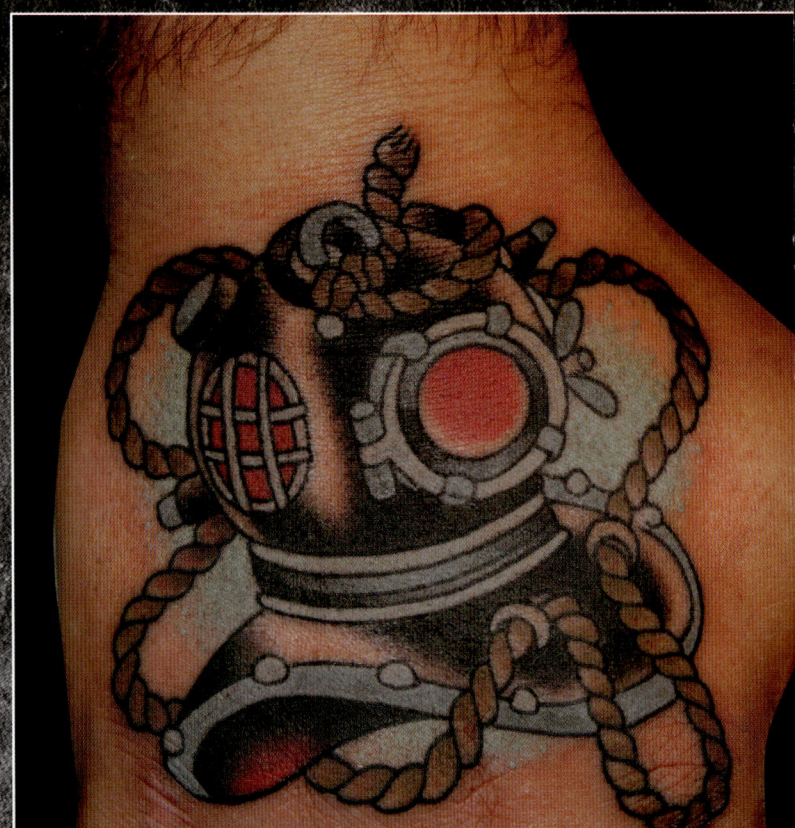

Dan Smith

Dan Smith

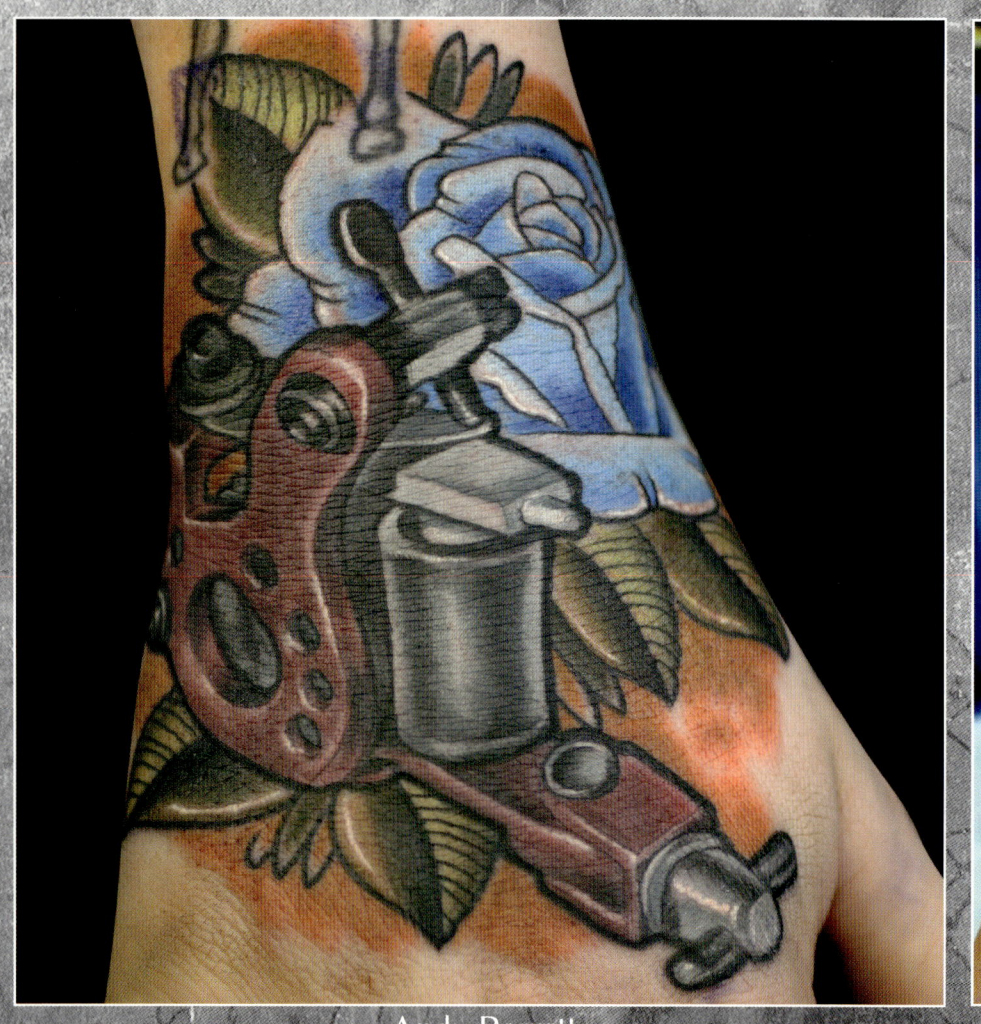

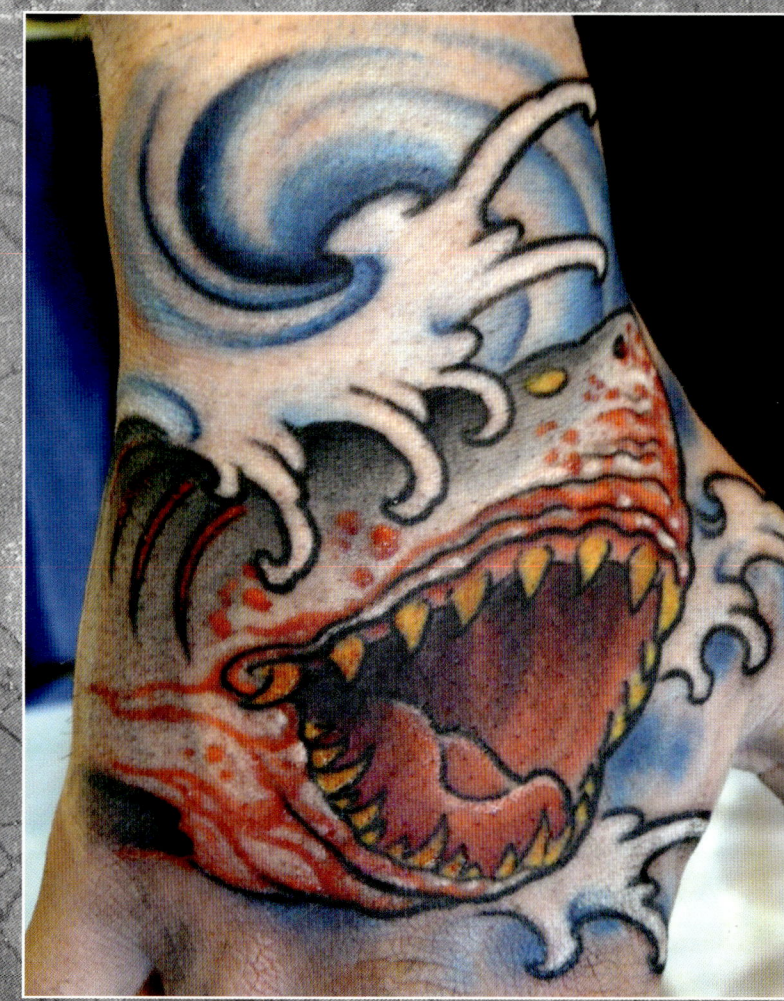

Andy Barrett

Scott Kelly

Dave Barton

Oddboy

Aric Taylor

Oddboy

Aric Taylor

Aric Taylor

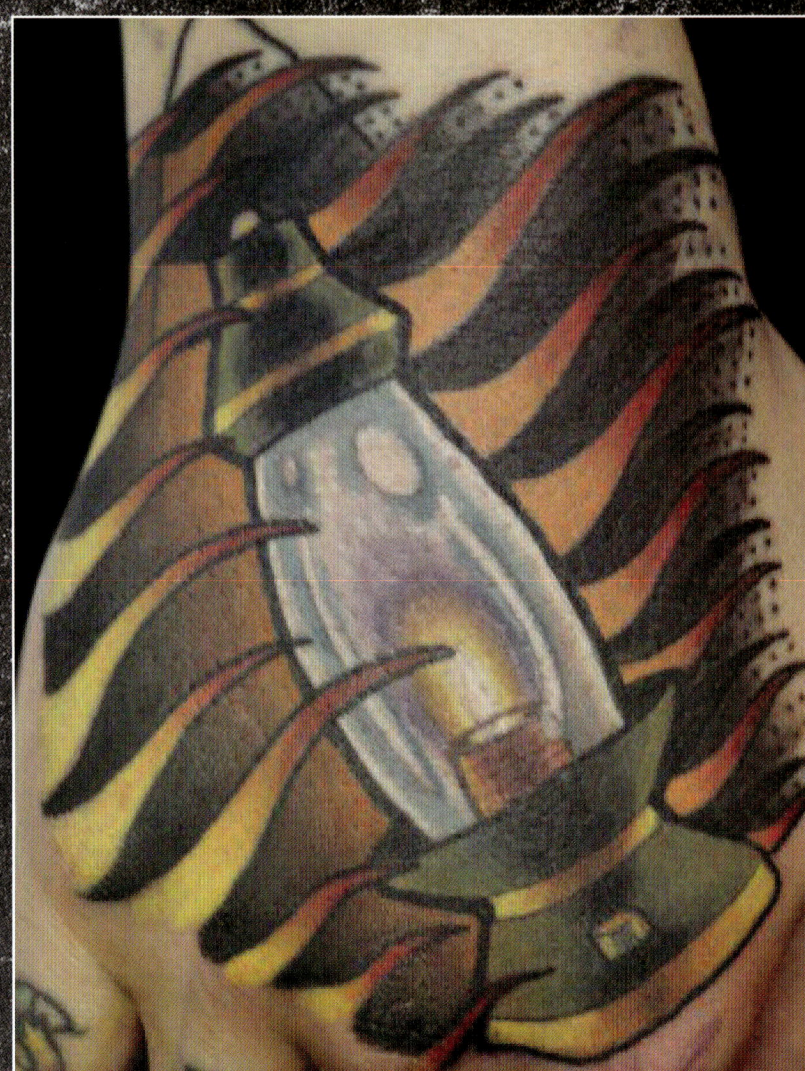

Jason Reeder

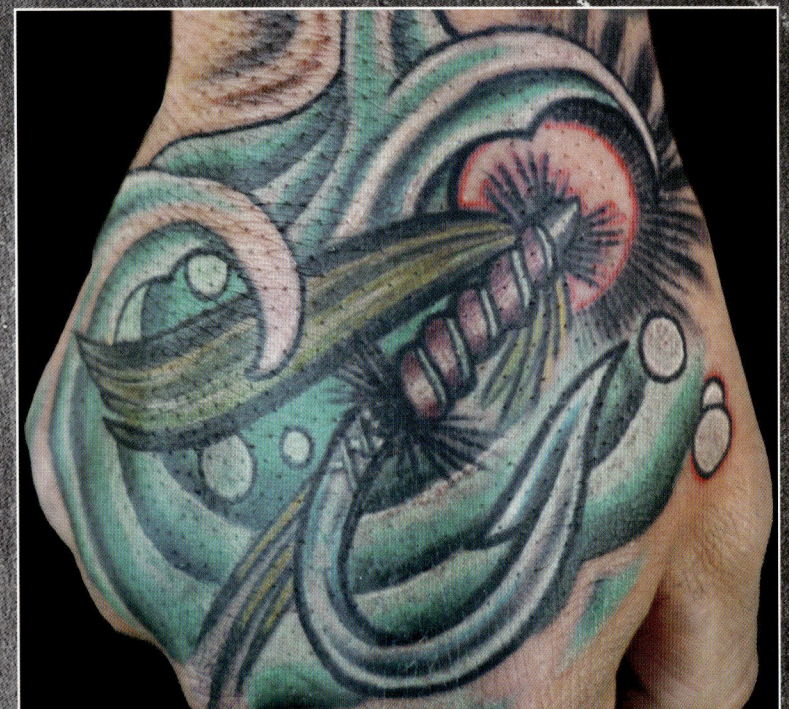

Becka MacDonald

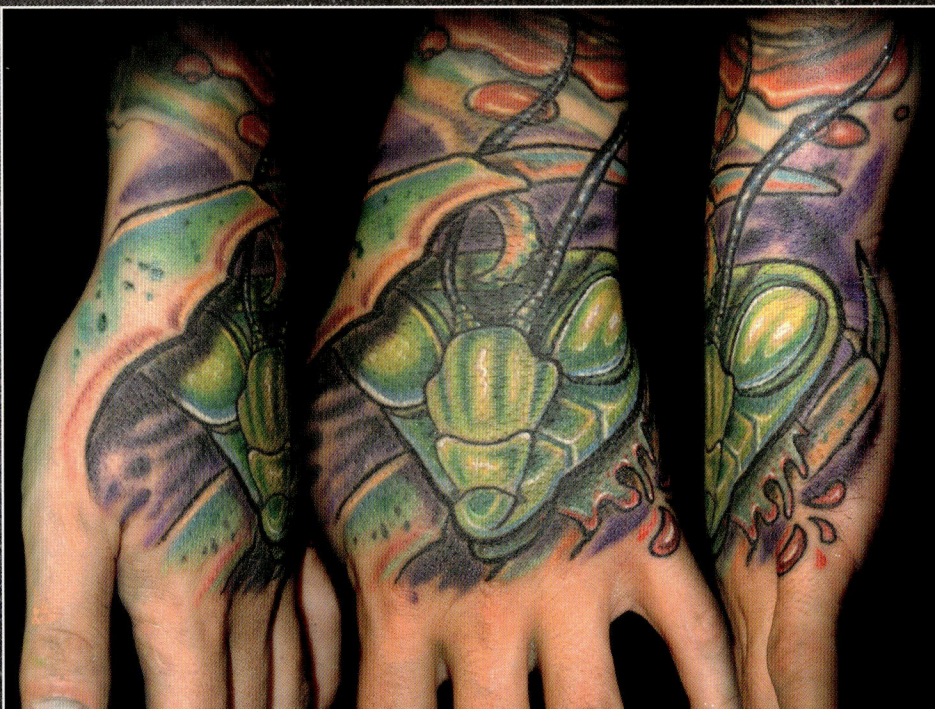

Dave Barton

Dave Barton

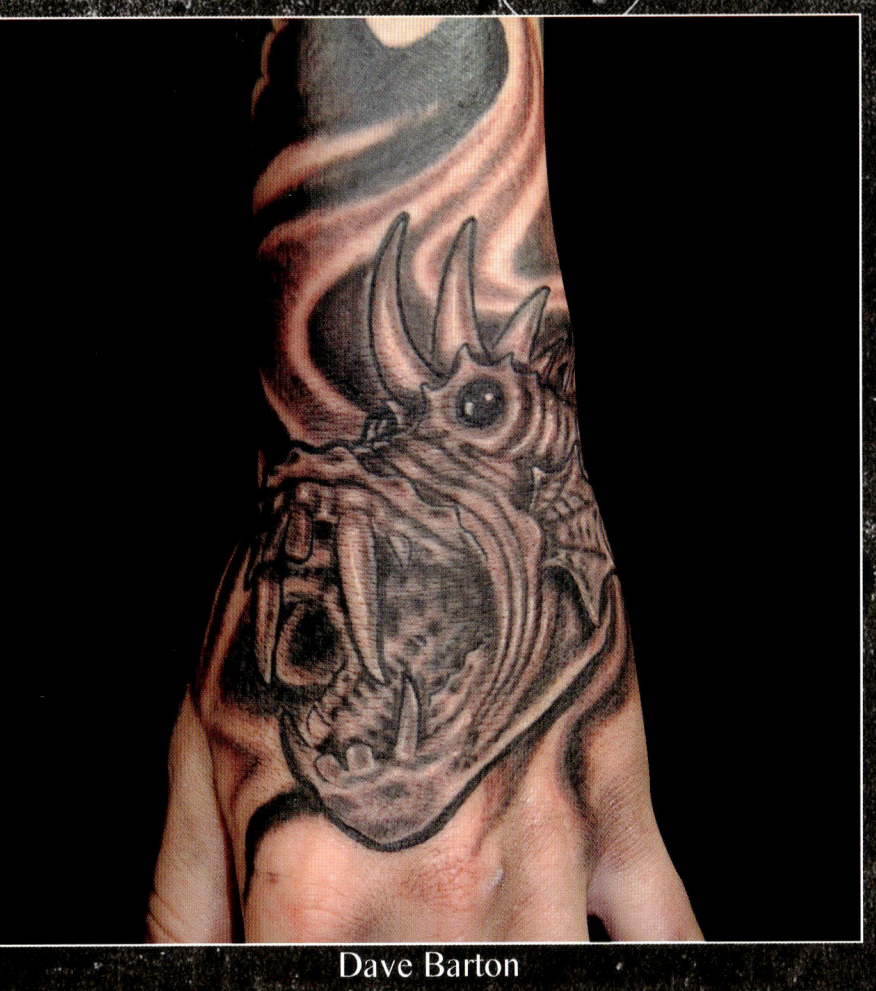

Bugs

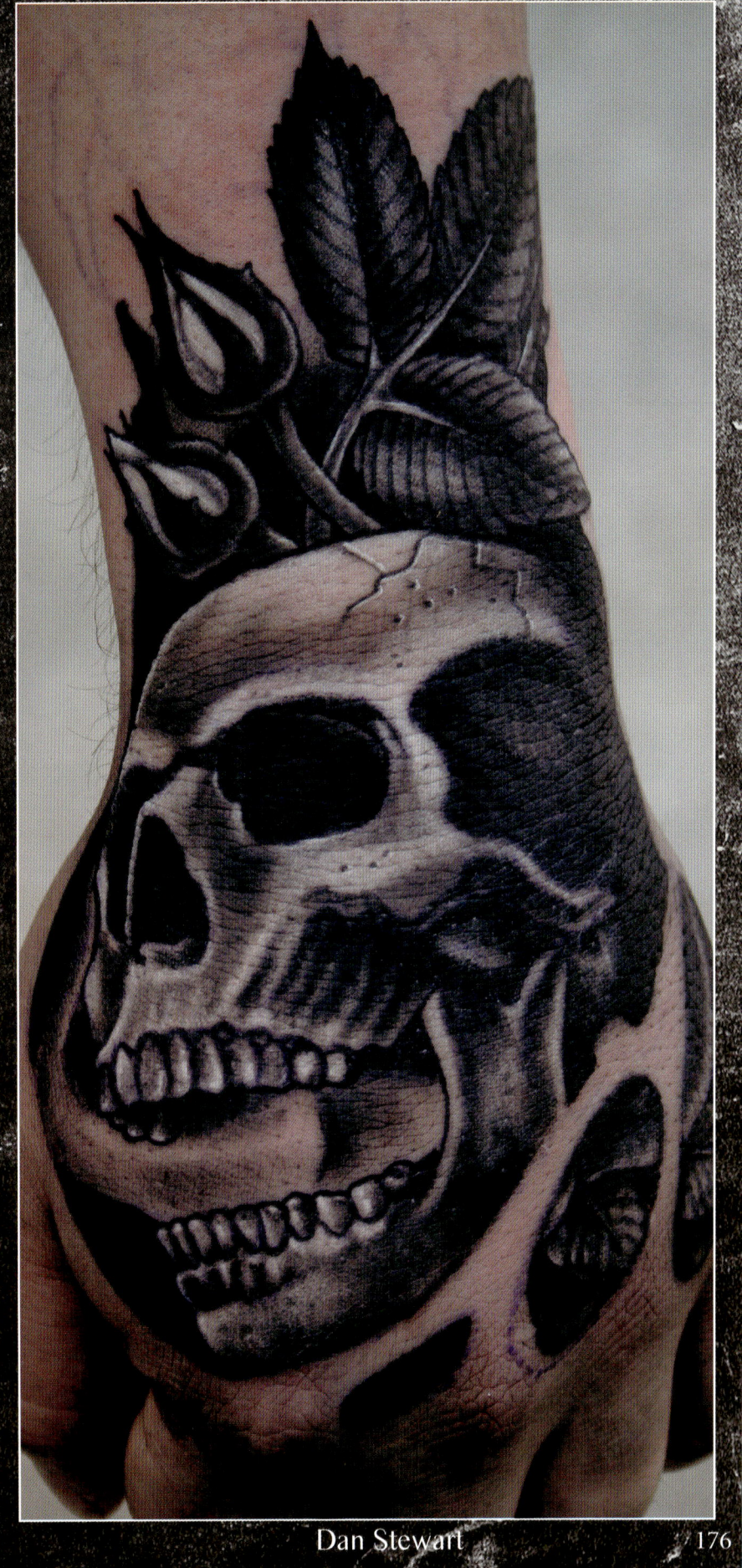

Dan Stewart

176

Jace Masula

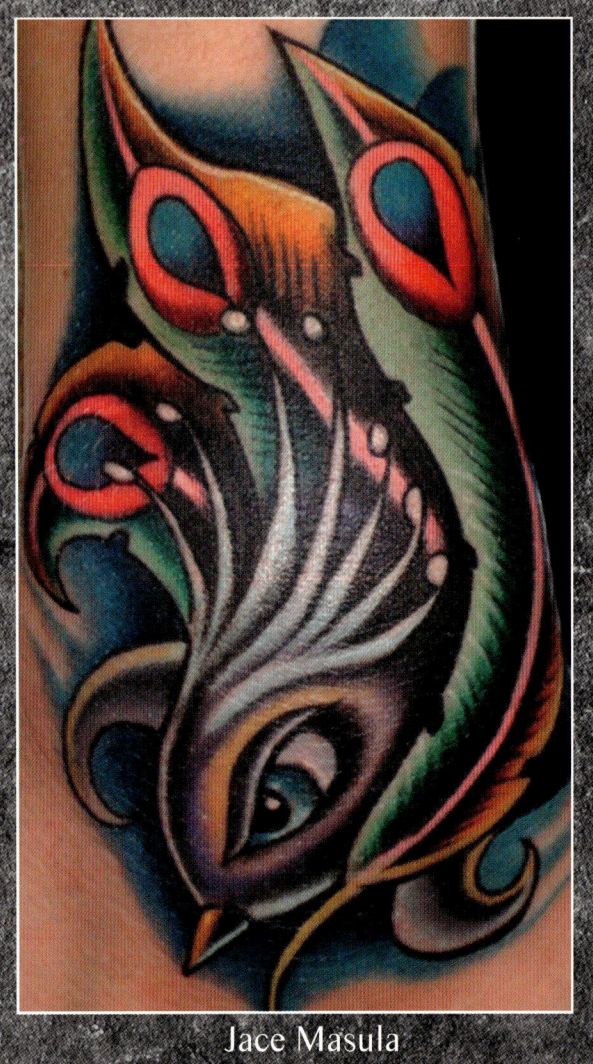

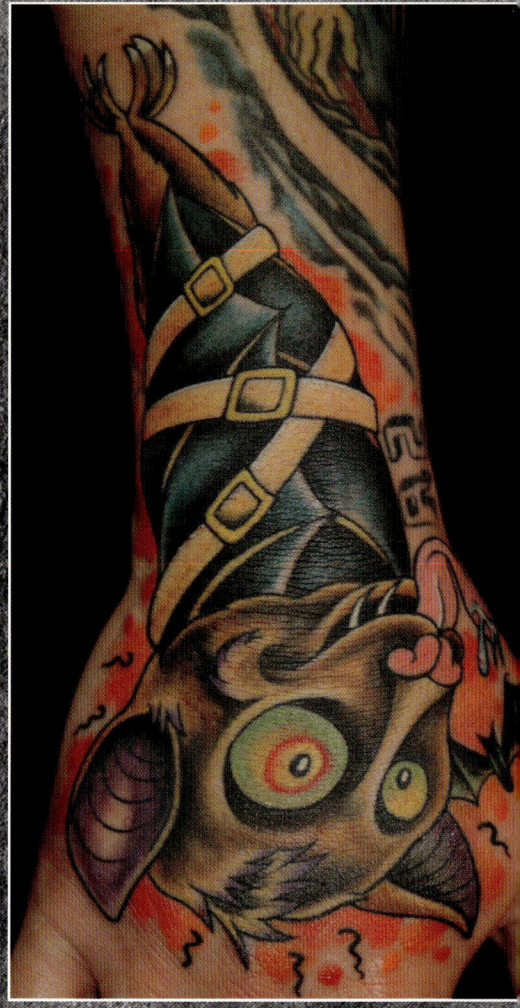

Dave Tedder

Holly Azzara

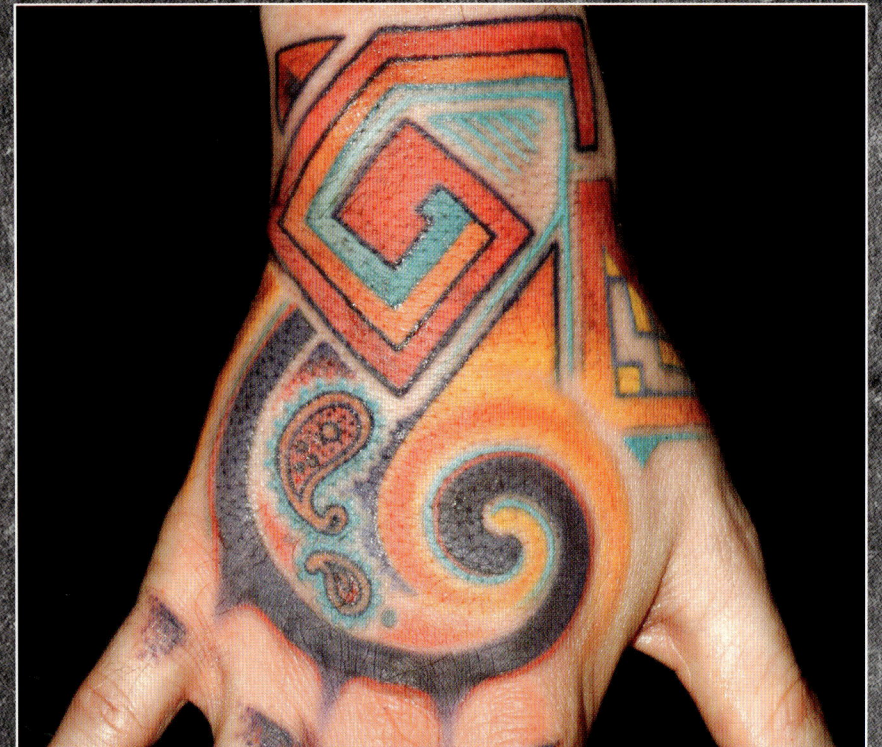

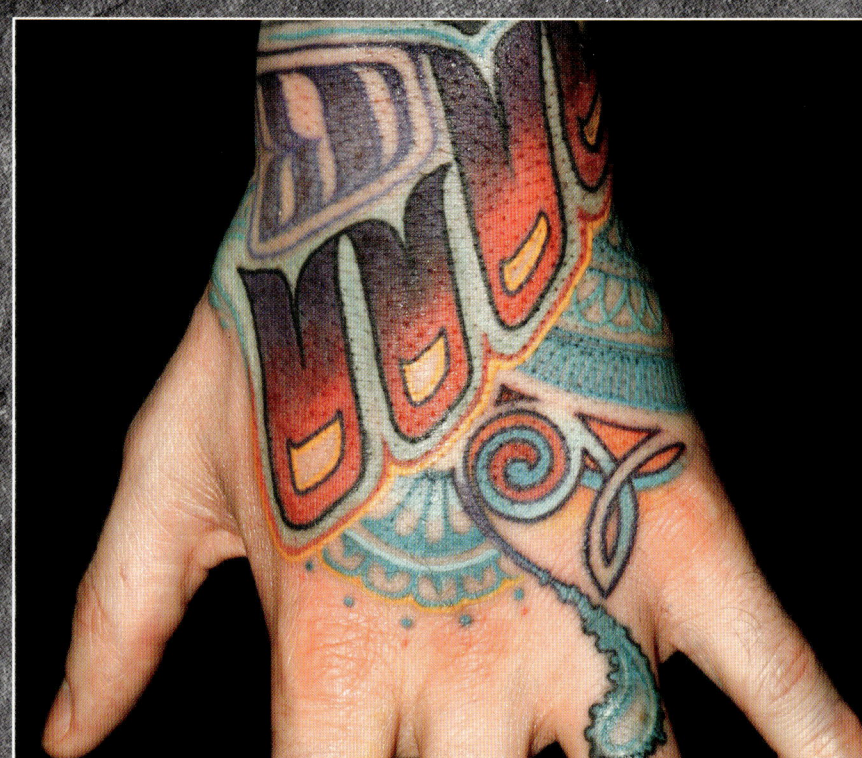

David Glantz

David Glantz

Marvin Silva

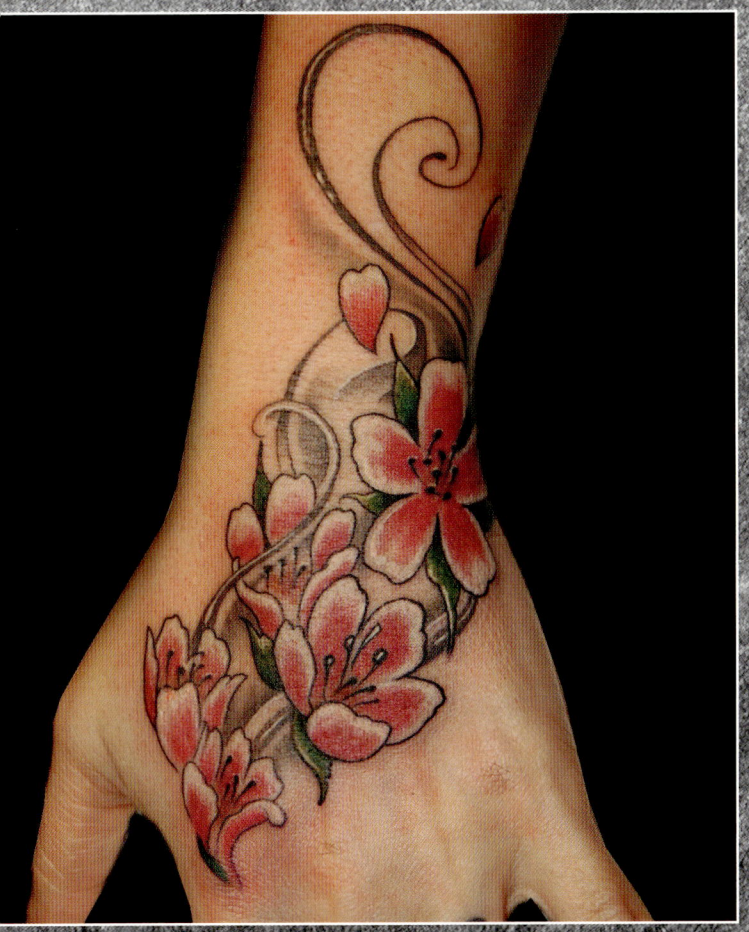

Matteo Cascetti

John Anderton

THE FEET

"Be sure you put your feet in the right place, then stand firm."- Abraham Lincoln

A third of all the bones in our bodies are located in our feet; in fact, there are 26 bones to be exact. Each foot also has 33 joints, 19 muscles, 10 tendons and 107 ligaments, making it quite an essential limb. When you consider how intricate the inner workings of the feet are, it seems only fitting that this body part should be used as a canvas for some wondrous tattoo artwork.

While most of the tattooed extremities featured in this book are difficult to conceal on a day-to-day basis, the feet are the exception to that rule, as they can easily be covered up with socks and shoes. On the flipside, tattooed feet can be brazenly displayed when barefoot, or while wearing sandals or open shoes. This transformative quality when it comes to revealing tattooed feet, make this body part choice a very popular location for tattoo collectors to place their designs.

While the skin on the feet can be more thin and tender than other areas of the body, and the protrusion of bones must also be taken into consideration, the duality of the feet and their mirror image quality make them a great spot to construct some very creative tattoo pieces. As you can see from the photos in this chapter, there are a lot of lucky feet in this world, wearing some spectacular and imaginative tattoo designs.

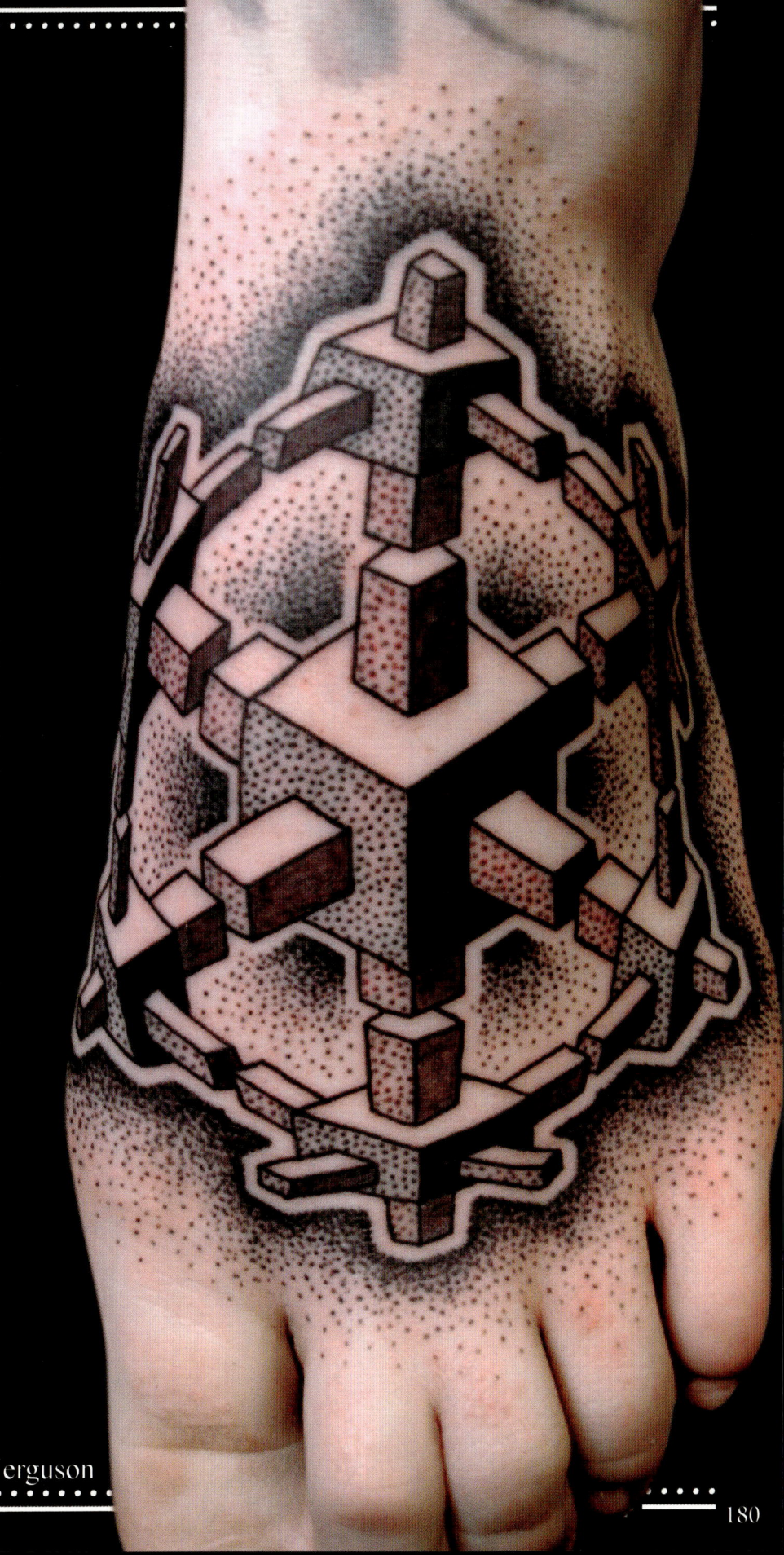

Cory Ferguson

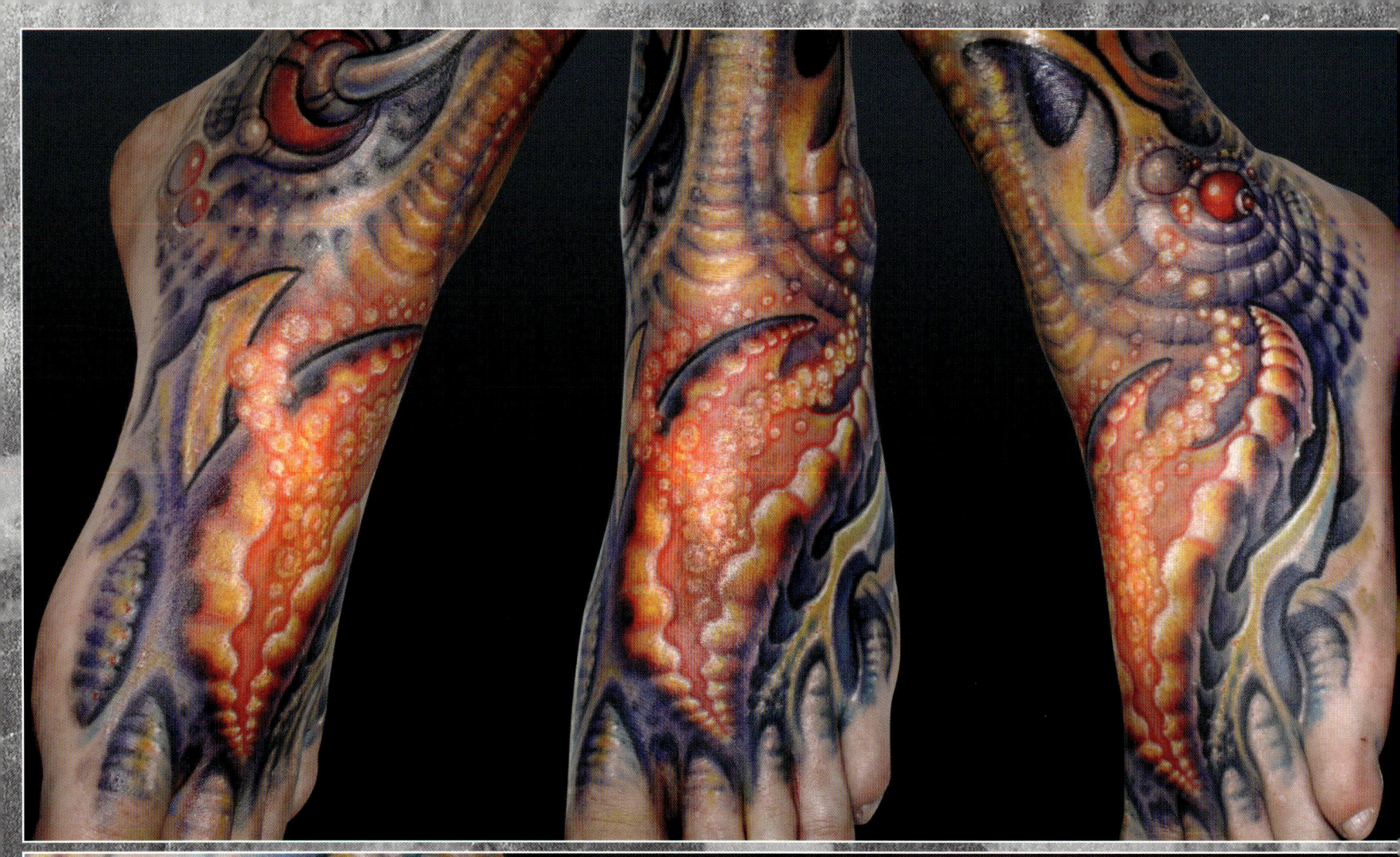

Guy Aitchison

Guy Aitchison

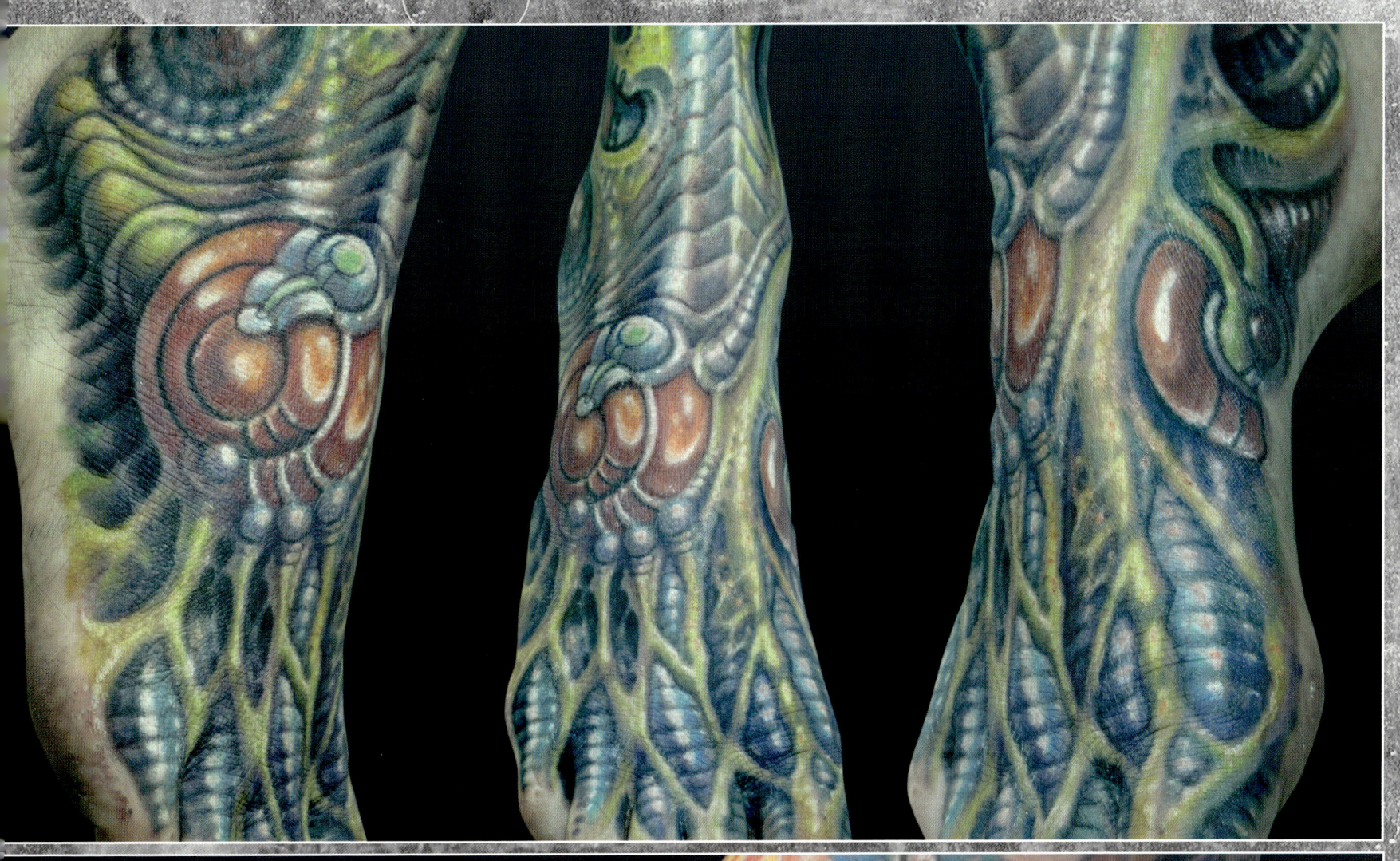

Guy Aitchison

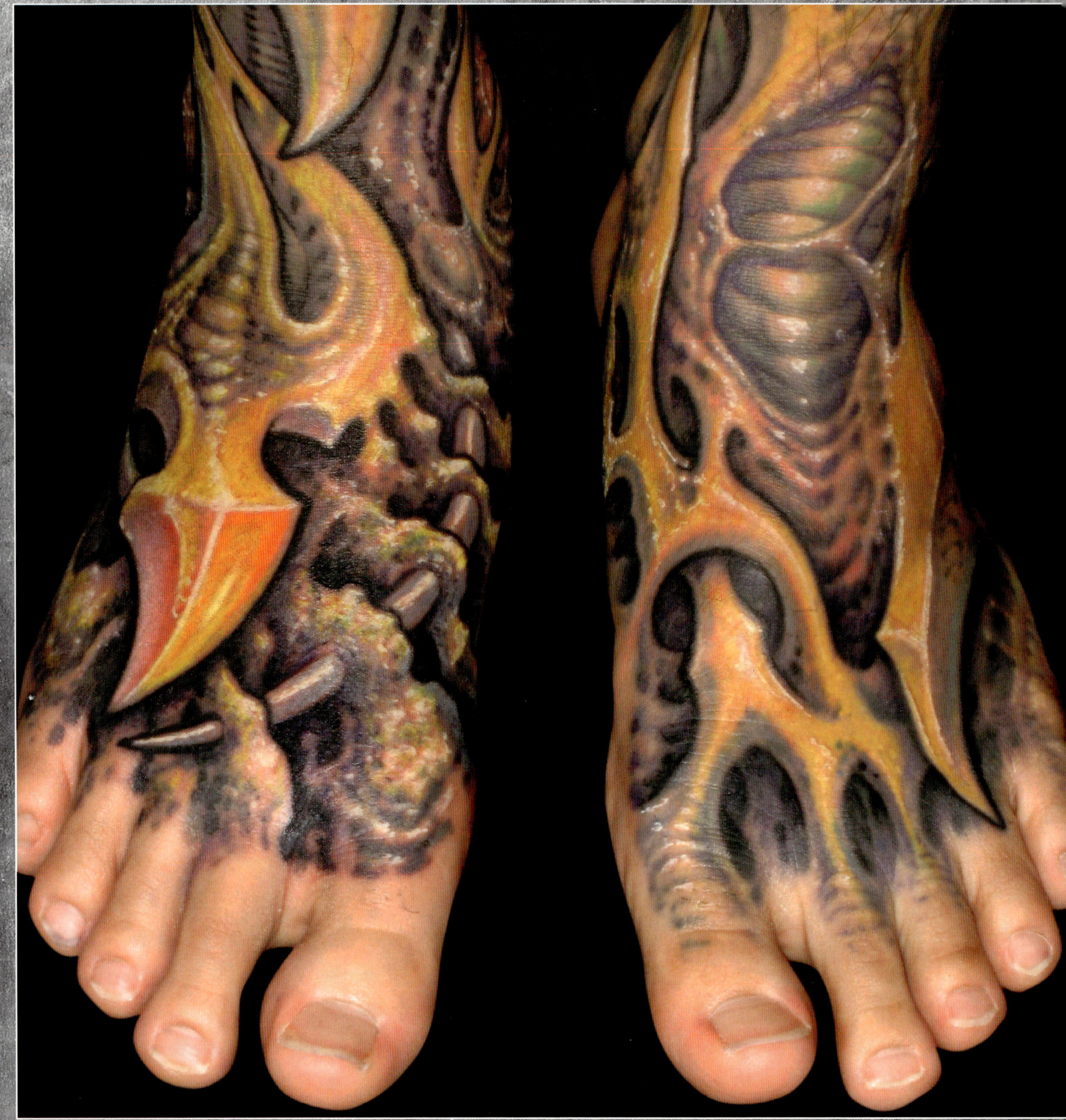

Andy Engel

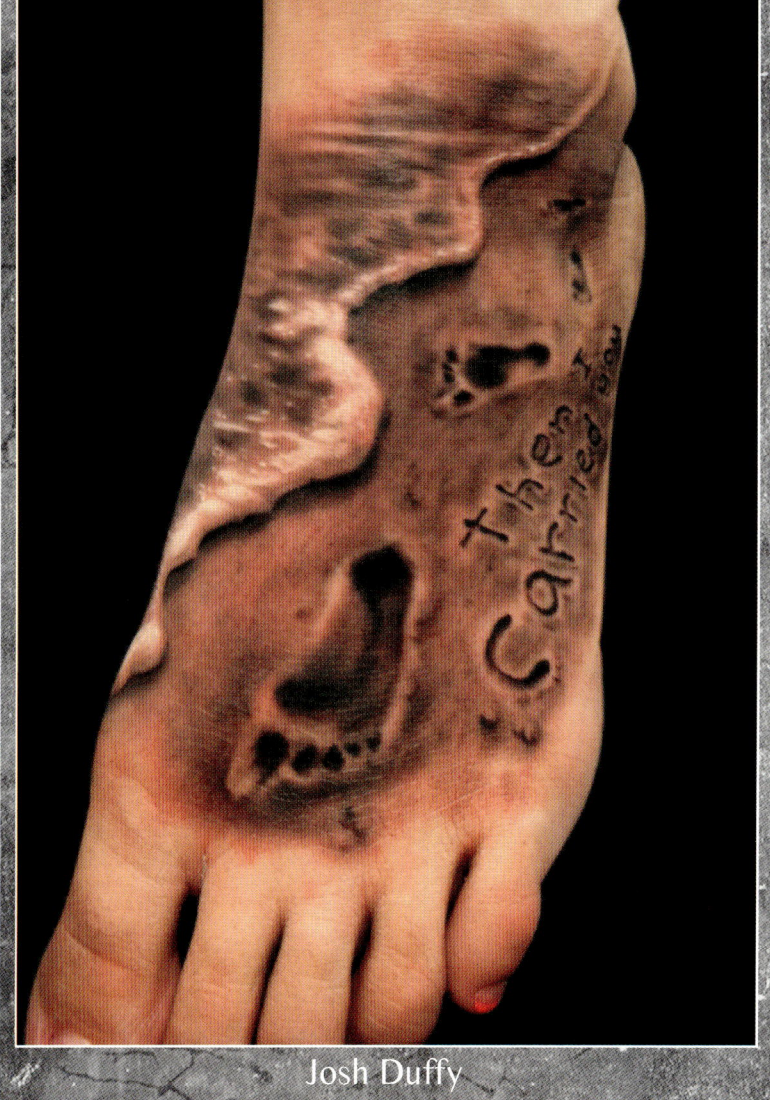

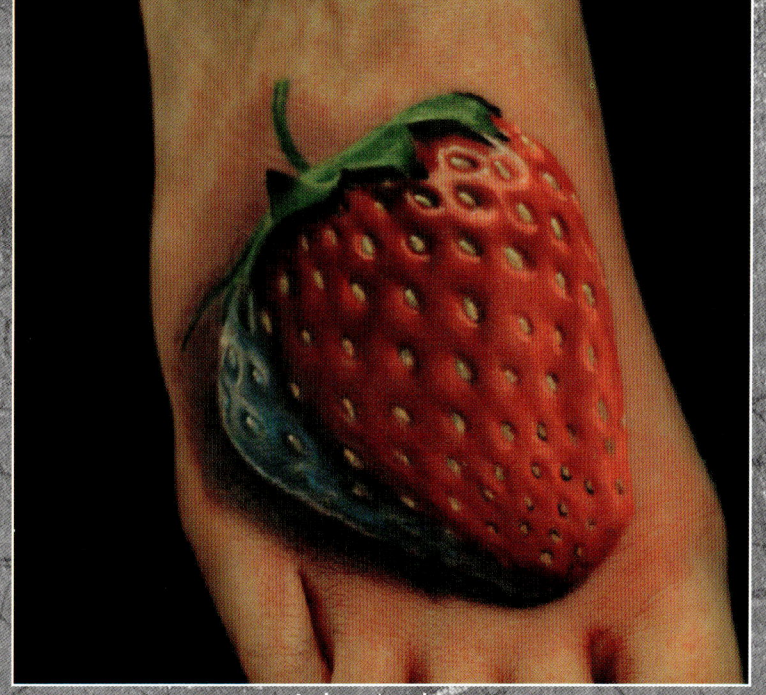

Josh Duffy

John Anderton

John Anderton

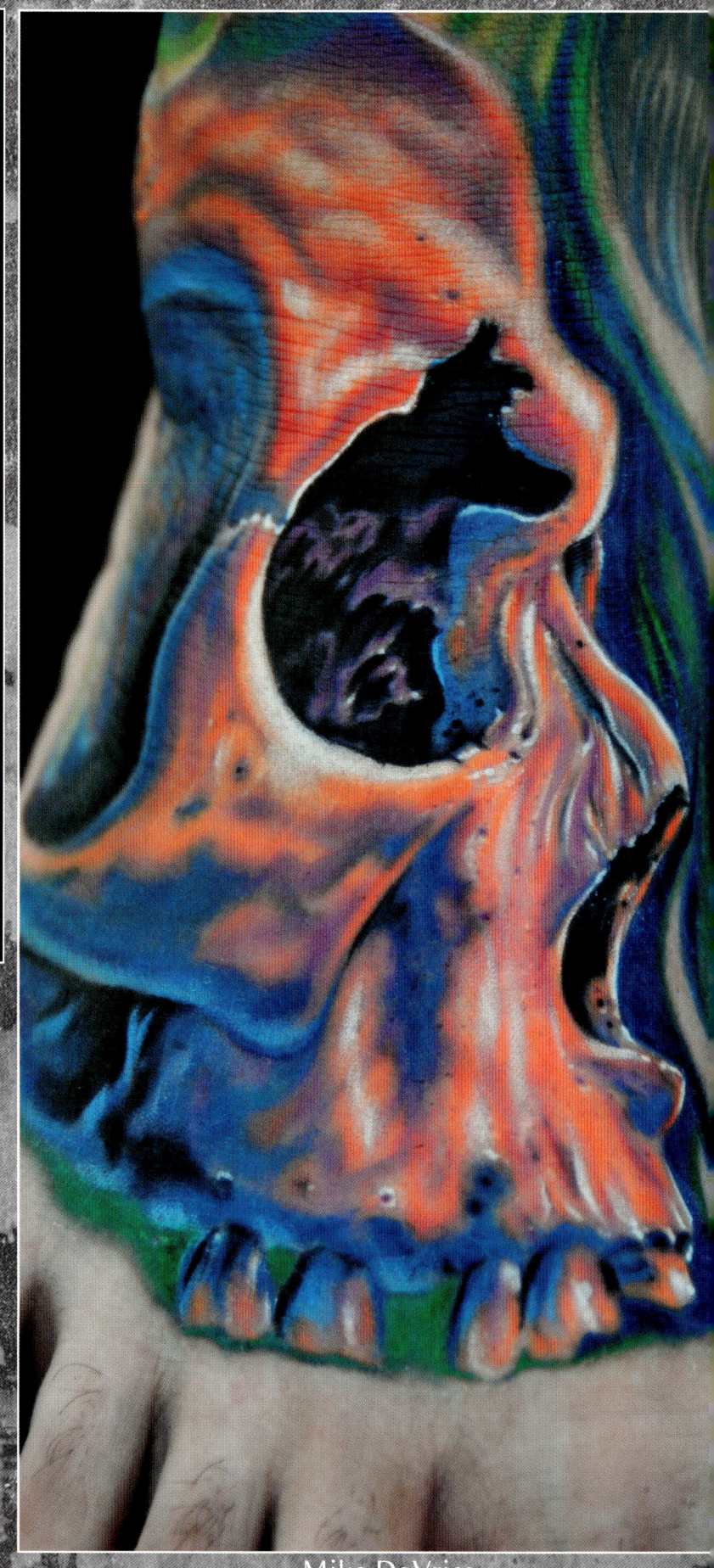

Rakel Nativdad

Mike DeVries

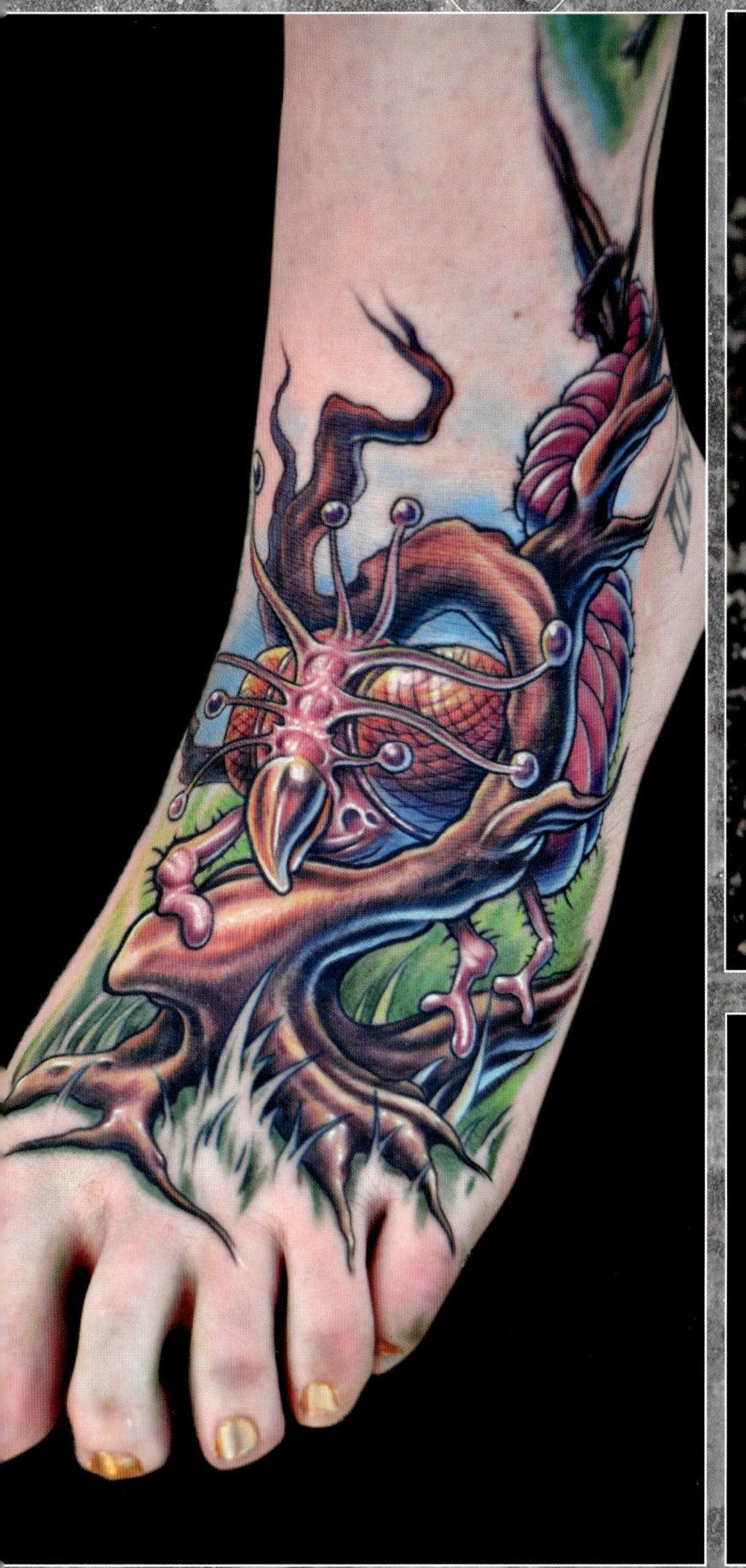

Mike DeVries

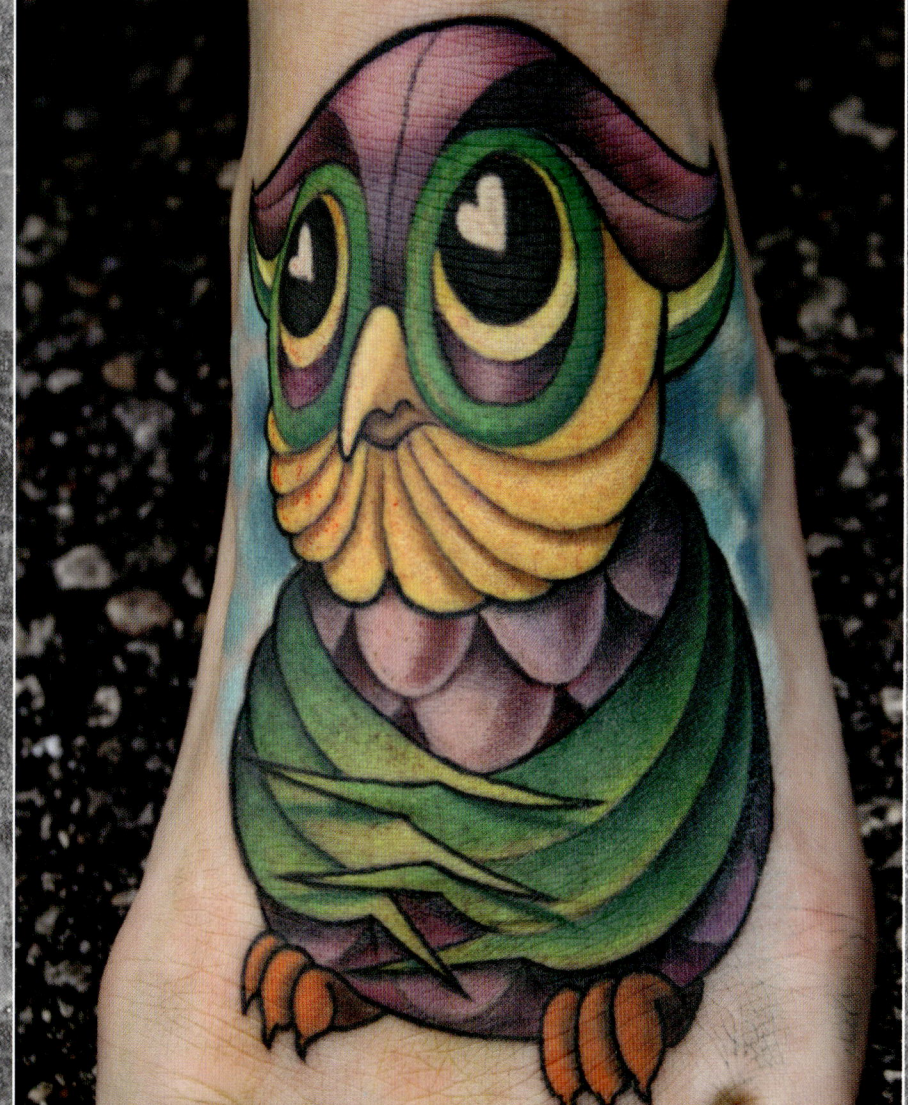

Jason Stephan

Shawn Barber

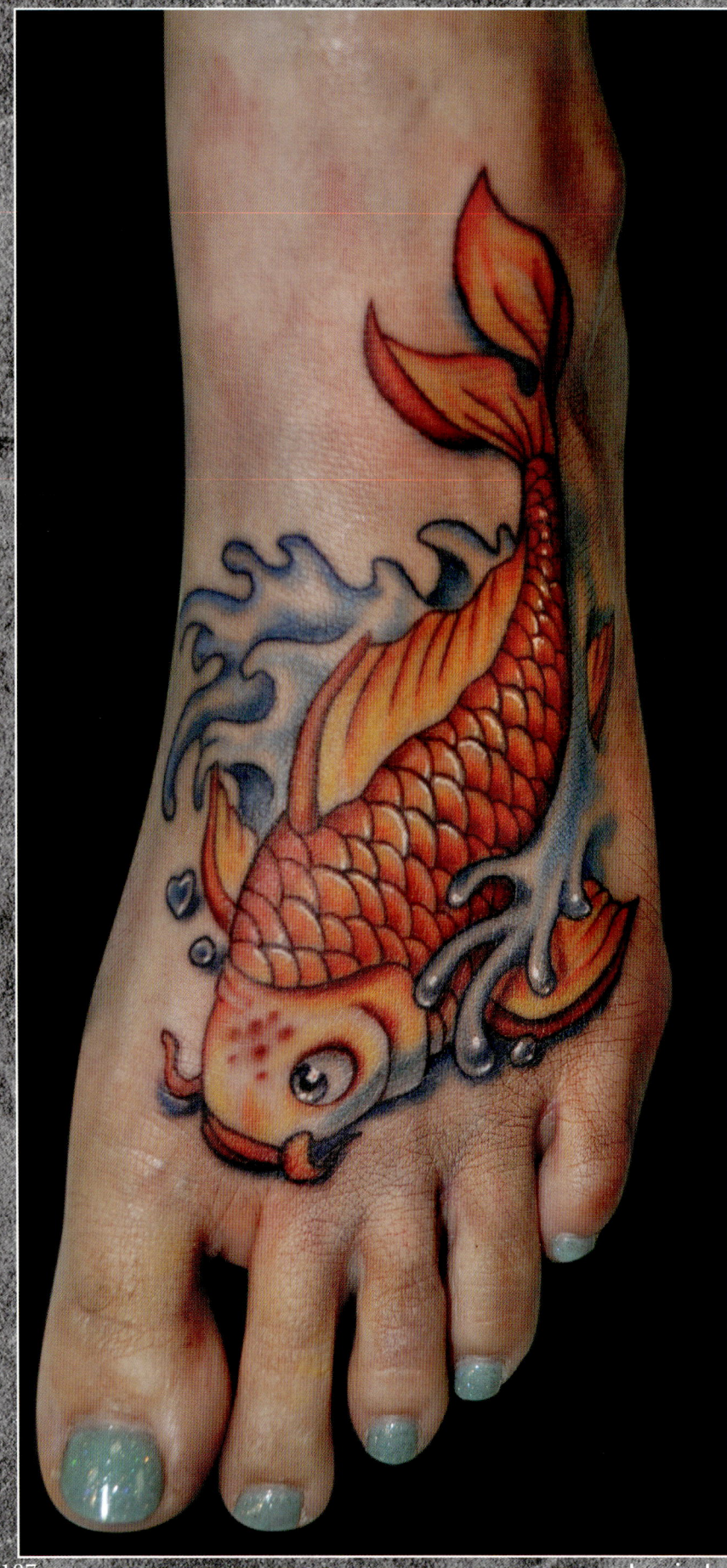

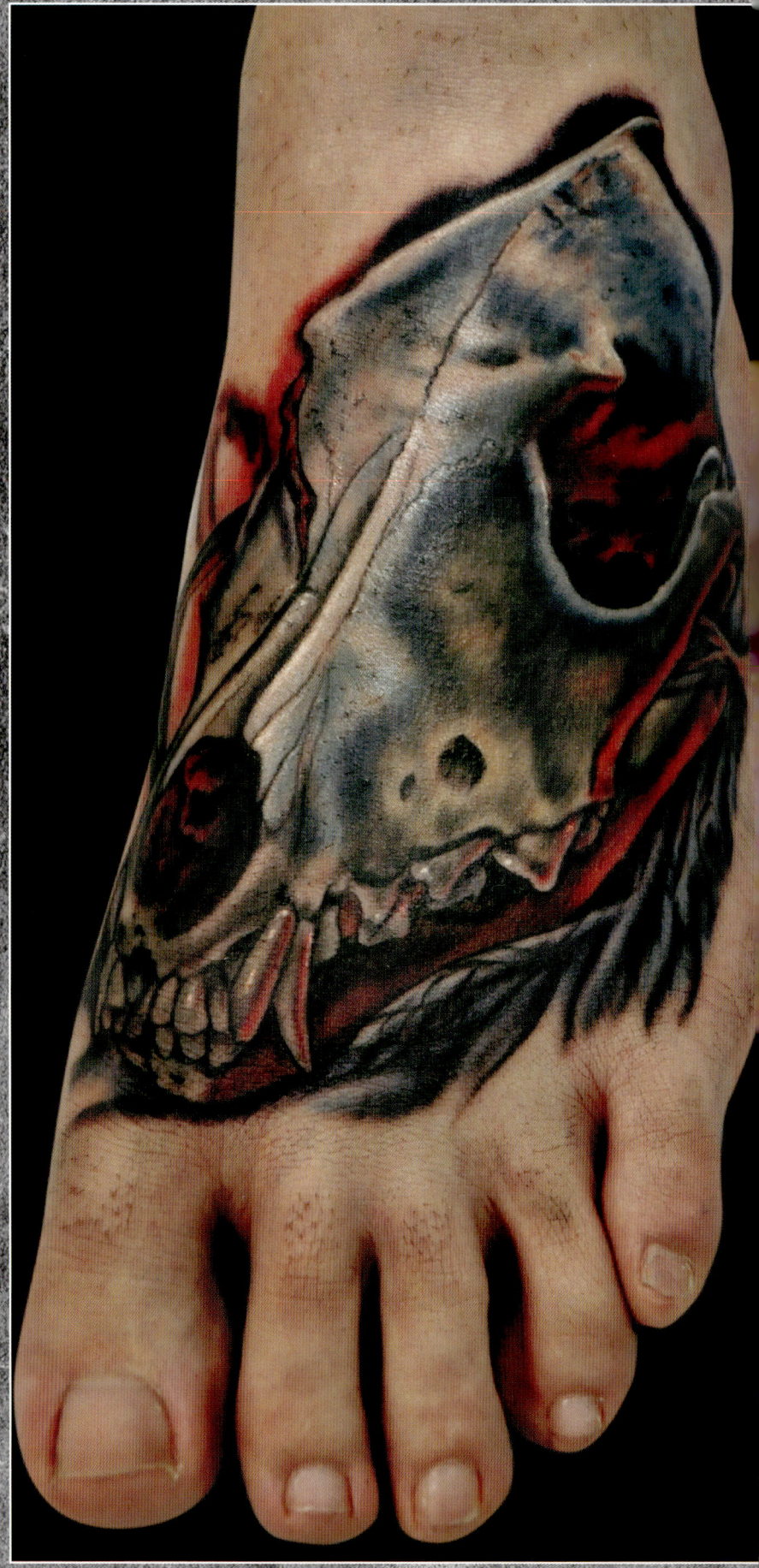

Jamie Lee Parker

Brent Olson

188

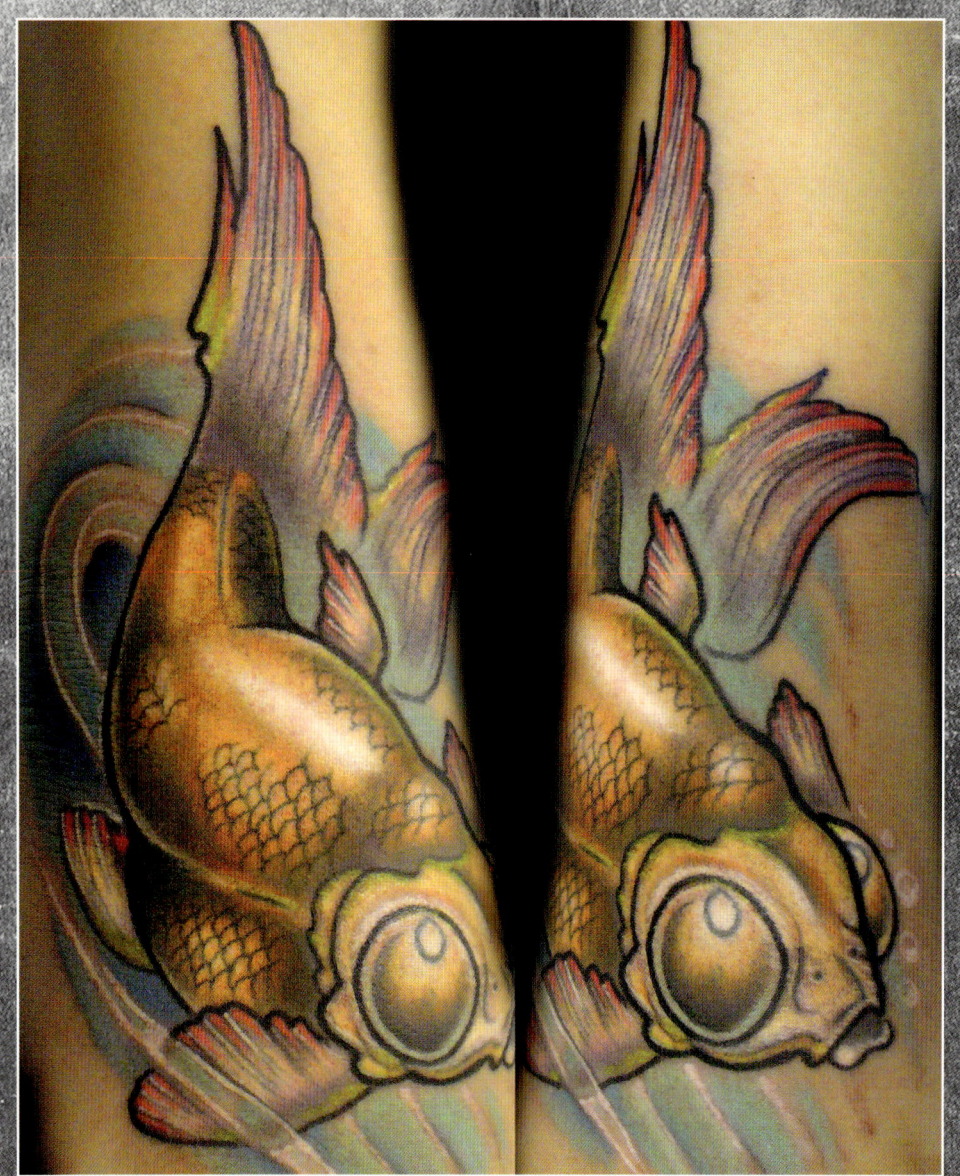

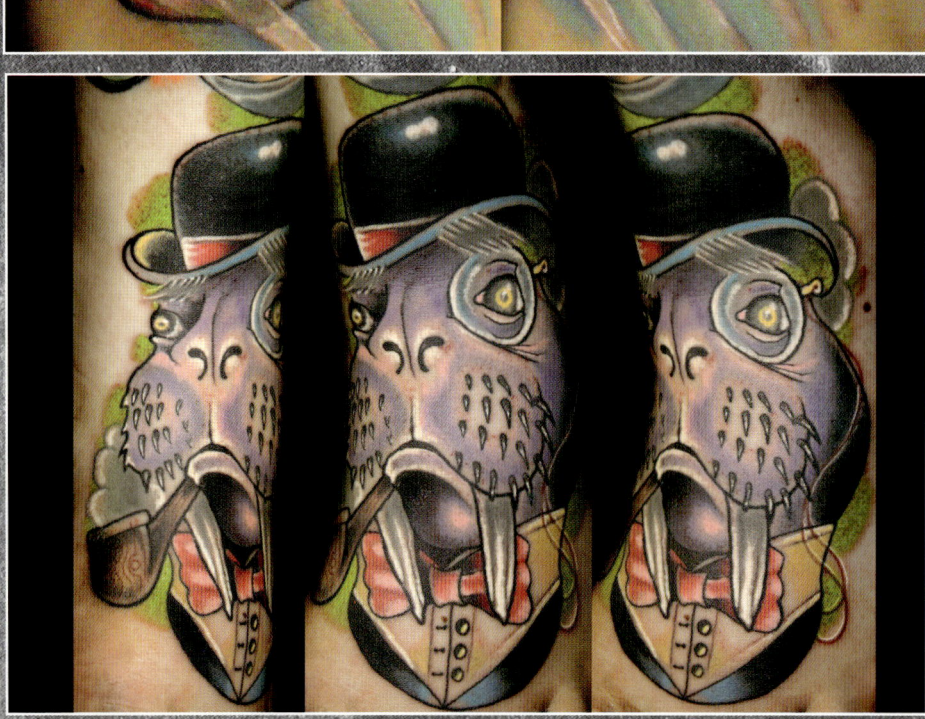

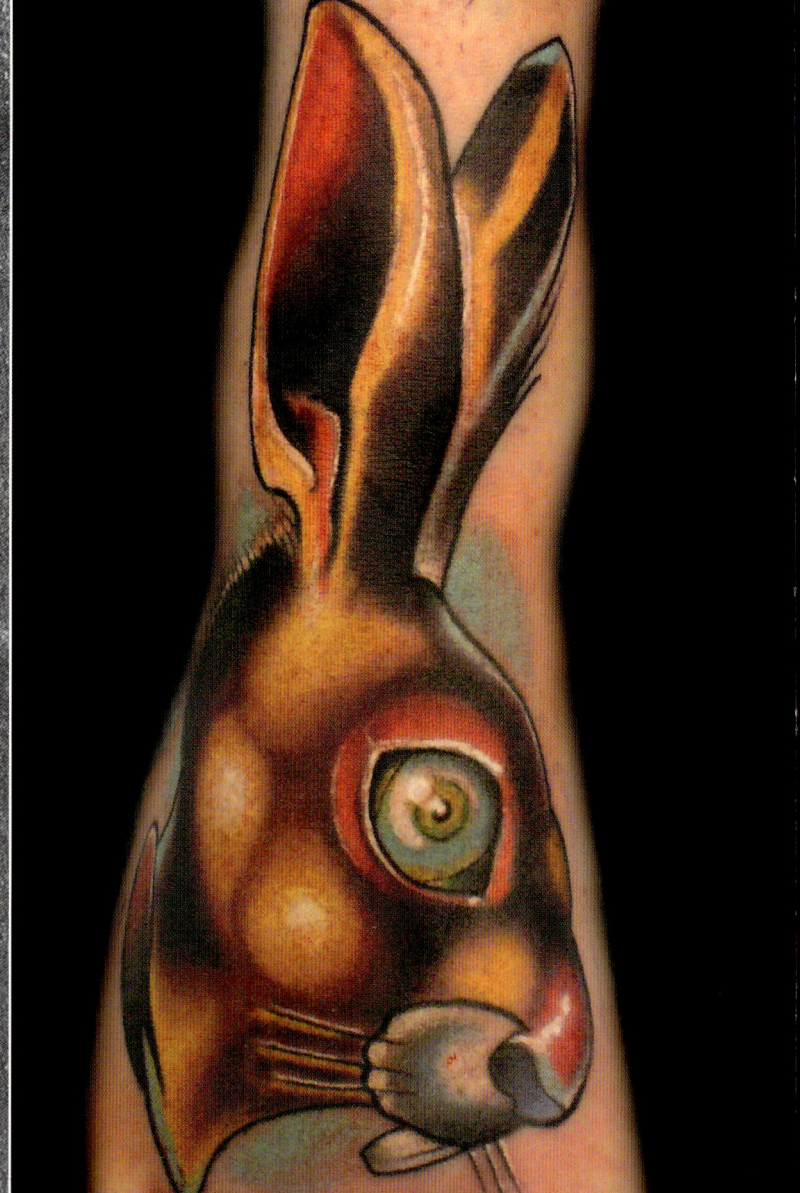

Sean Herman

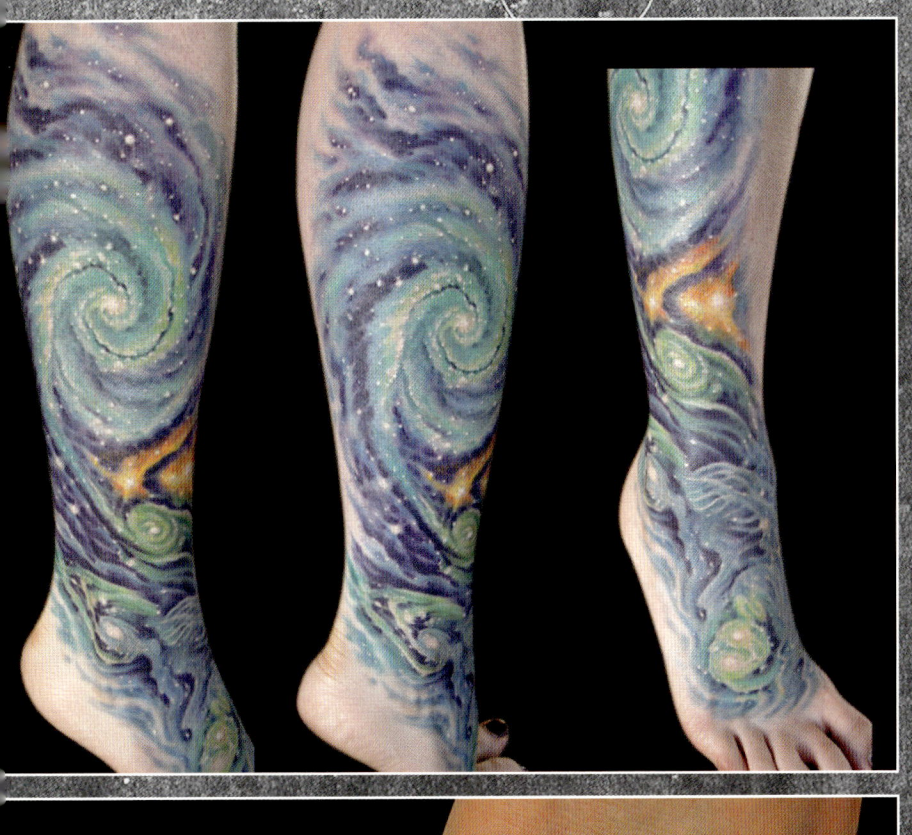

Steve Morris

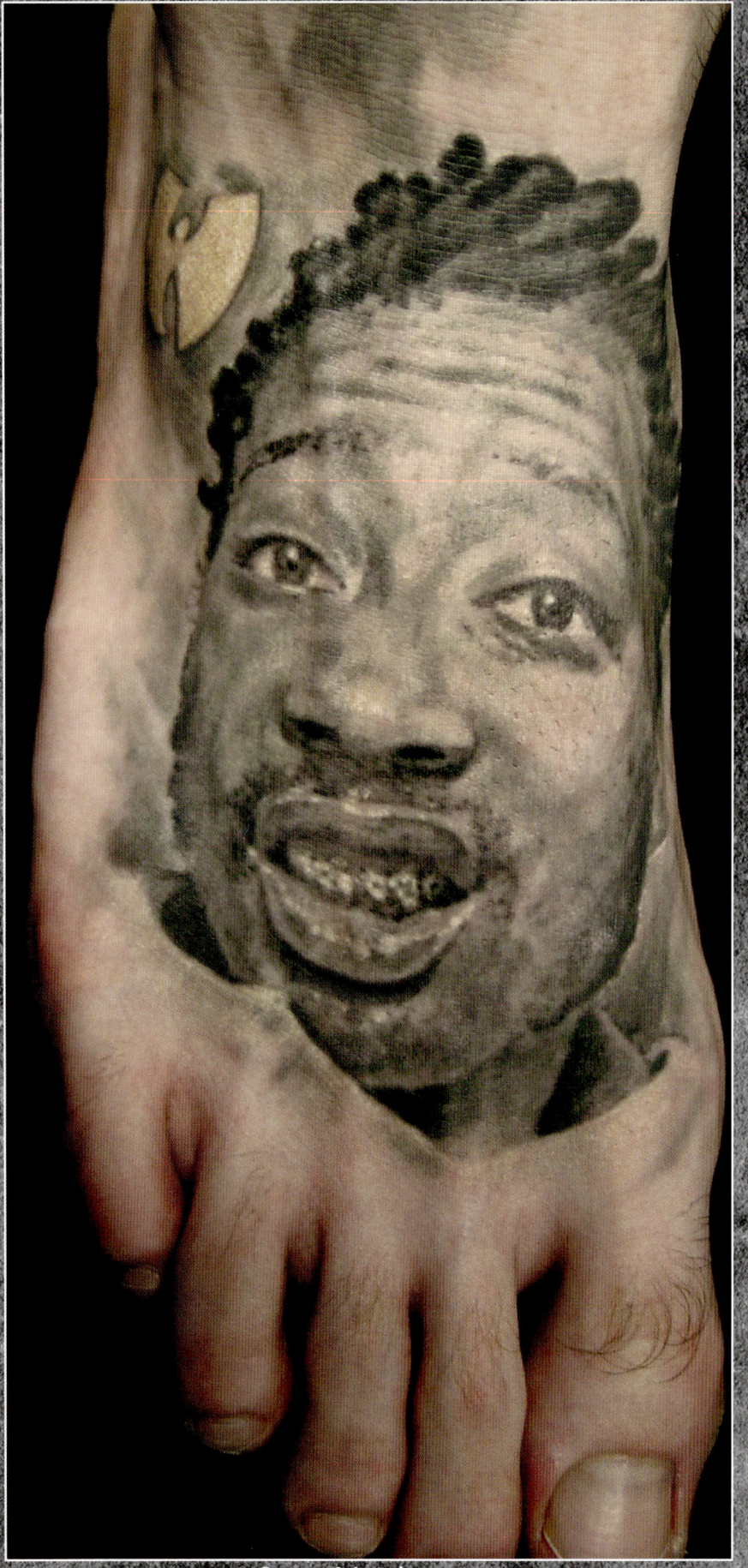

Steve Morris

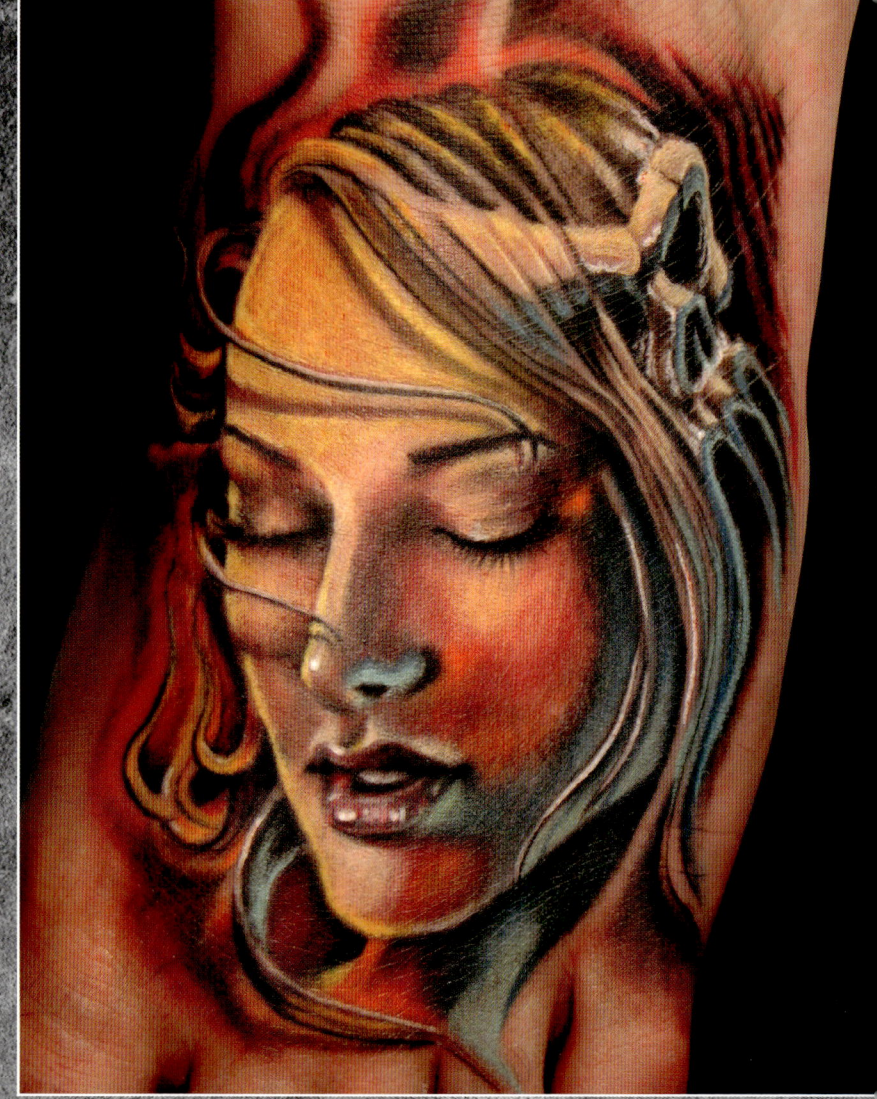

Steve Morris

Stefano Alcantara

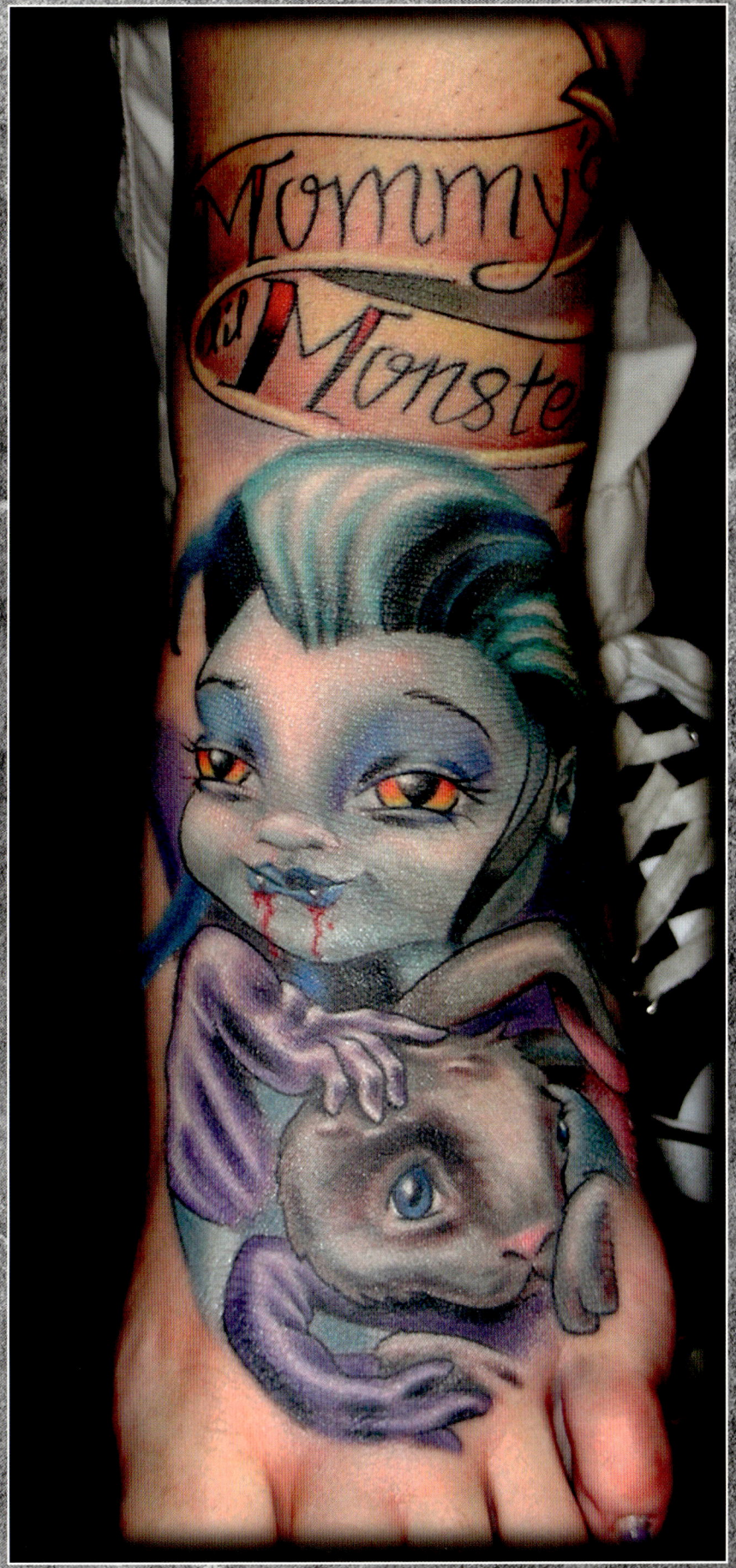

Sweet Laraine

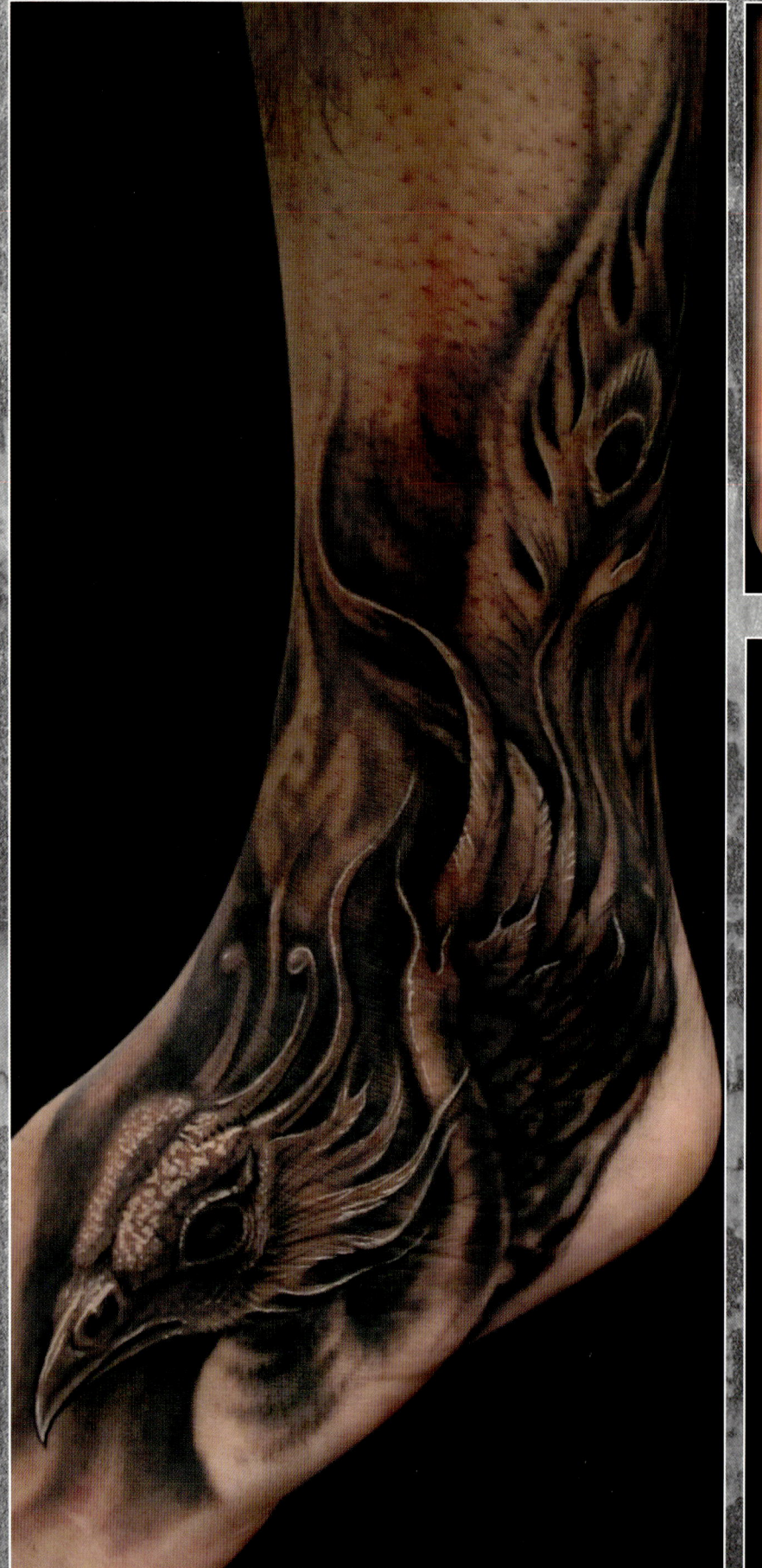

Stefano Alcantara

Sweet Laraine

Cory Ferguson

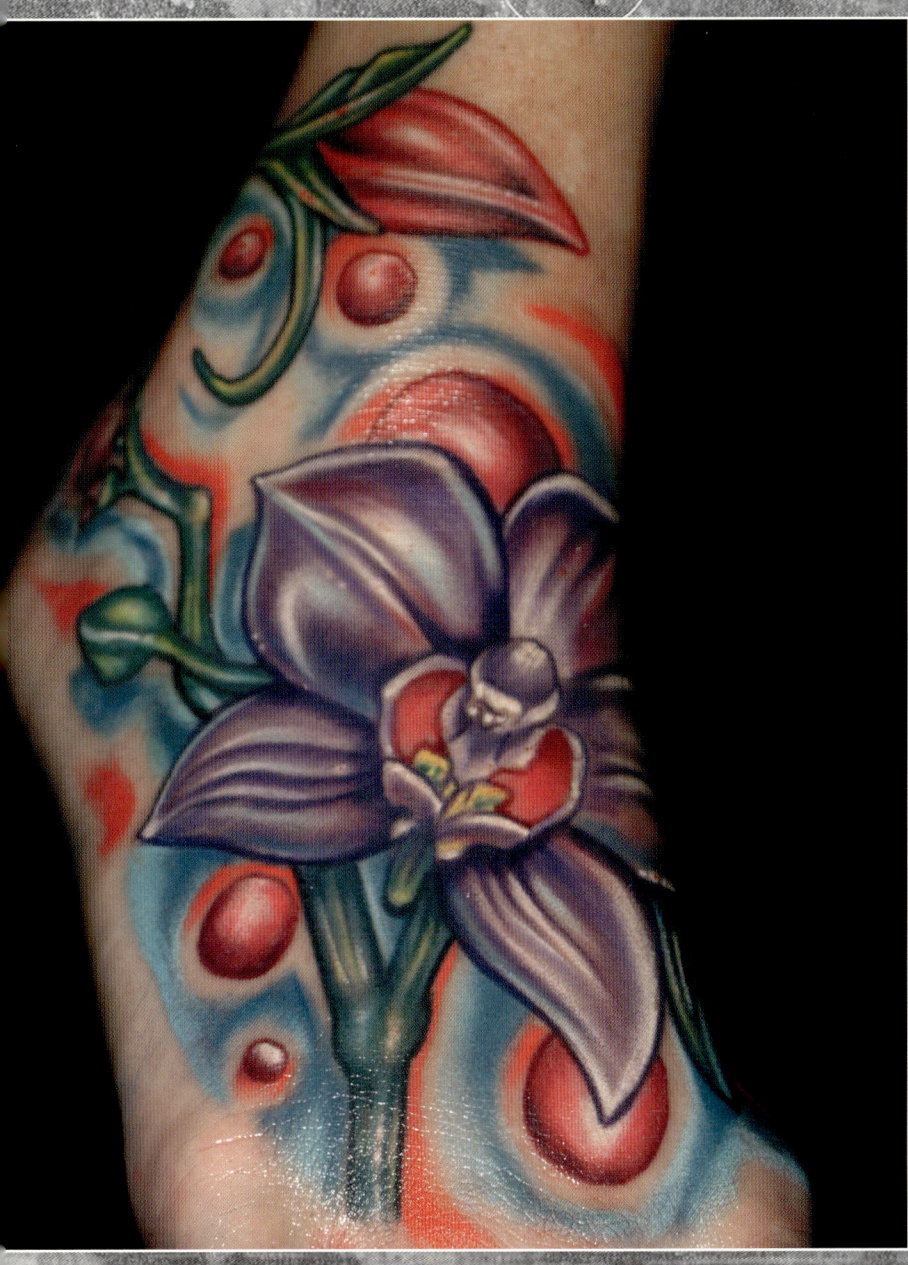

Roman Abrego

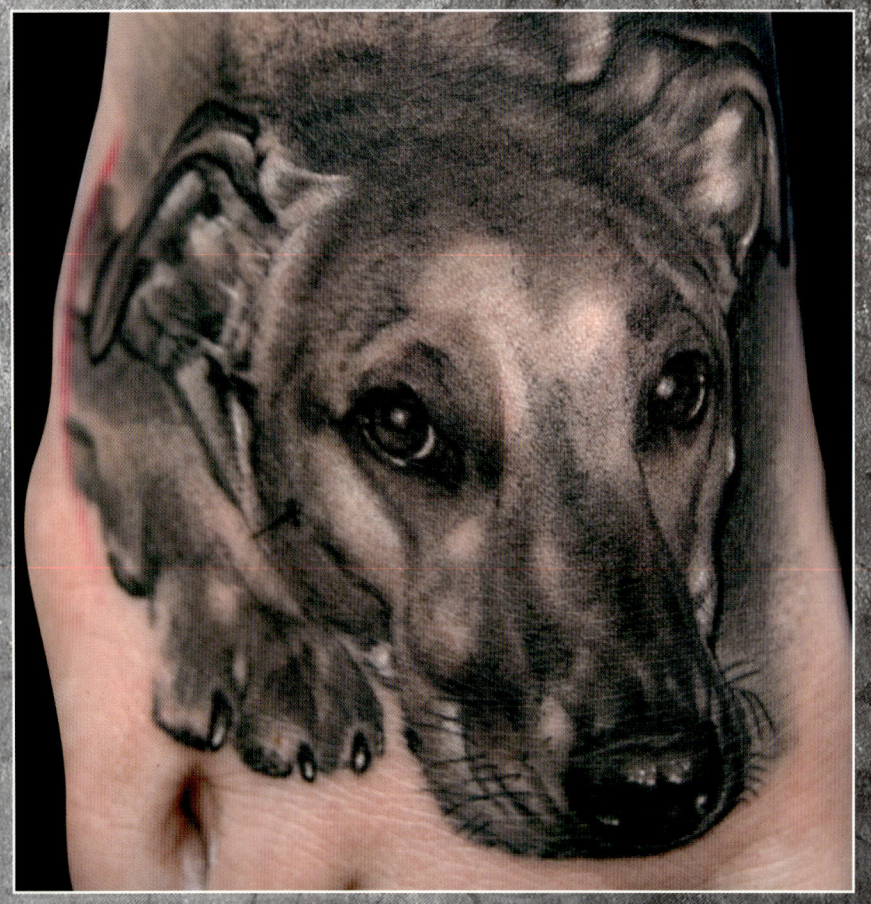

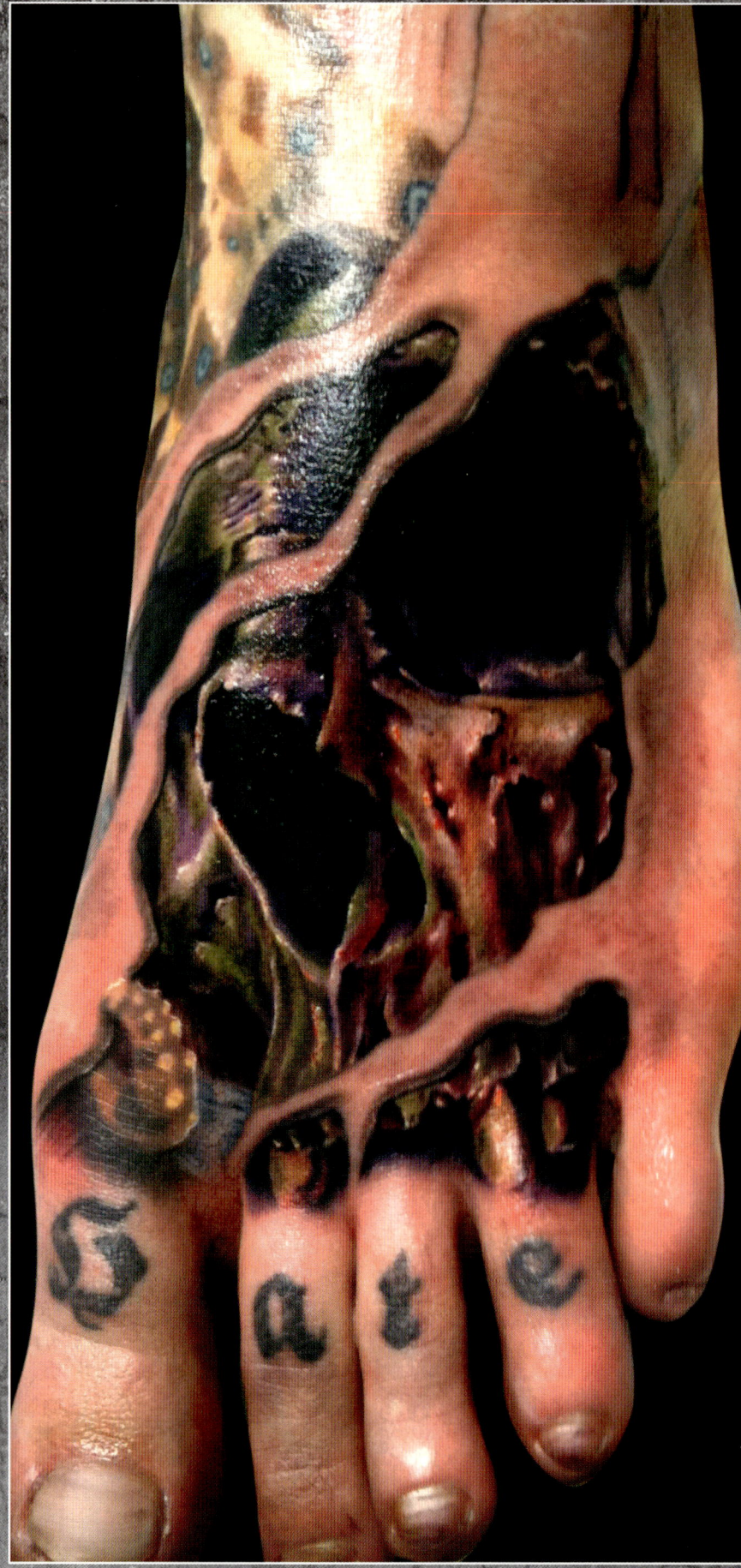

Andy Engel

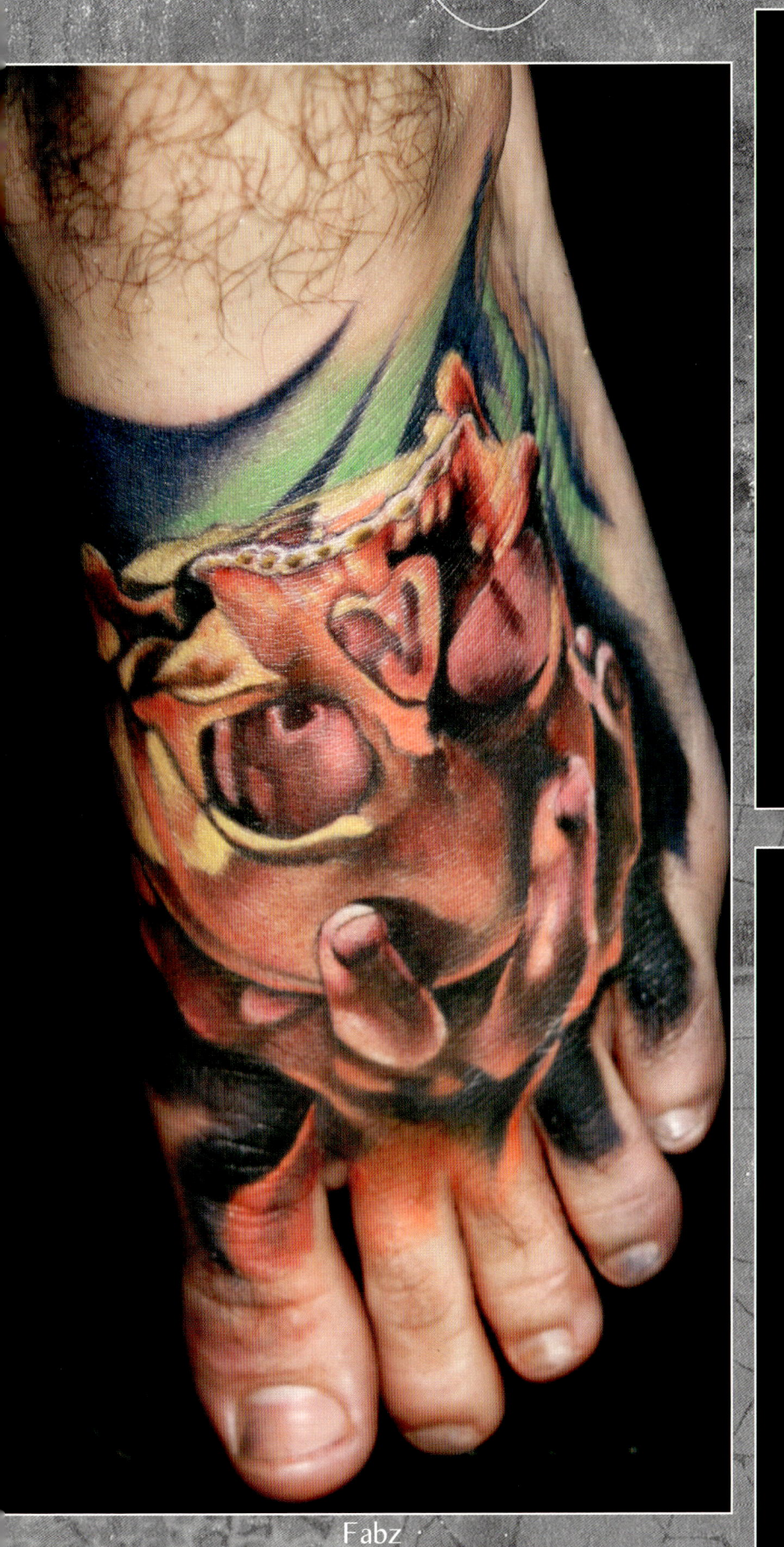

Fabz

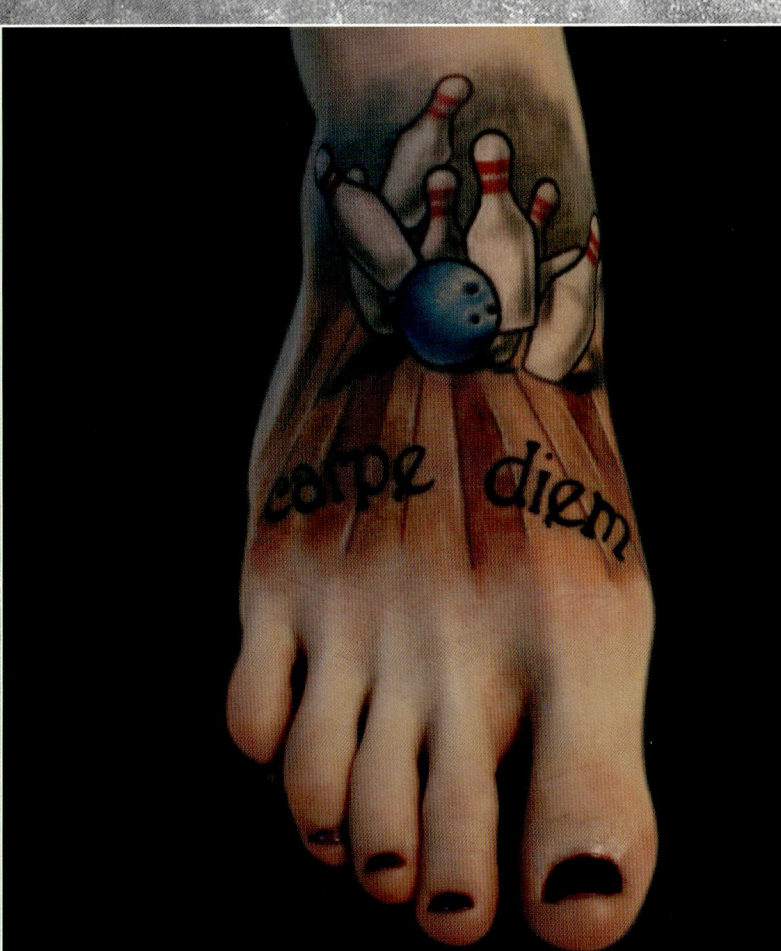

Ty McEwen

Ty McEwen

Ty McEwen

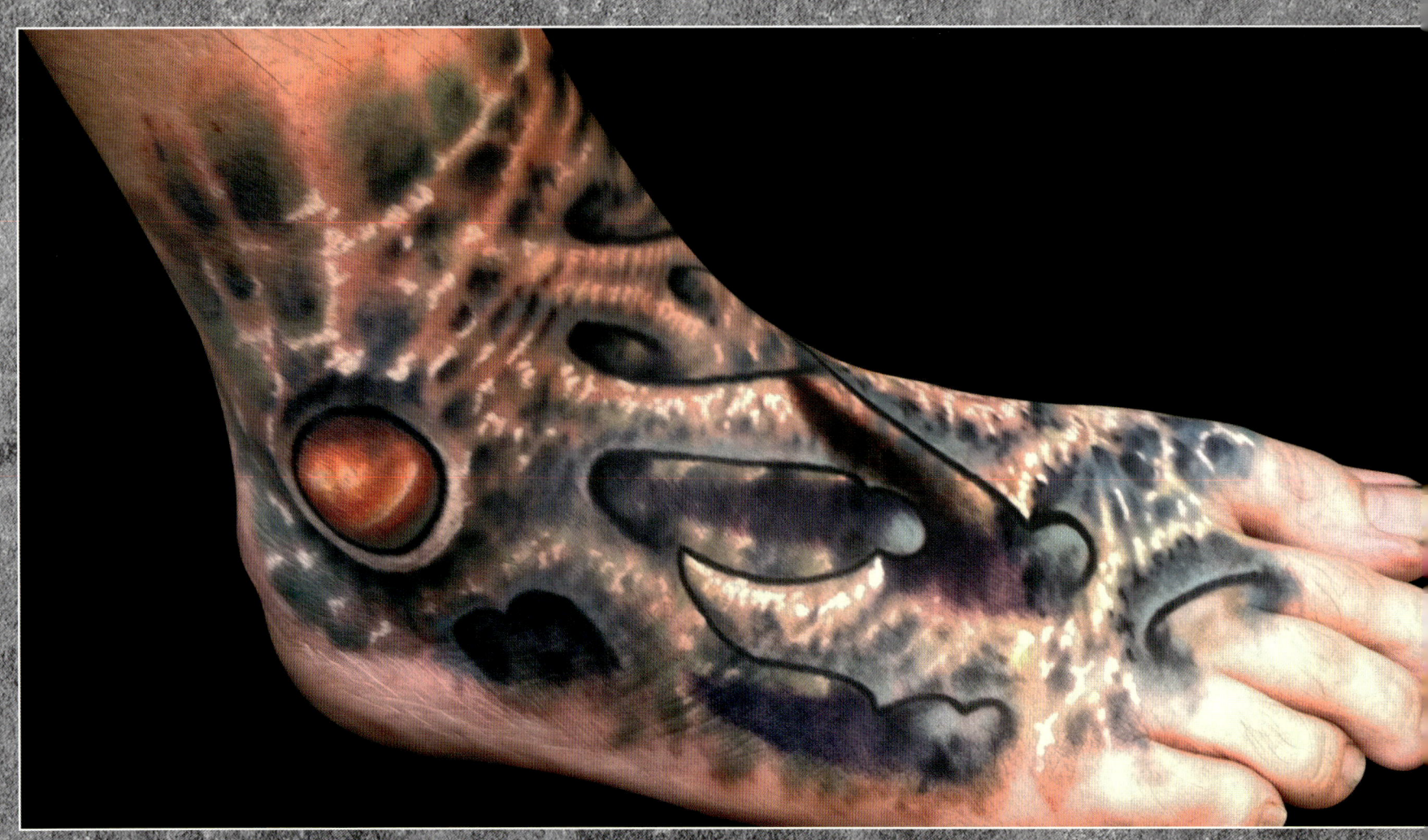

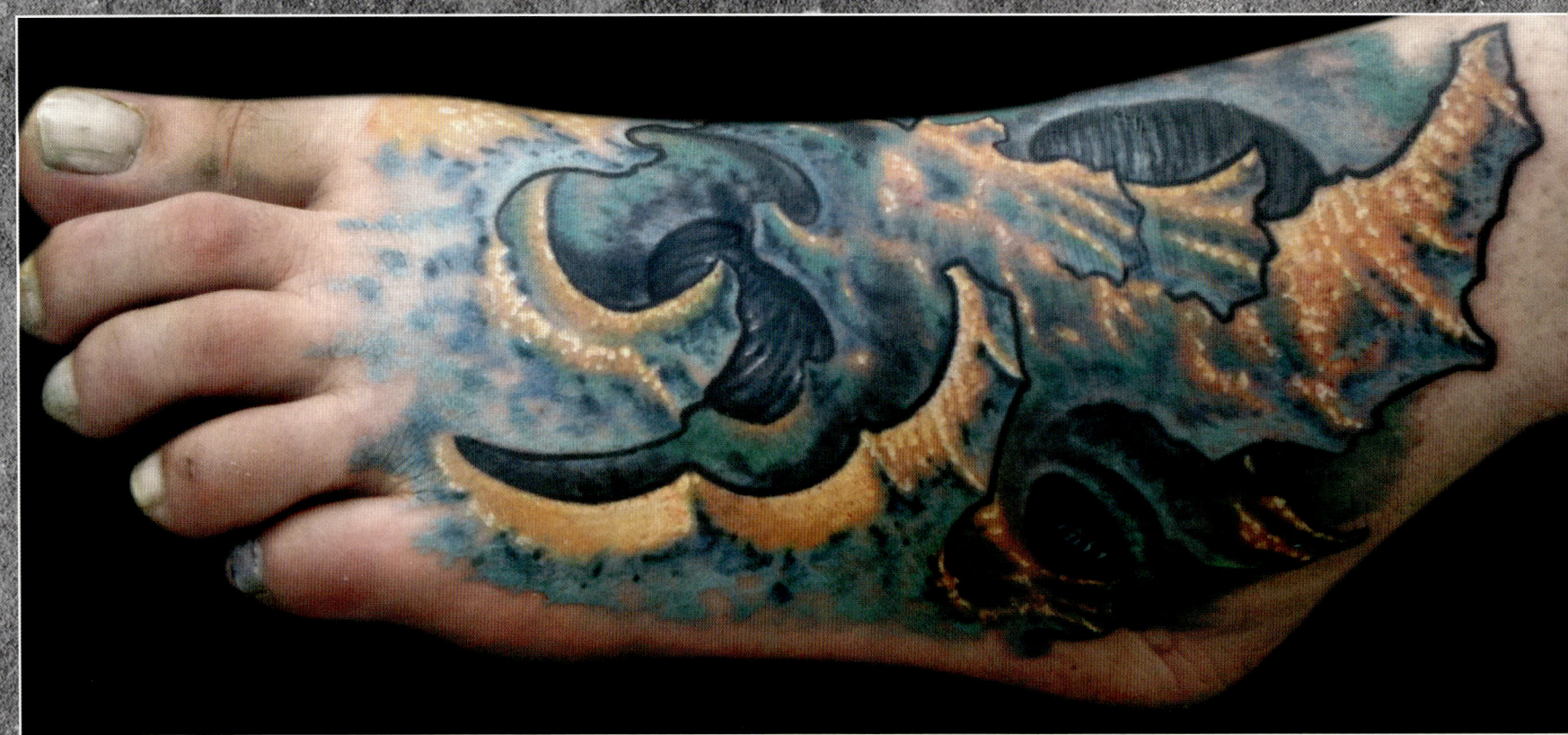

Kyle Cotterman

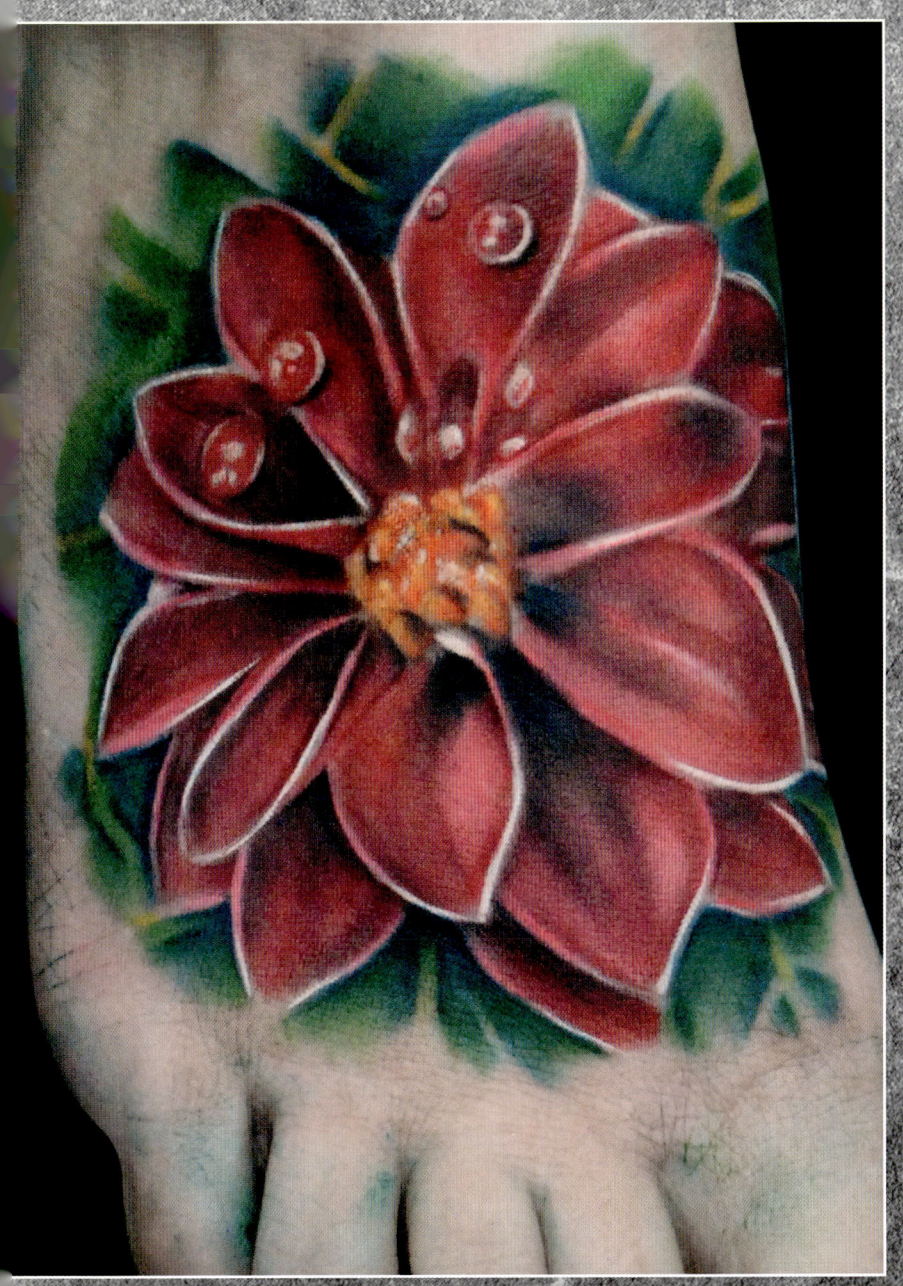

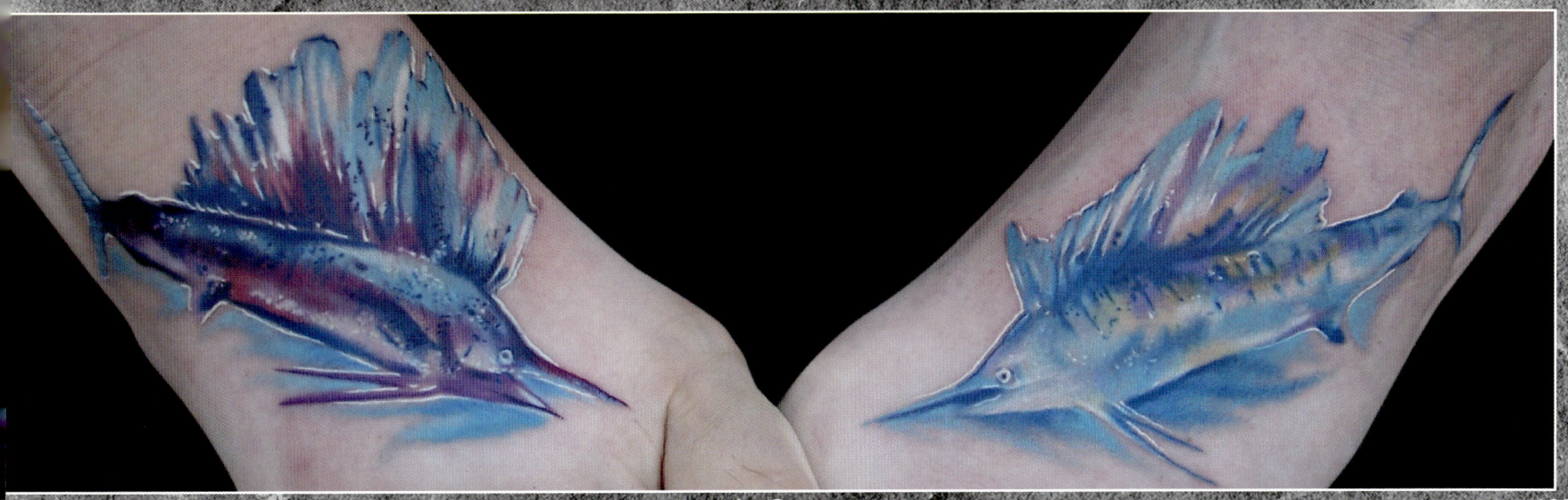

Ben Corn

Ben Corn

Marc Durrant

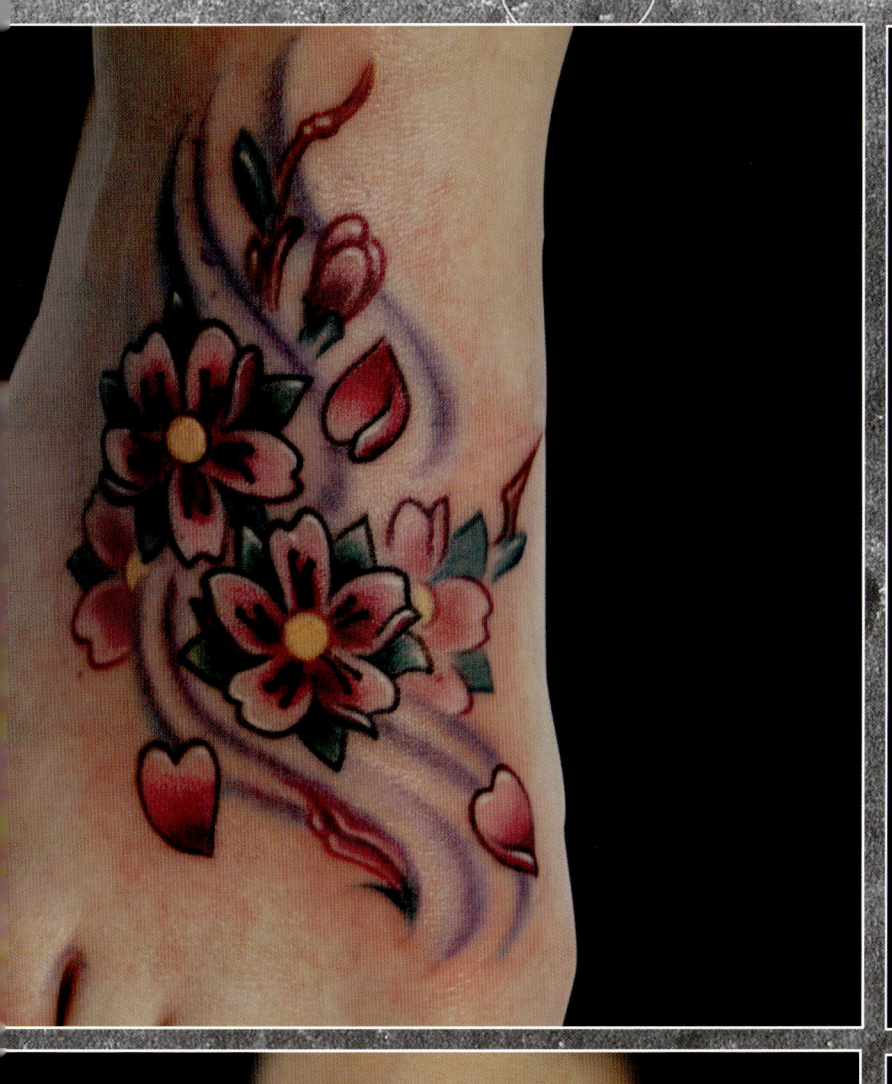

Marvin Silva

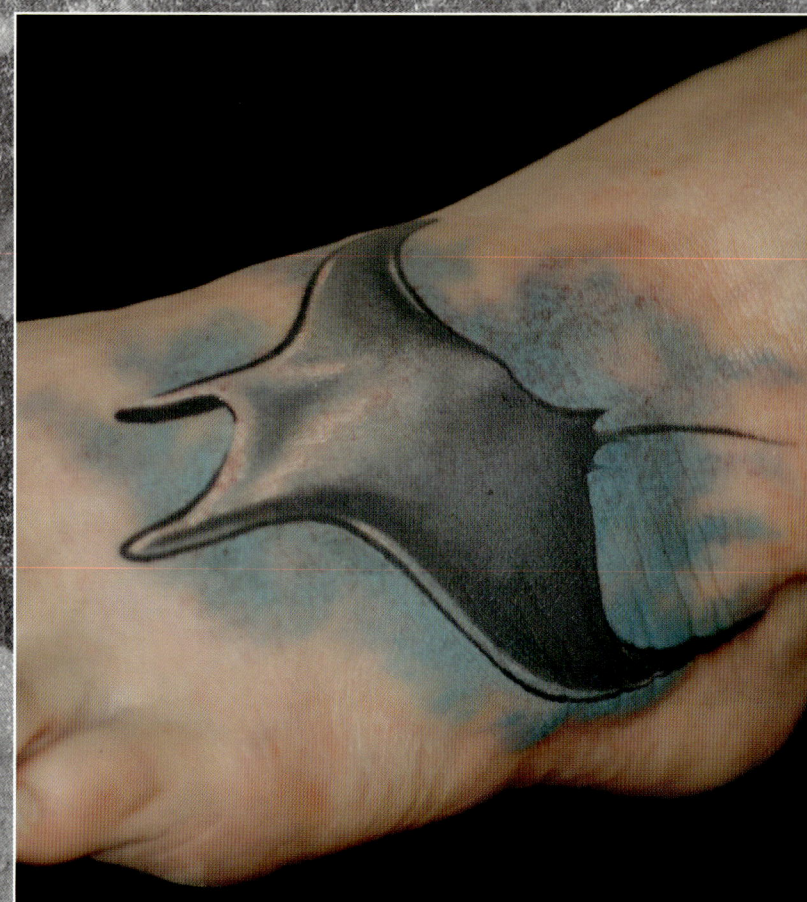

Leigh Oldcorn

Miro Tomas

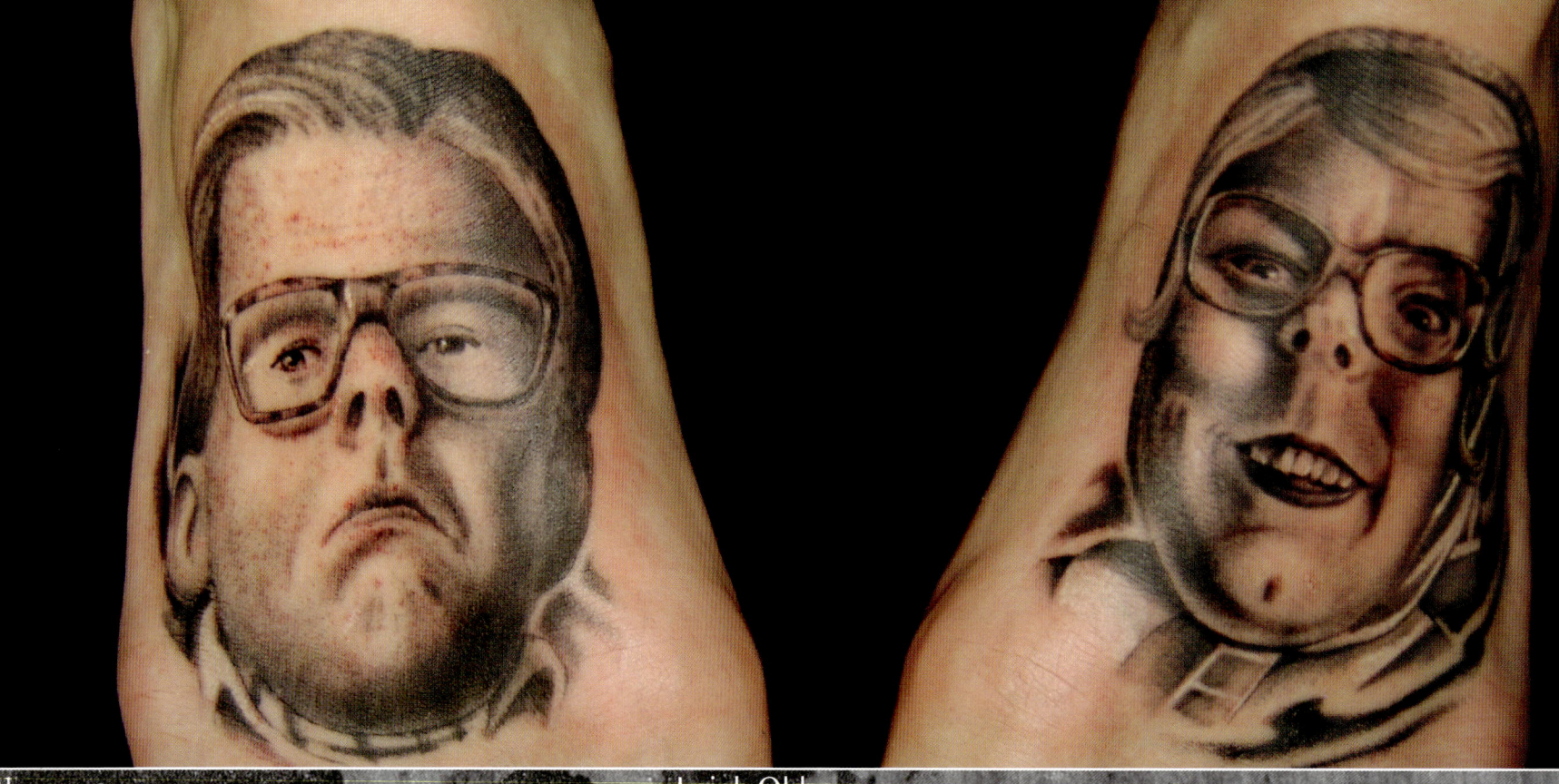

Leigh Oldcorn

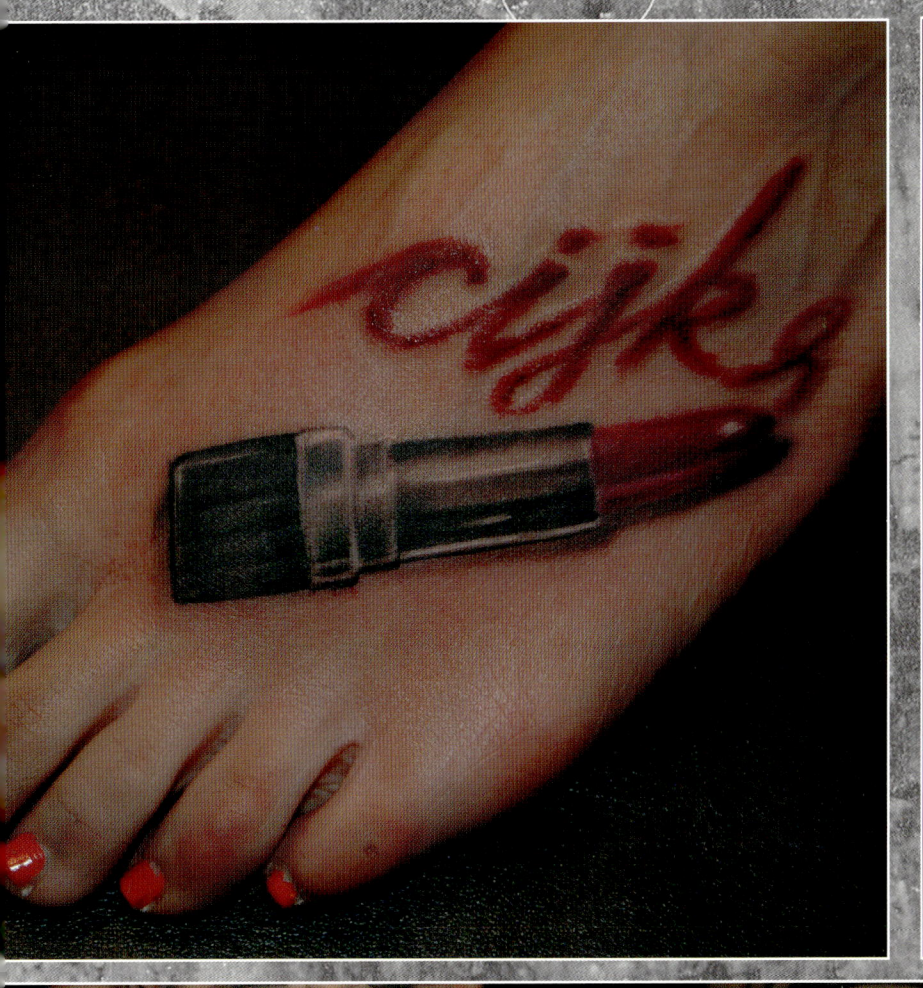

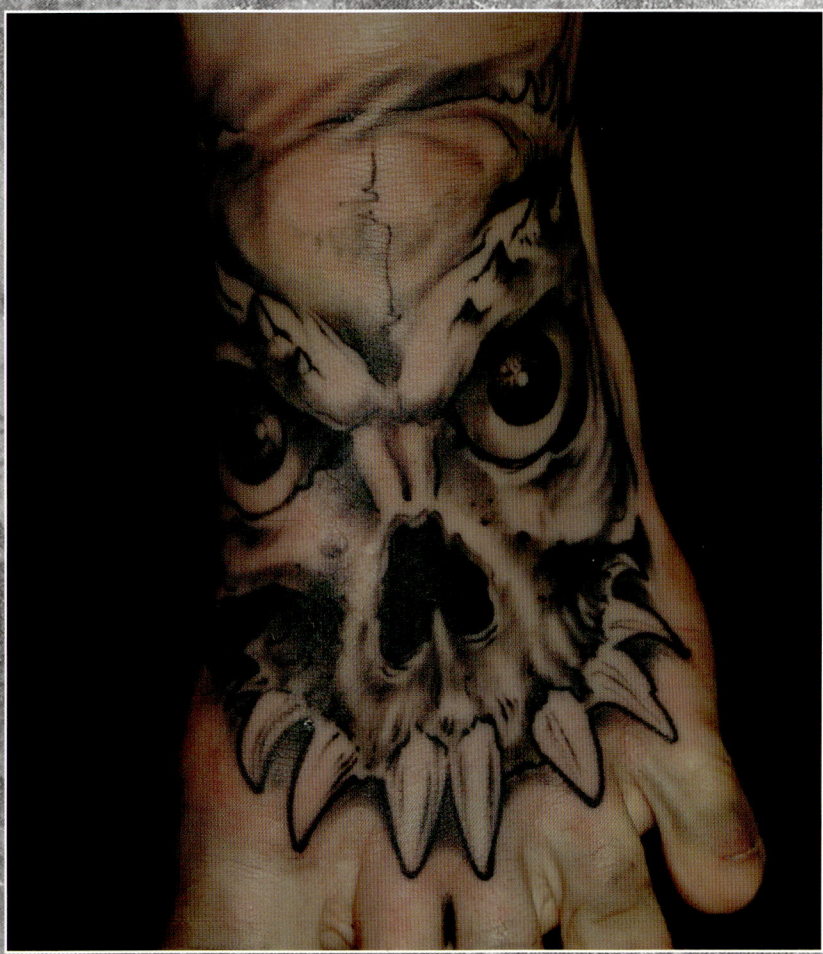

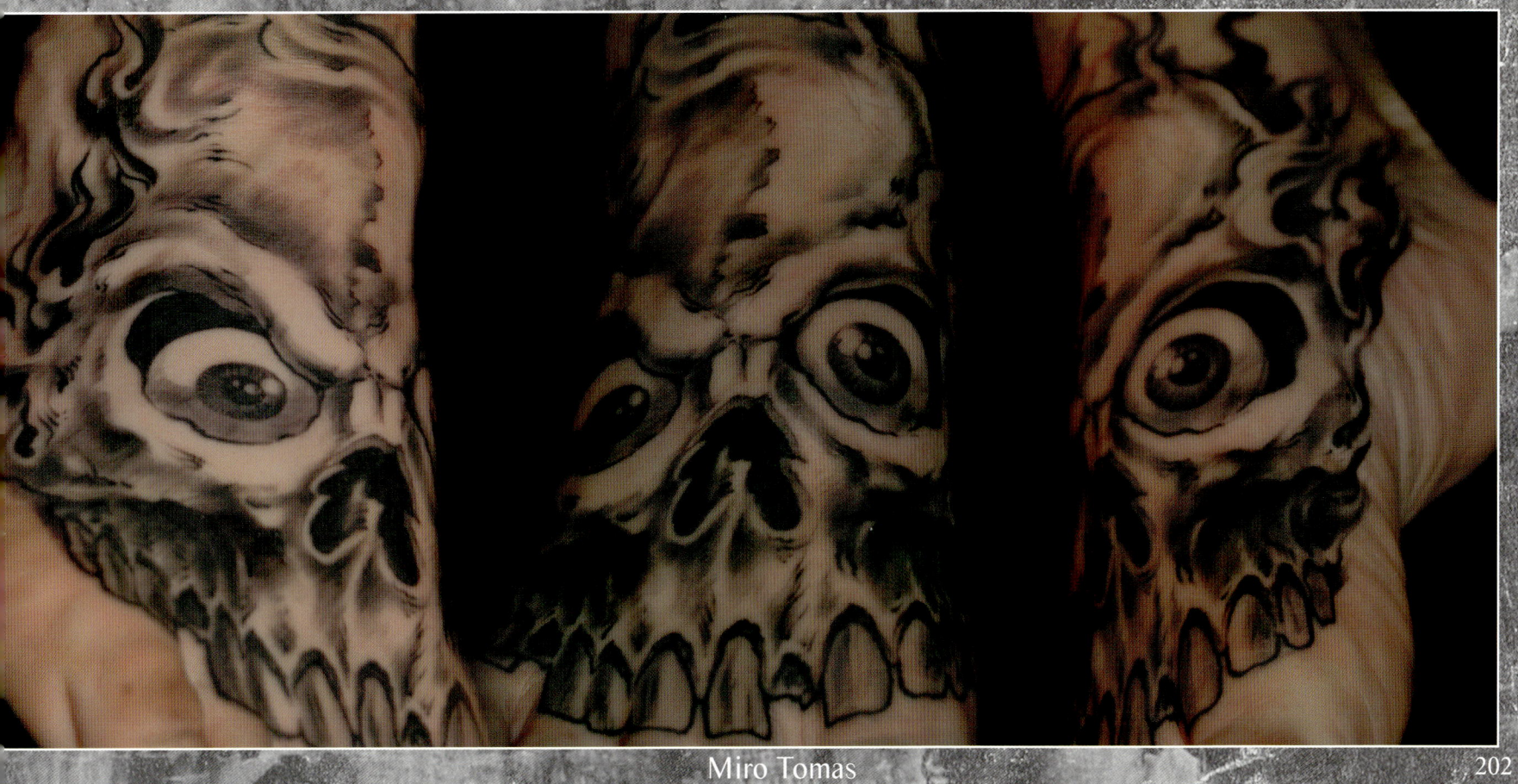

Miro Tomas

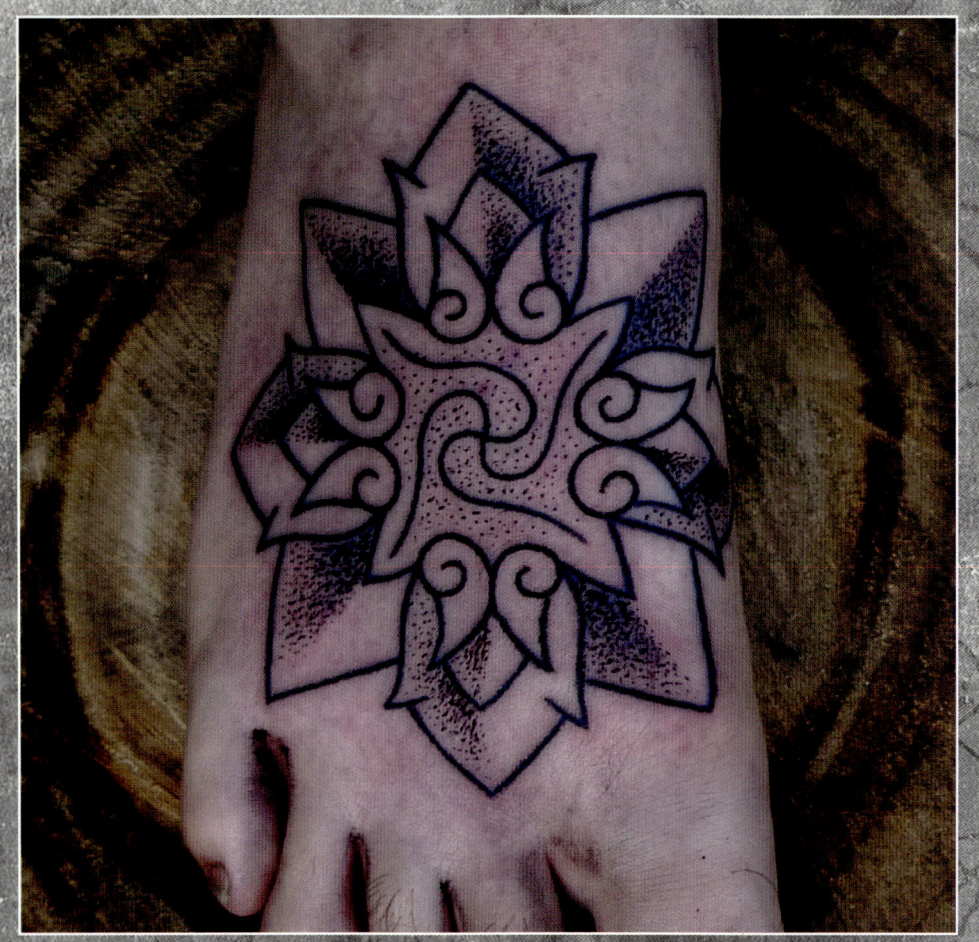

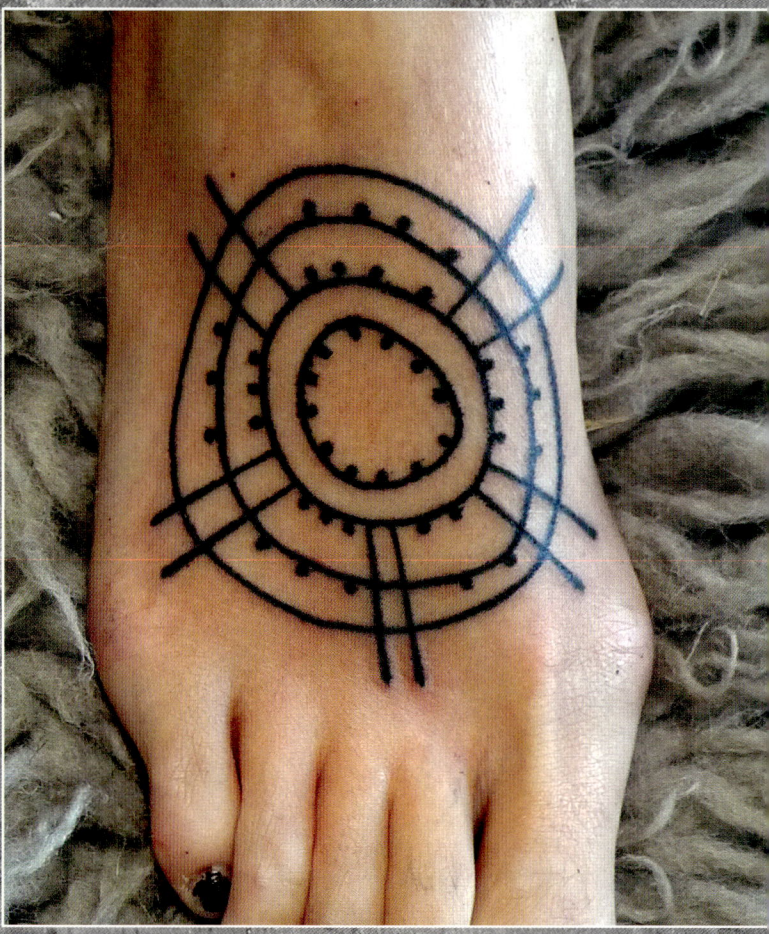

Colin Dale

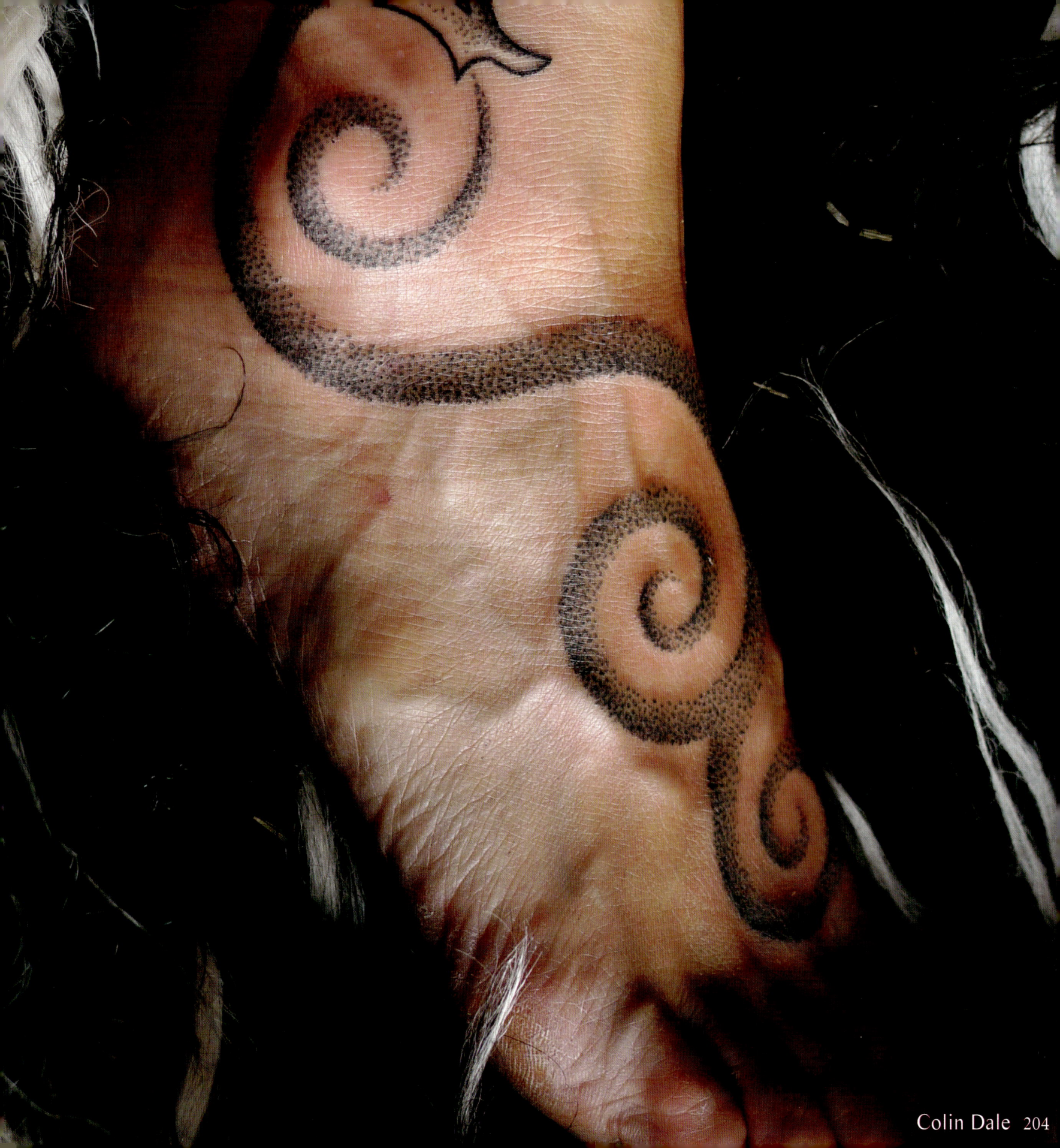

Colin Dale 204

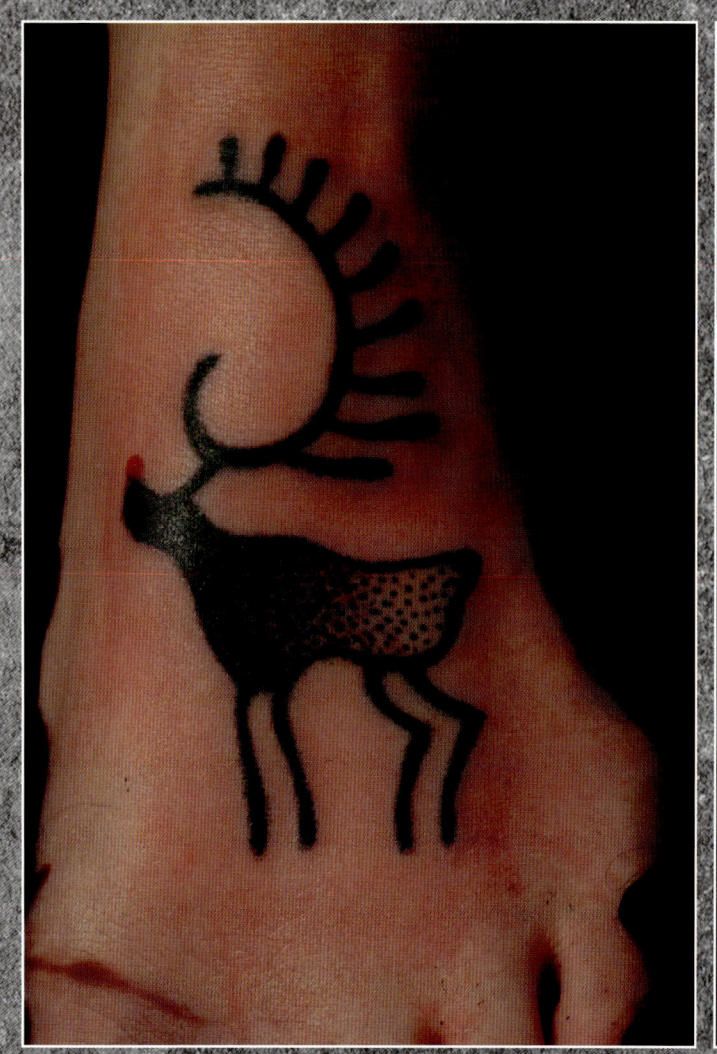

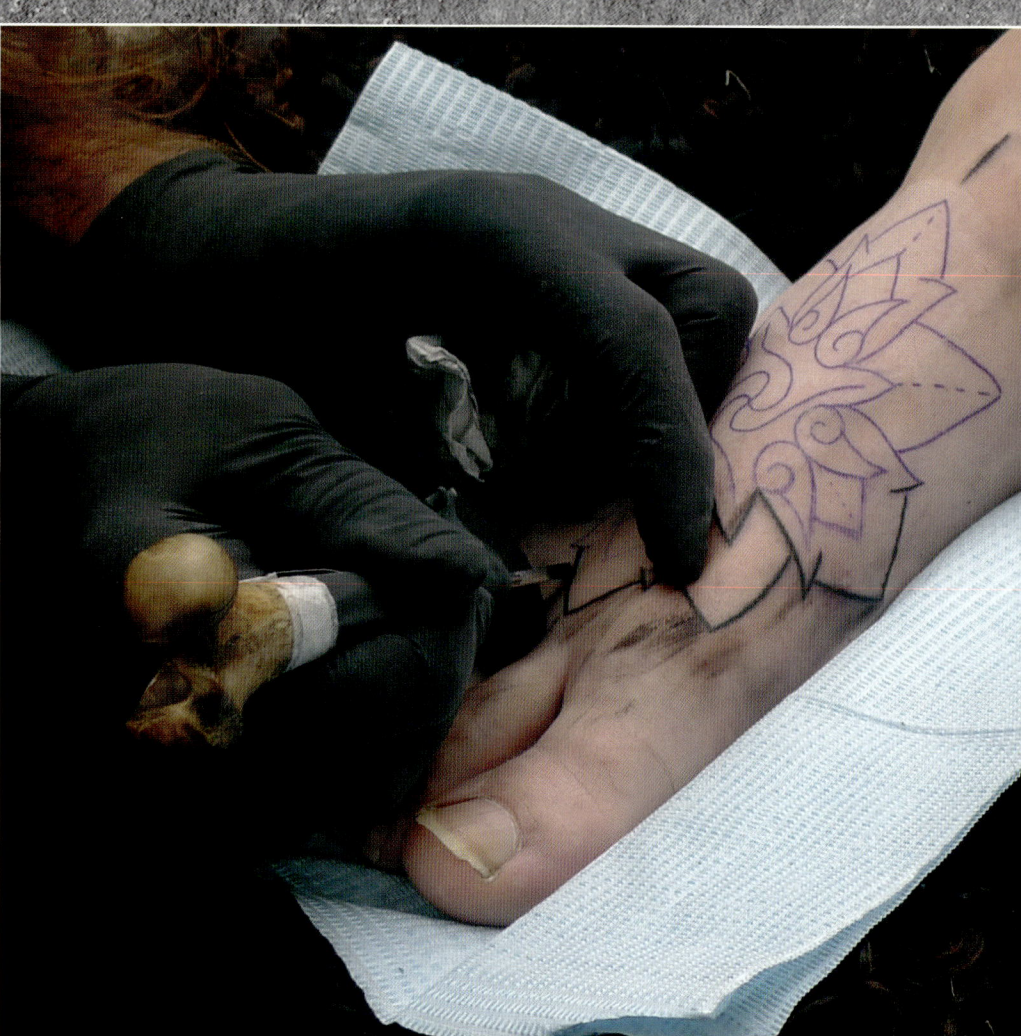

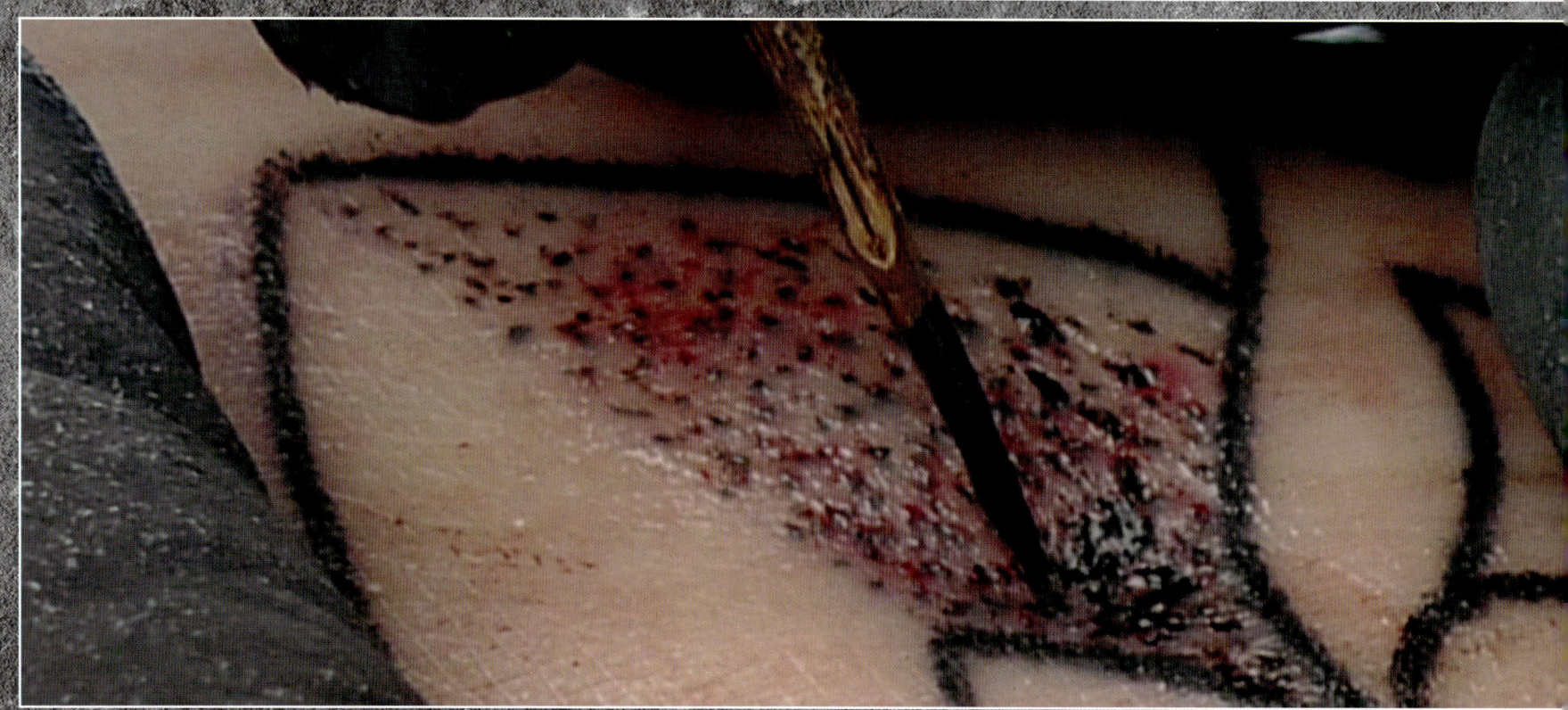

Colin Dale

Colin Dale

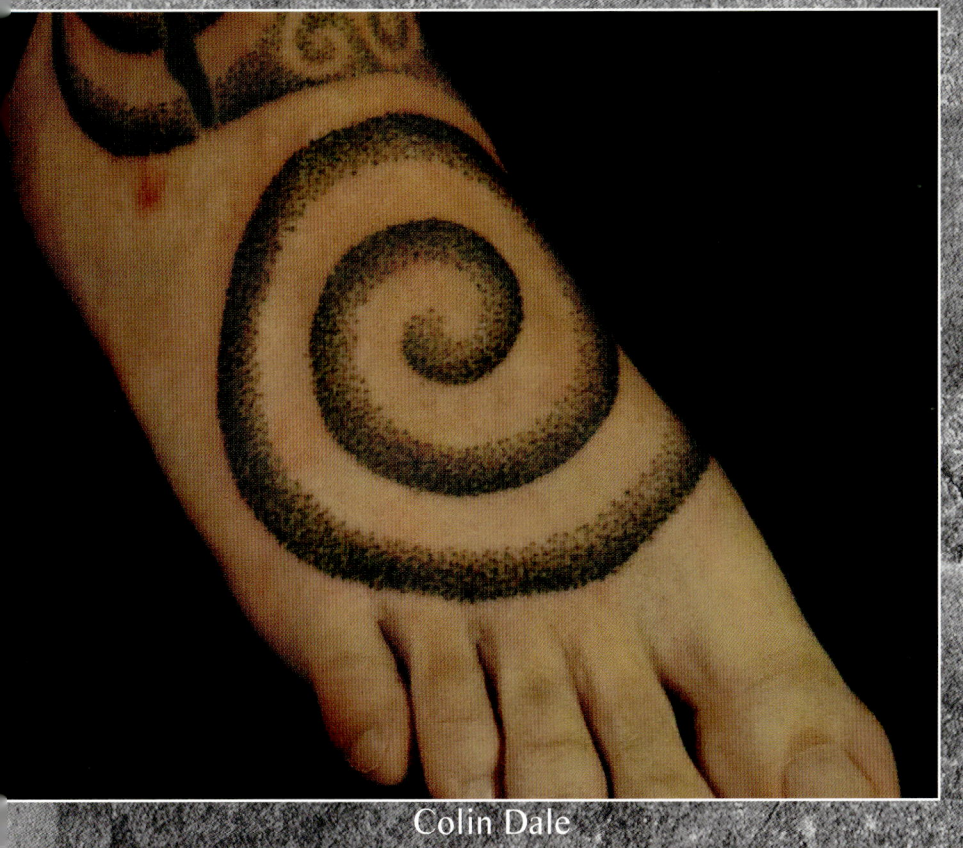

Dmitriy Samohin

Colin Dale

206

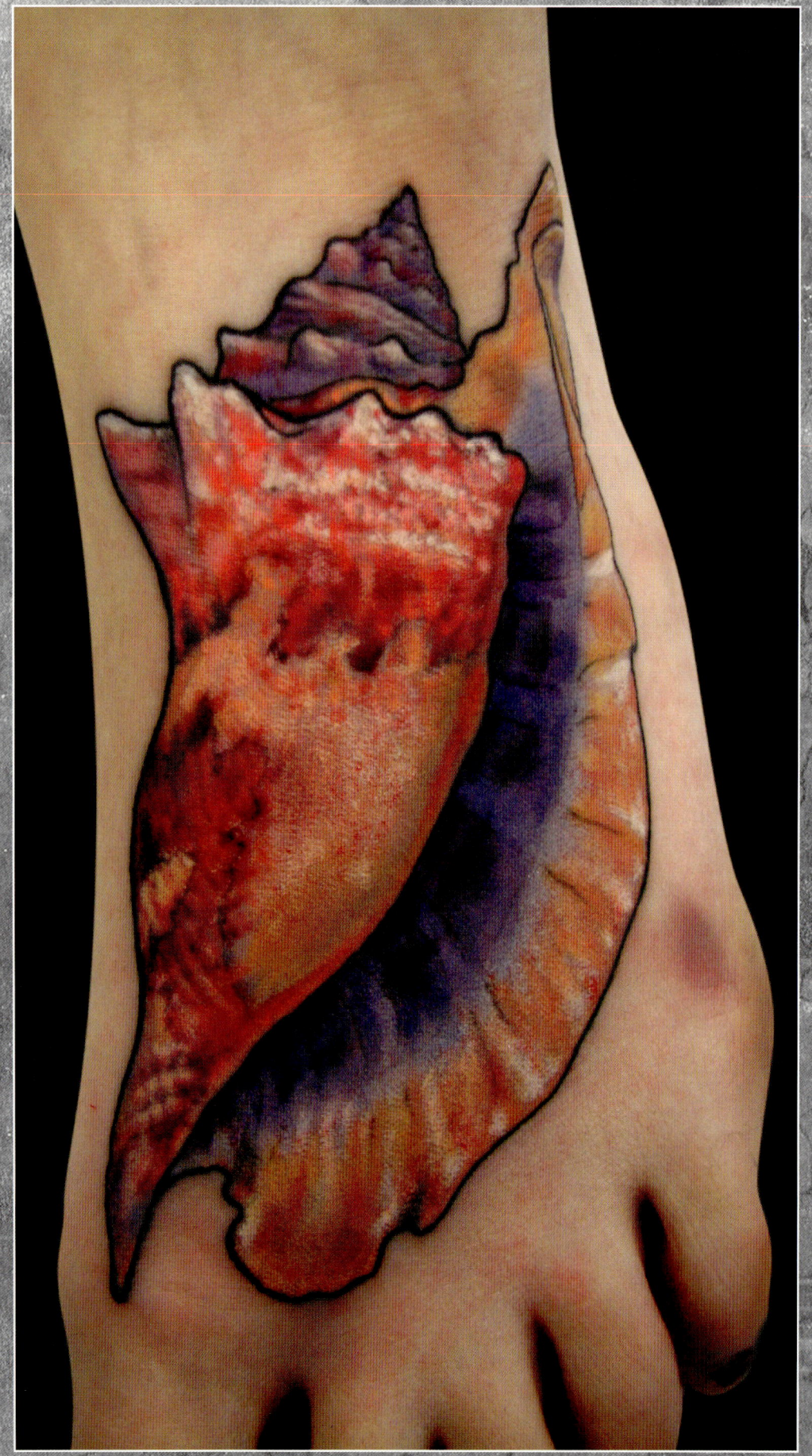

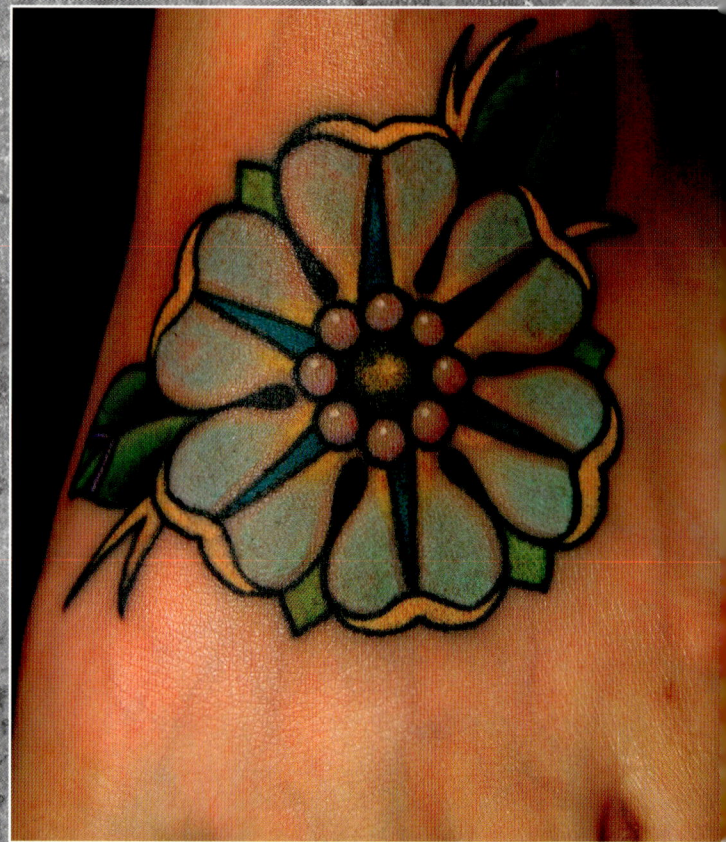

Jeff Croci

Jim Sylvia

Jim Sylvia

Jim Sylvia

Johan Velthuizen

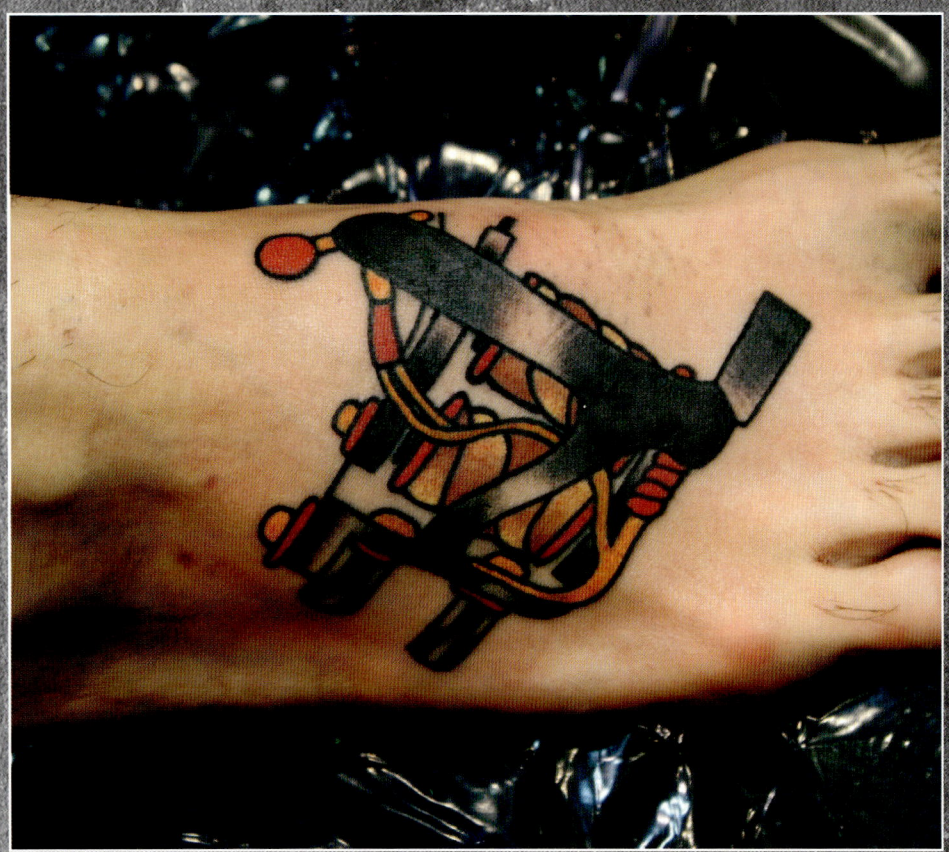

Myke Chambers

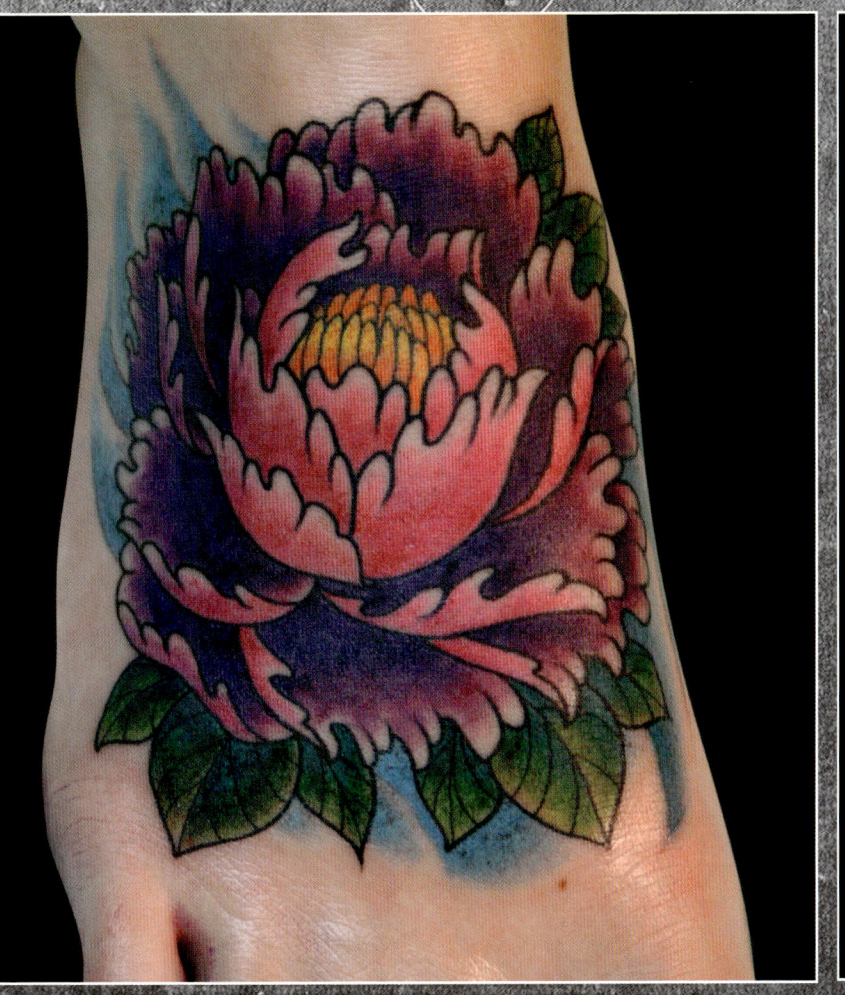

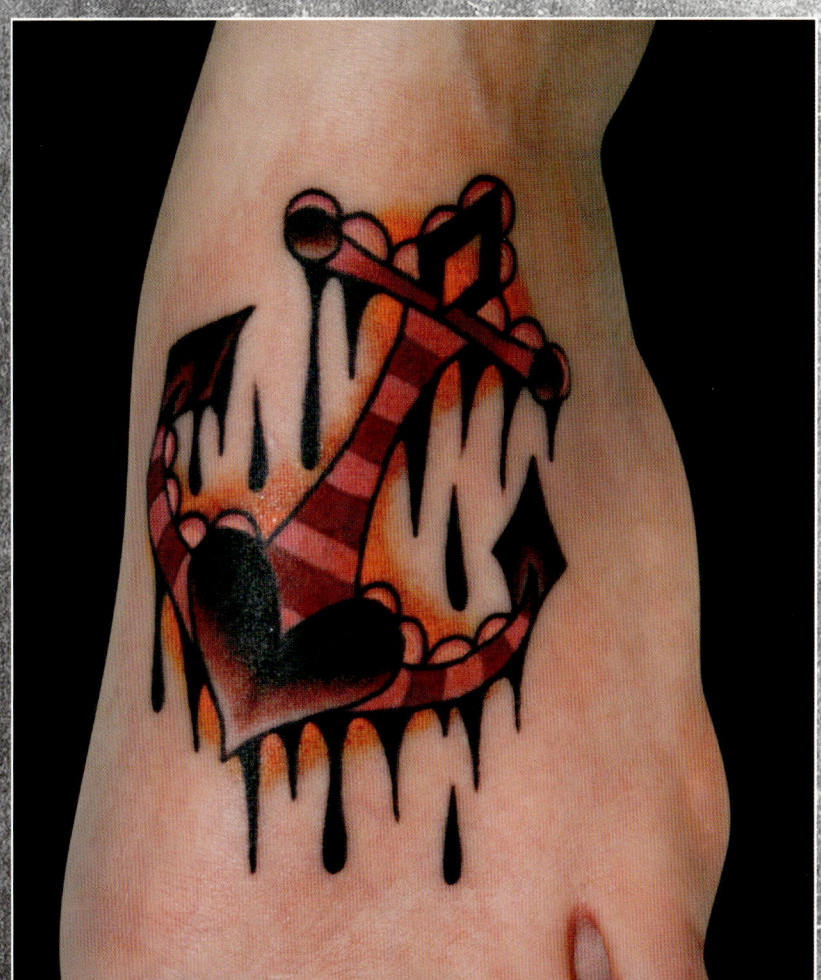

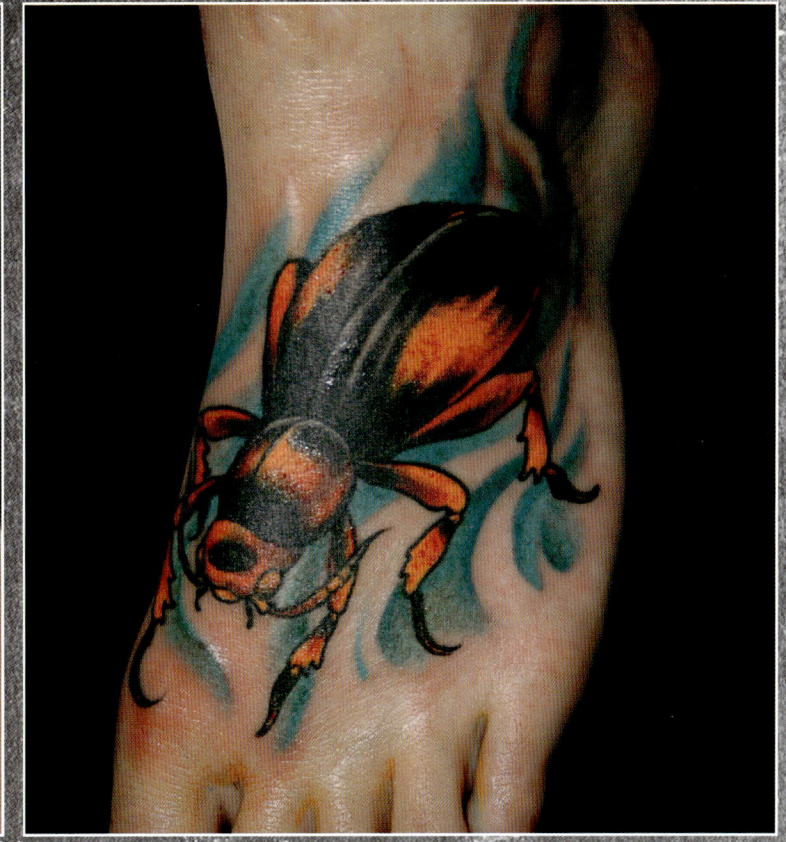

David Bruehl

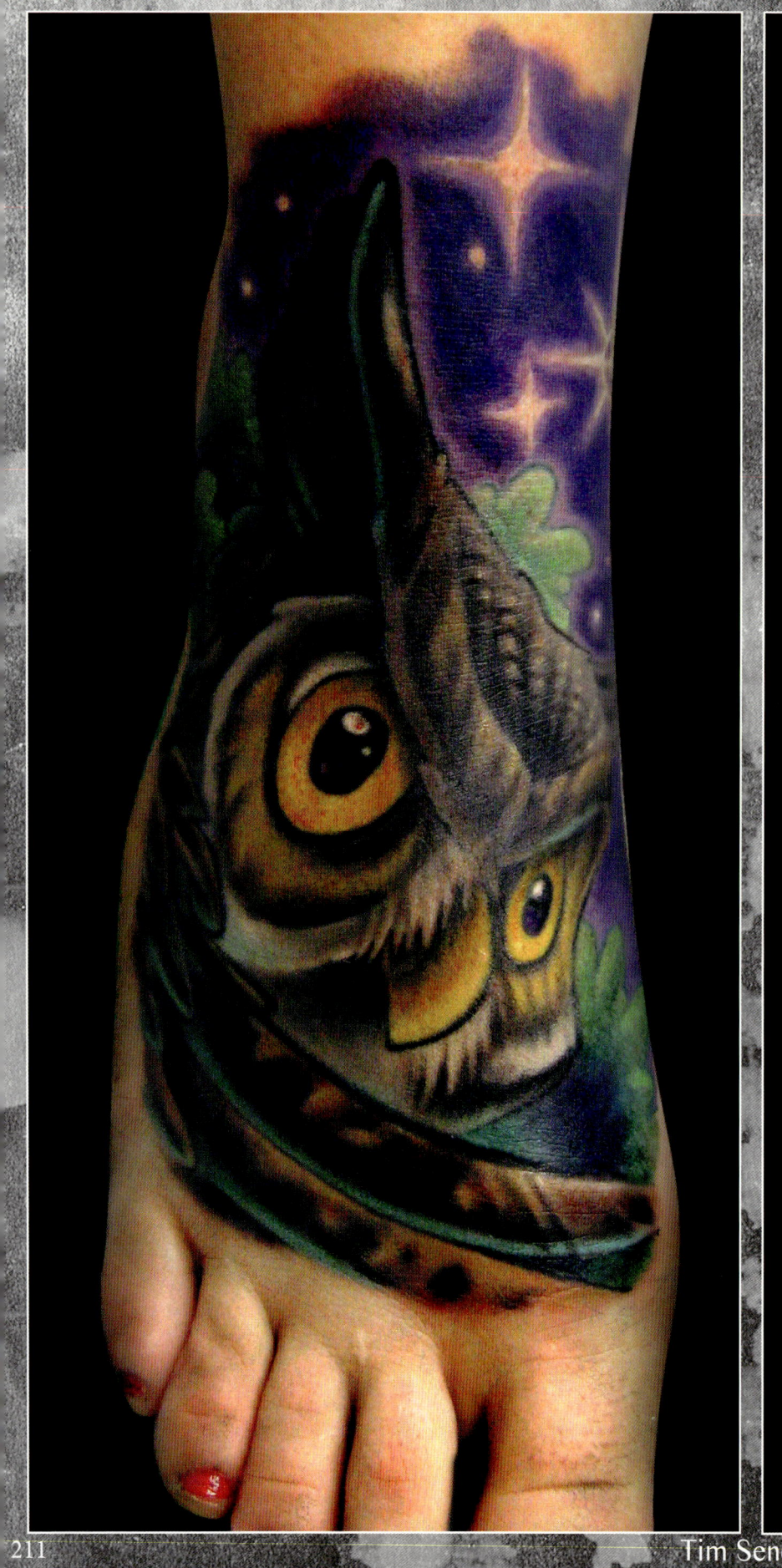

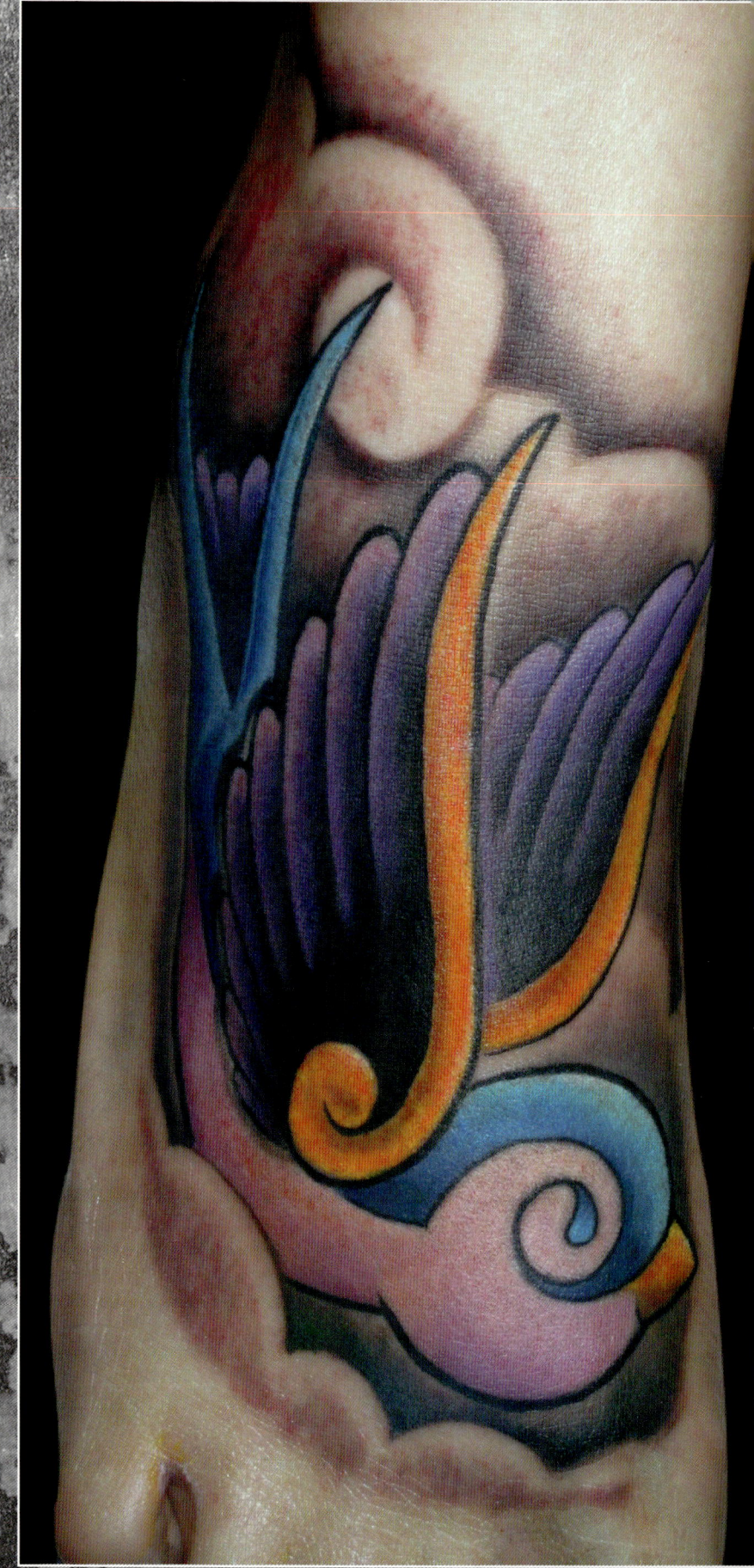

Tim Senecal

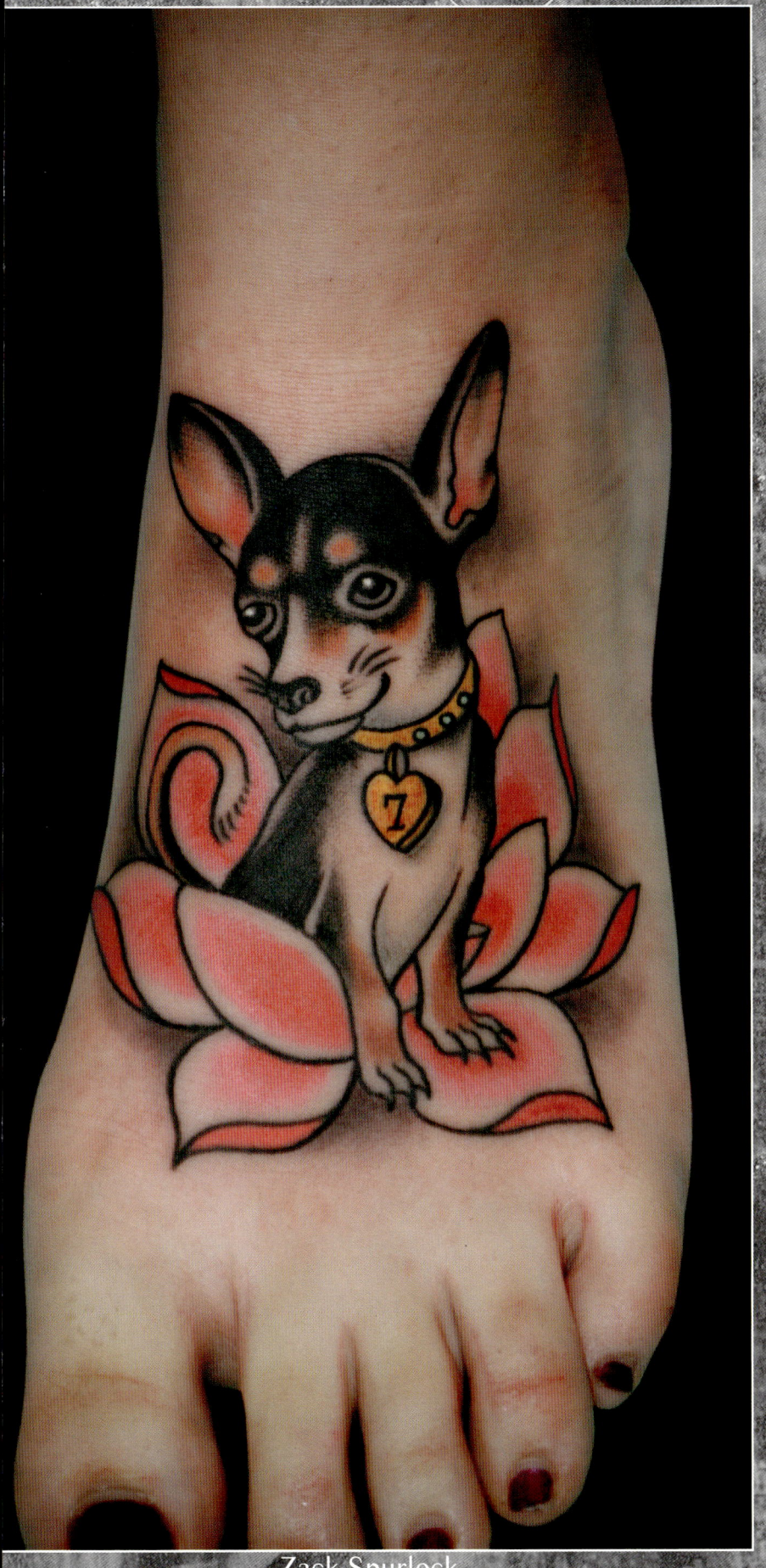

Zack Spurlock

Zack Spurlock

Ben Stubbs

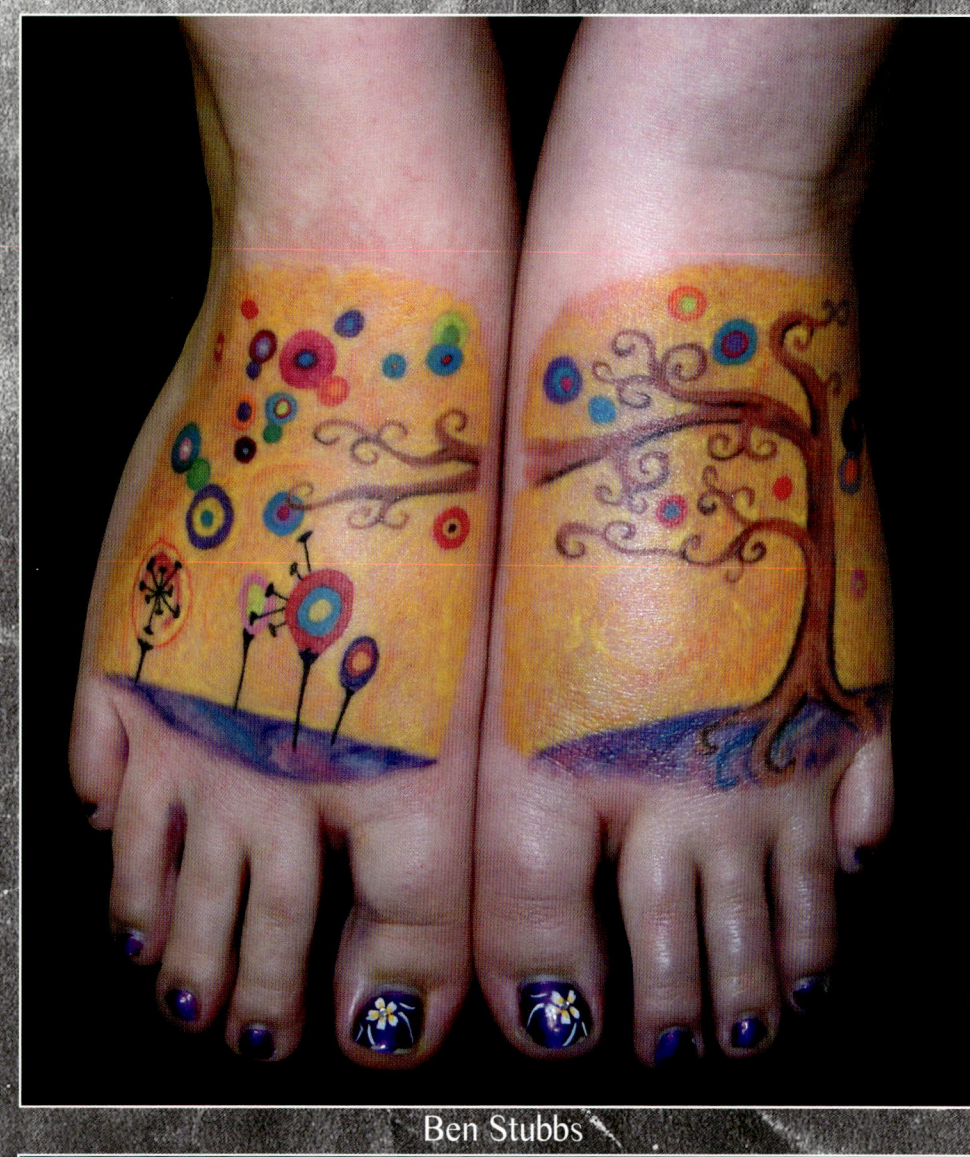

Ben Stubbs

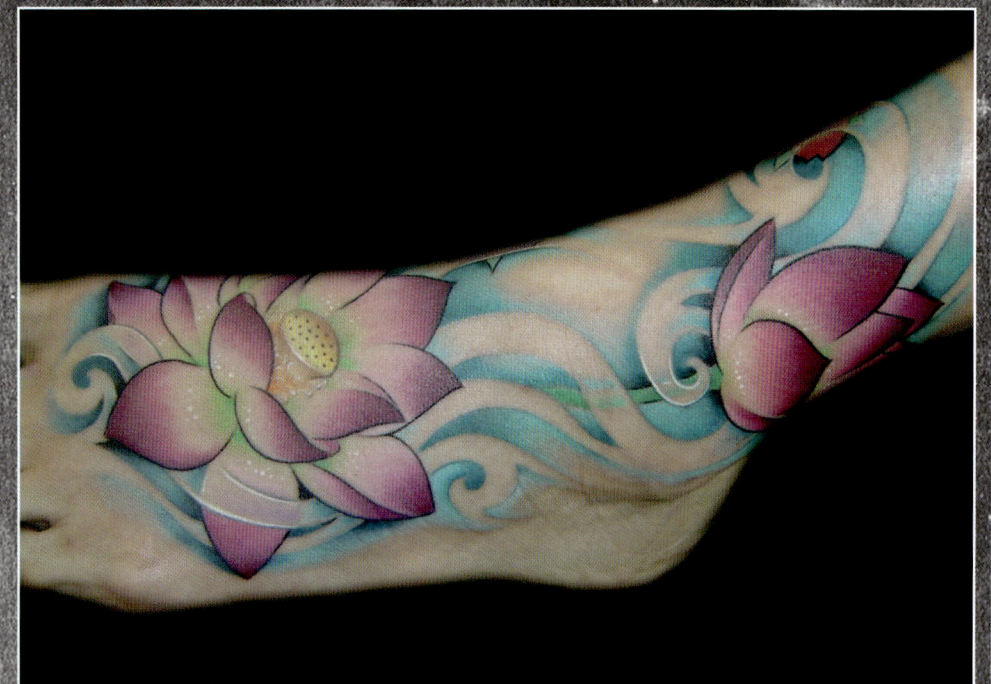

Juan Salgado

Juan Salgado

Paul Booth

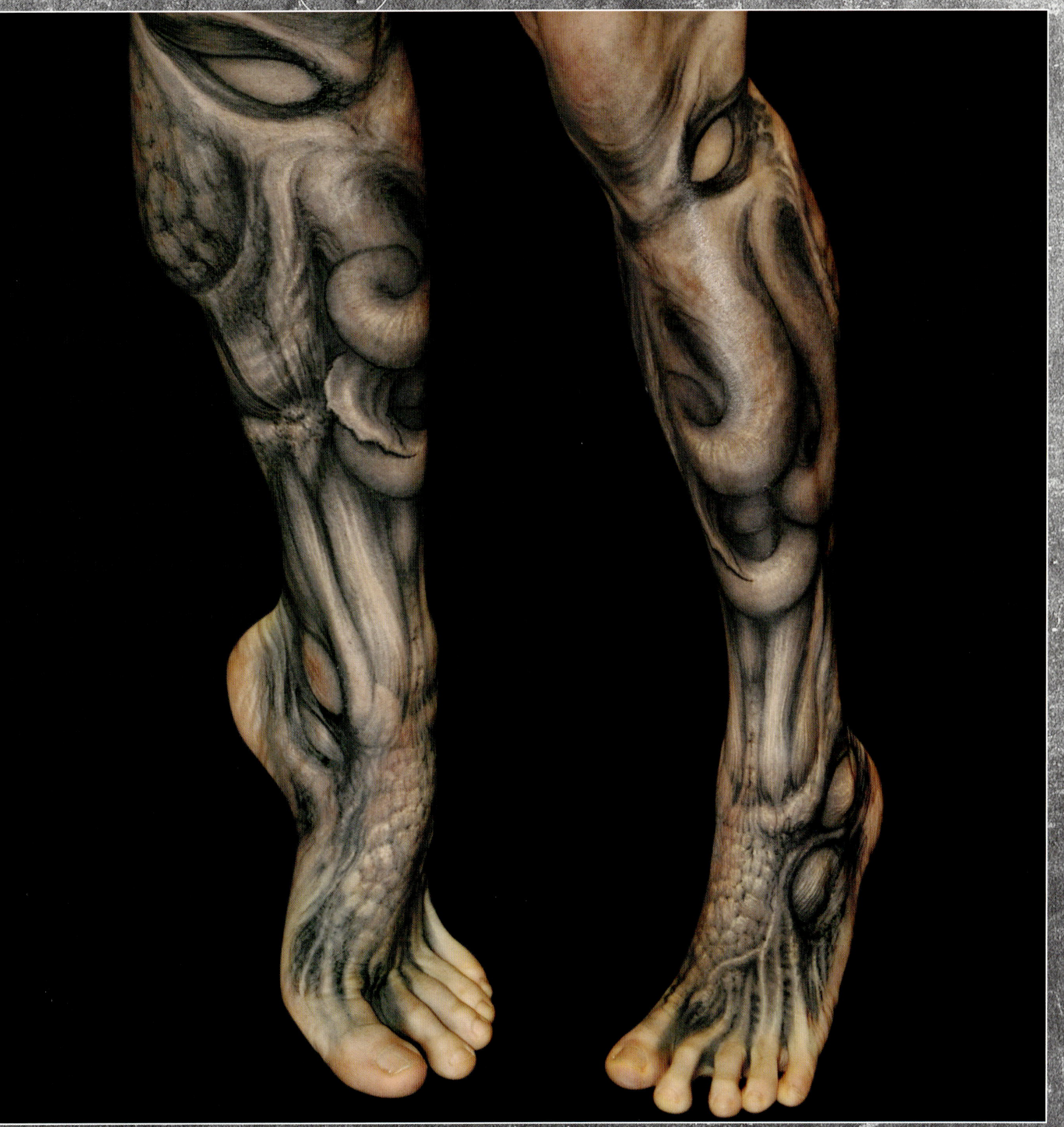

Billy Beans

Billy Beans

Clay Fraser

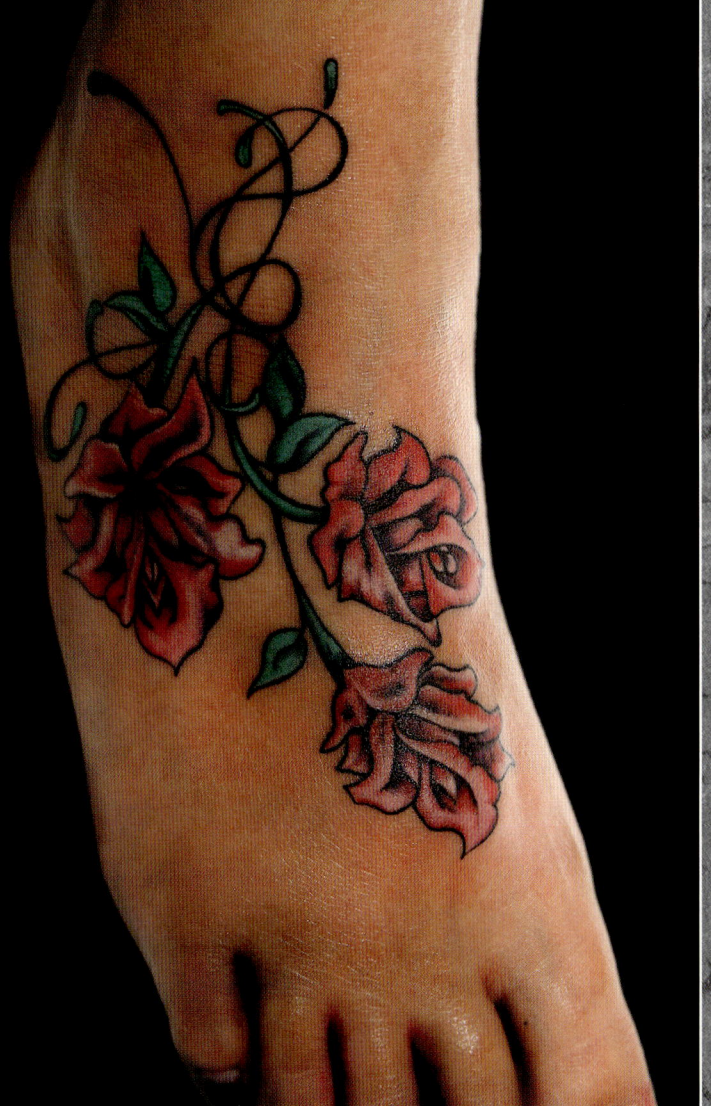

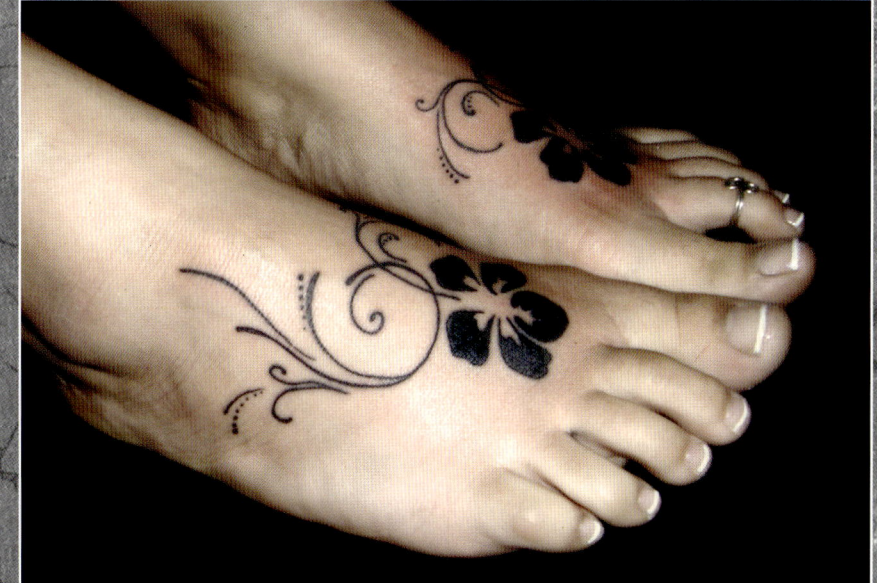

Clay Fraser

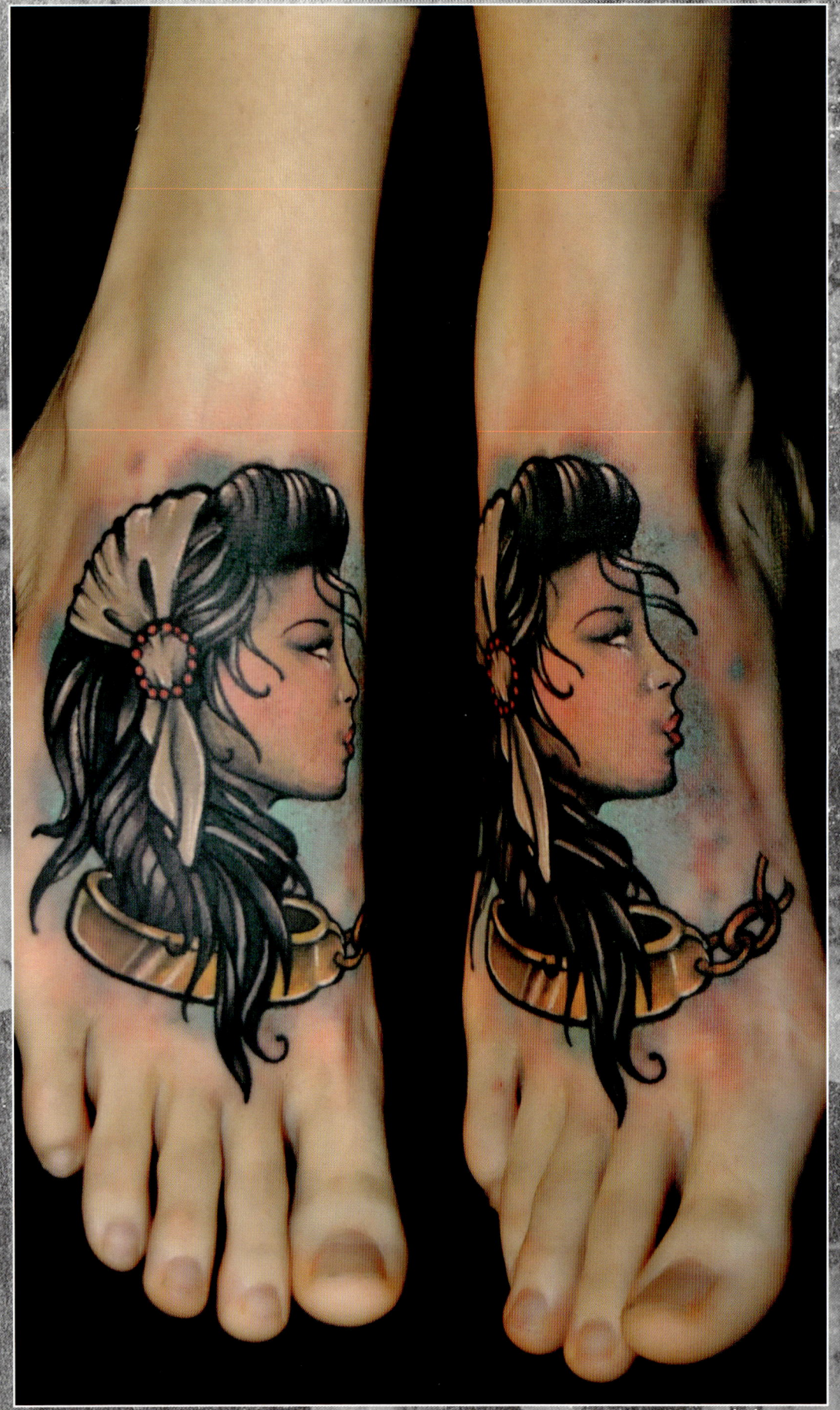

Roman Warwink

Short

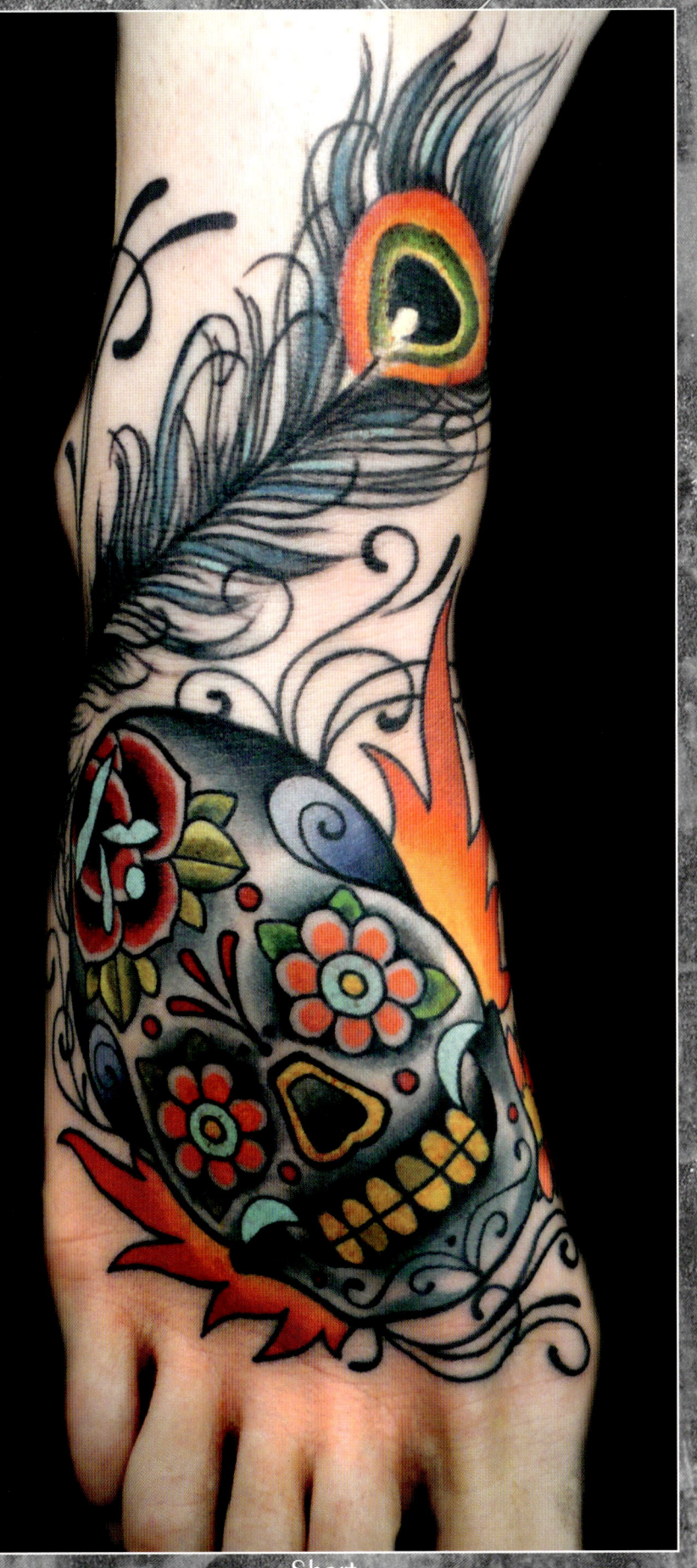

Short

Matt Griffith

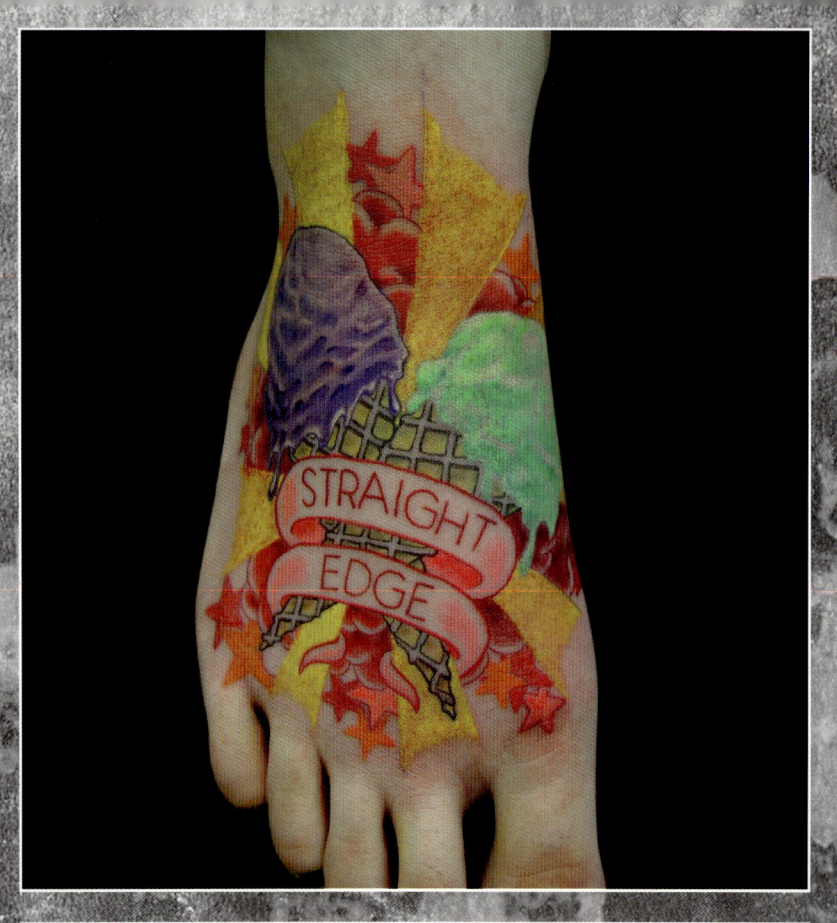

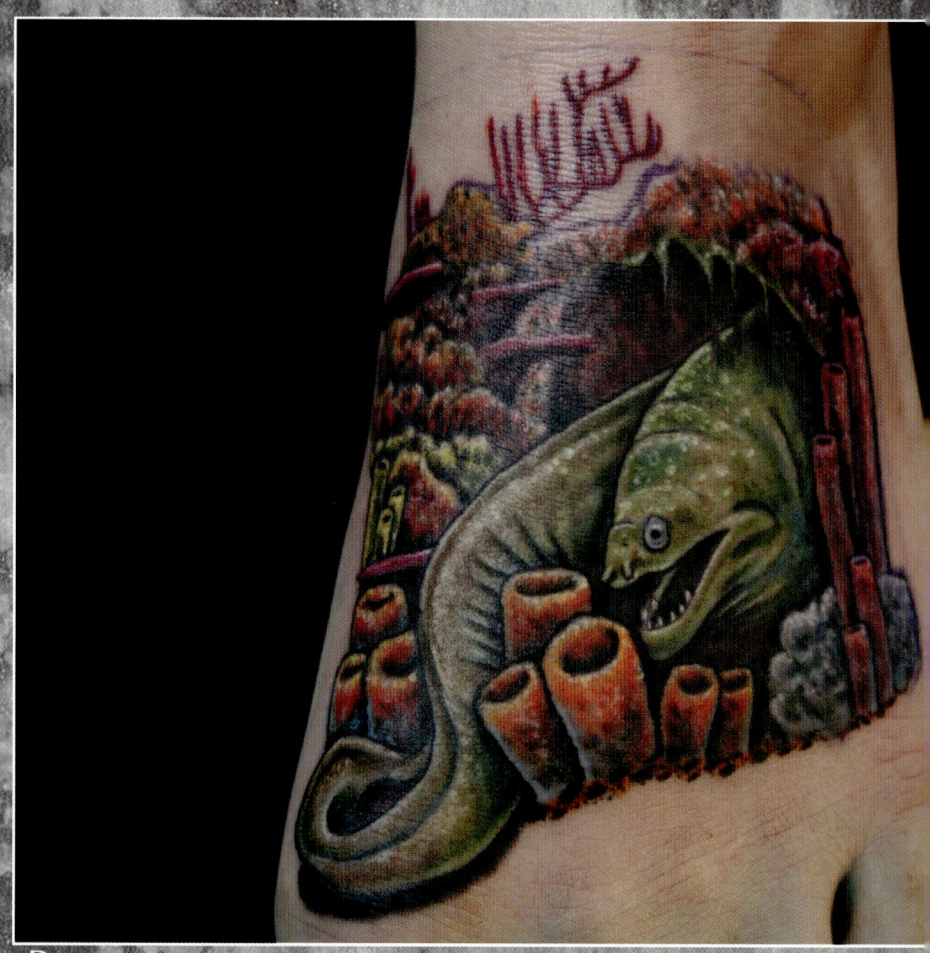

Larry Brogan

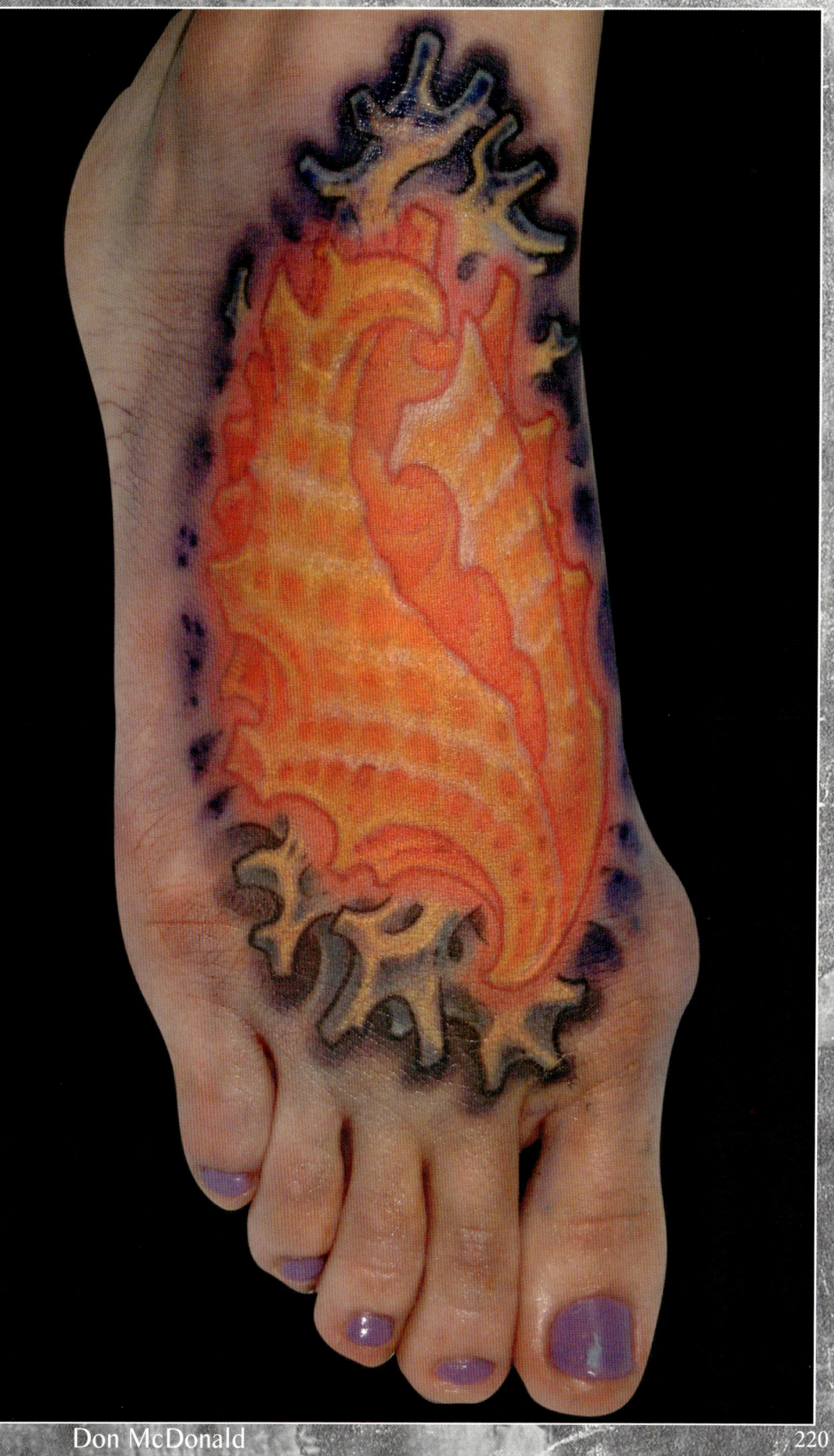

Don McDonald

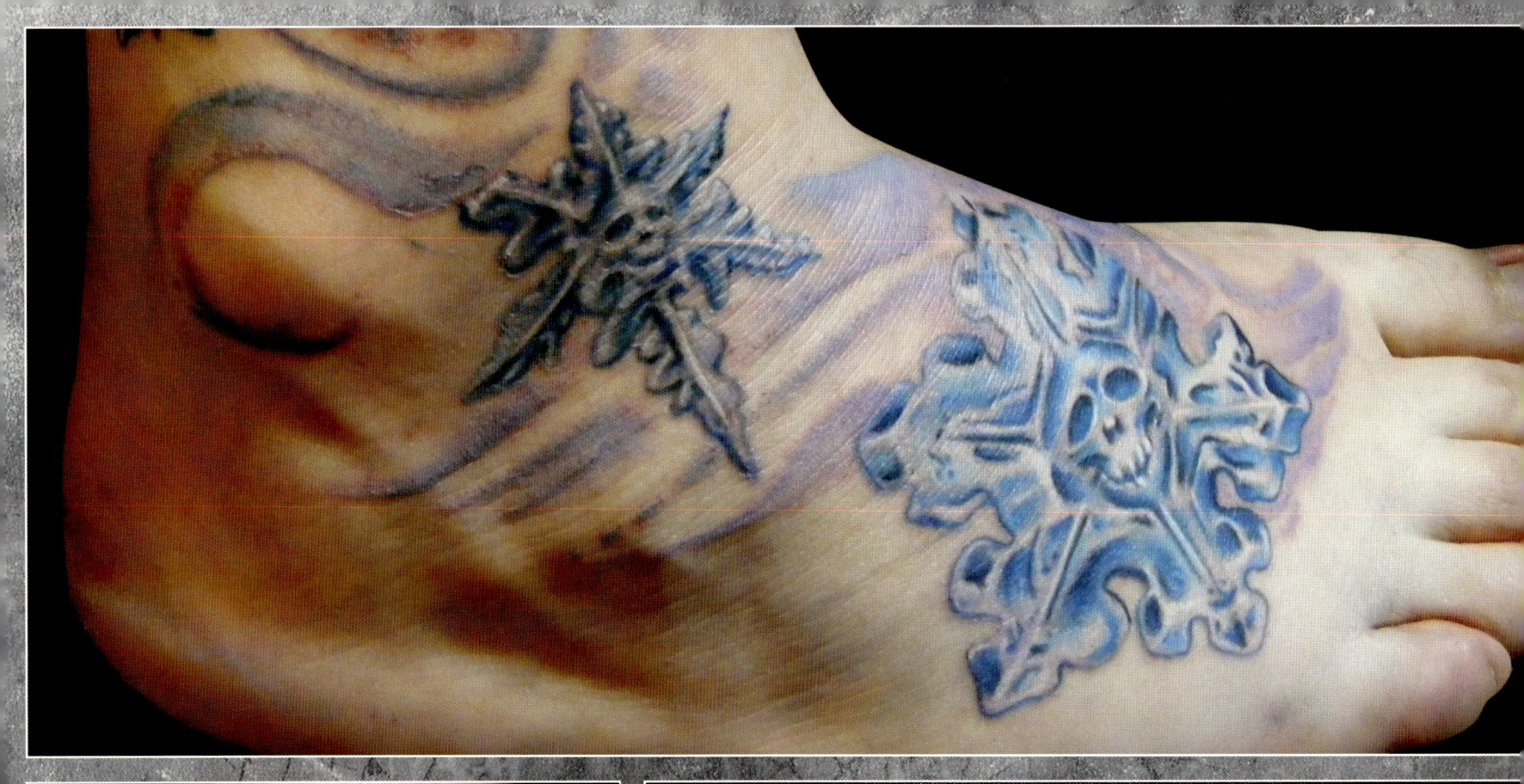

Endre Szabo

Endre Szabo

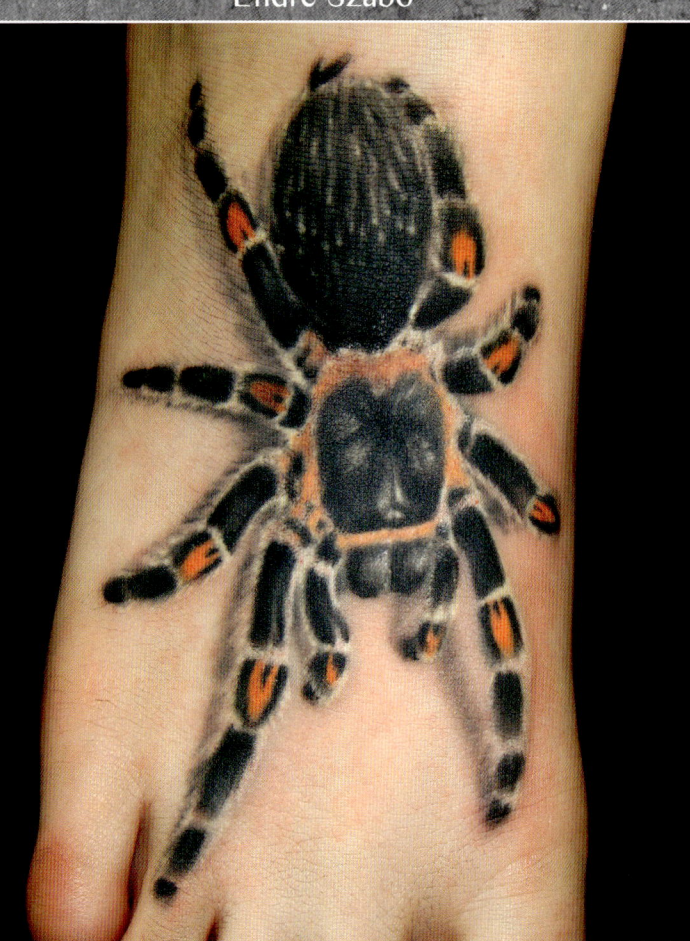

Daksi Smithy

Jee Sayalero

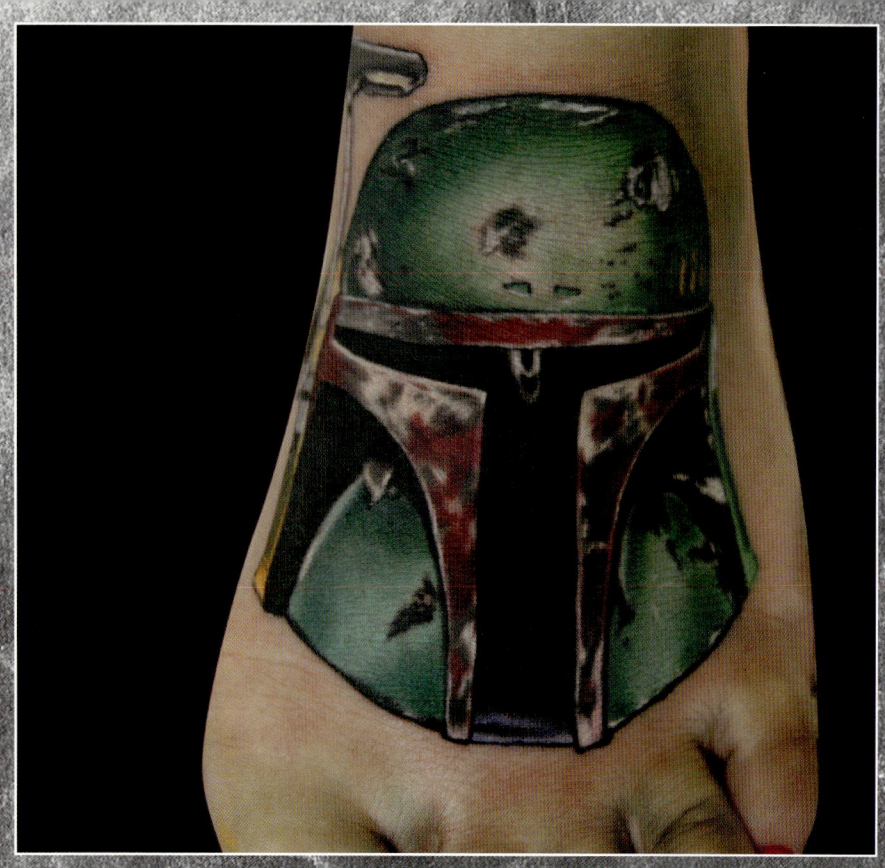

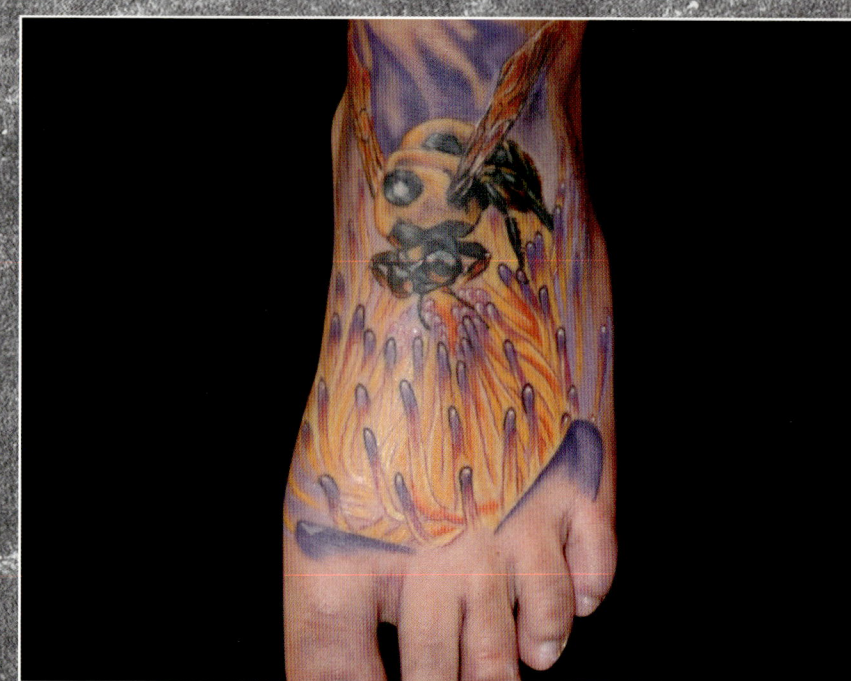

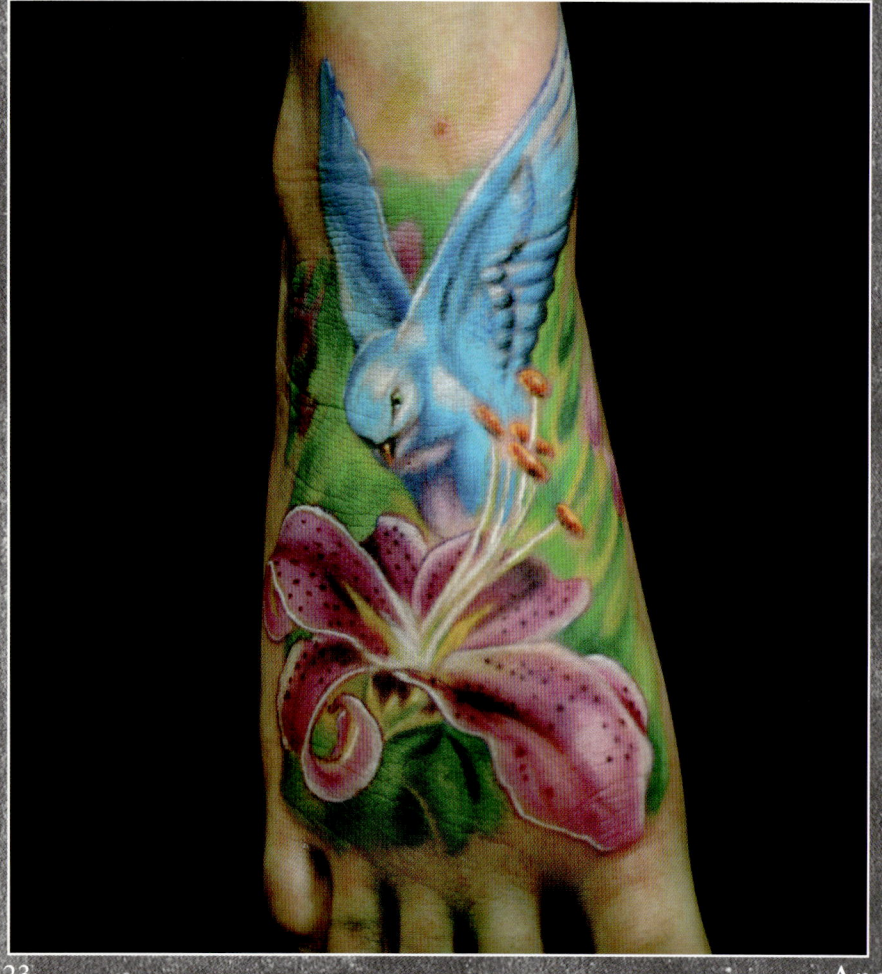

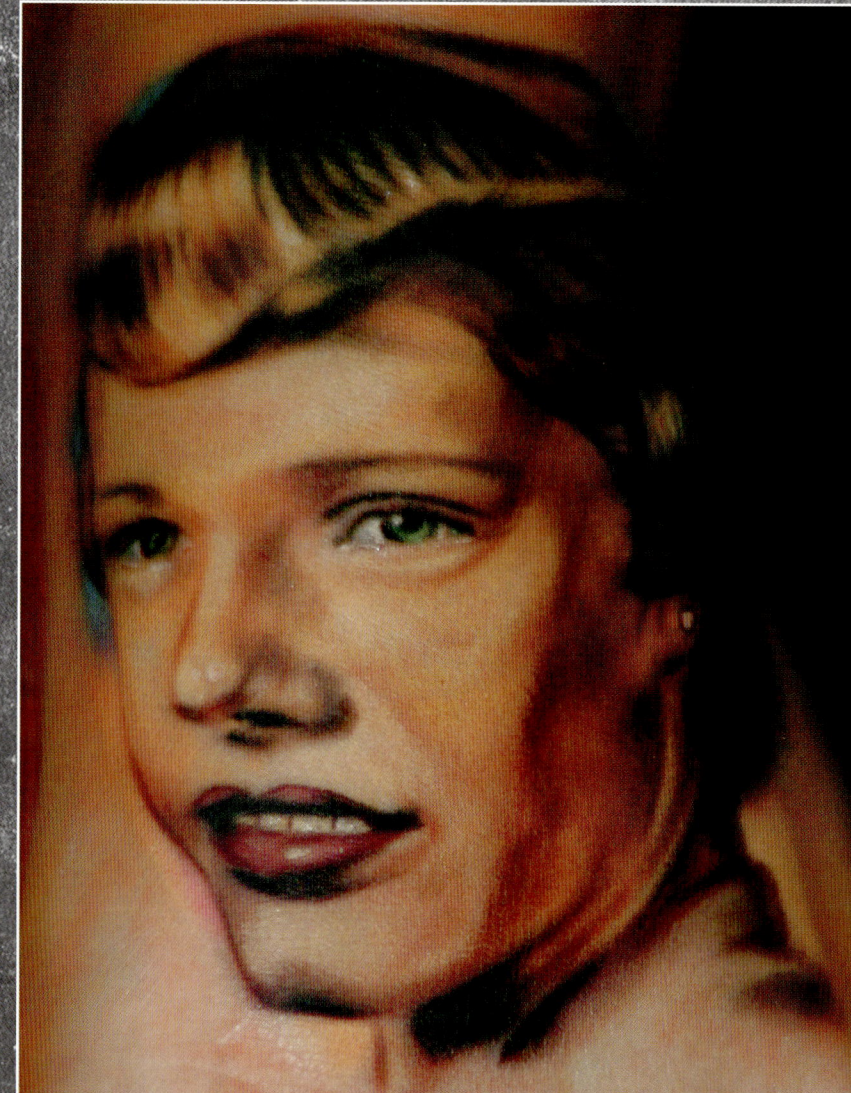

Andy Barrett

Daksi

Jasmin Austin

Dan Smith

Dan Smith

Christian Perez

Christian Perez

Steve Byrne

Byron Drechsler

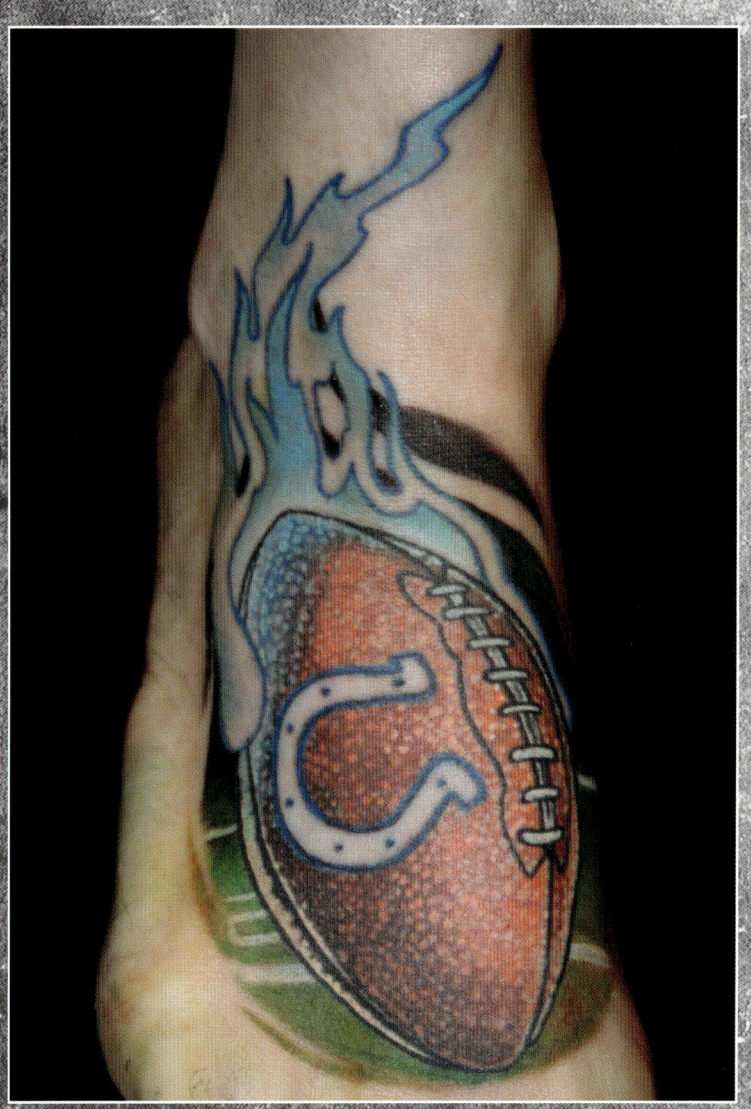

Dan Stewart

Dan Stewart

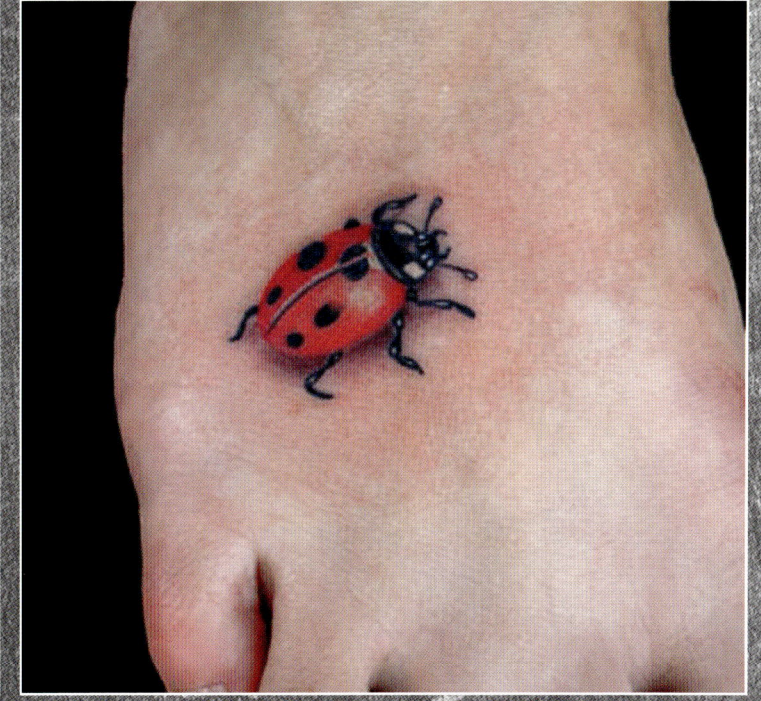

Sean Fletcher

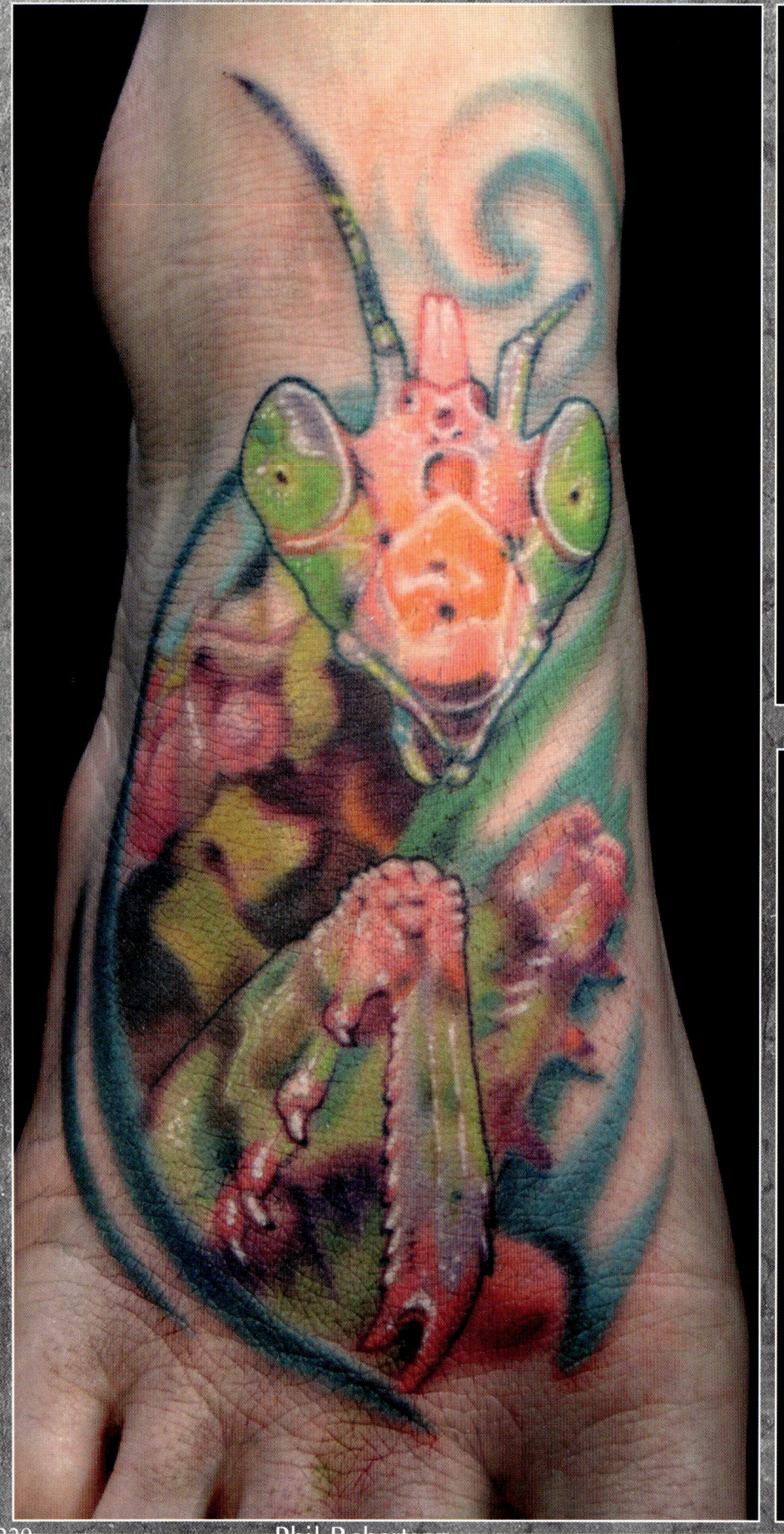

Phil Robertson

Jason Vaughn

Jason Vaughn

Vince Villalvazo

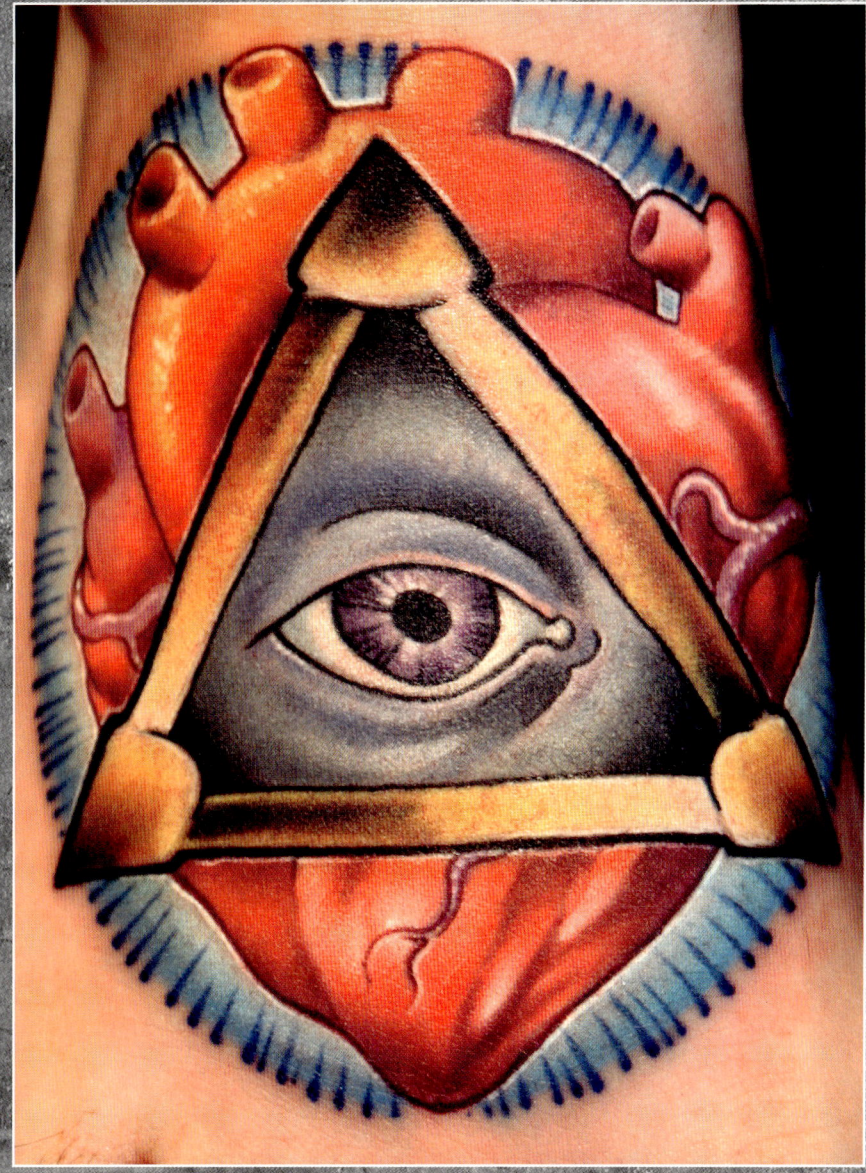

Tim Pangburn

Tim Pangburn

Robert Hendrickson

230

Jeremiah Barba

Deno

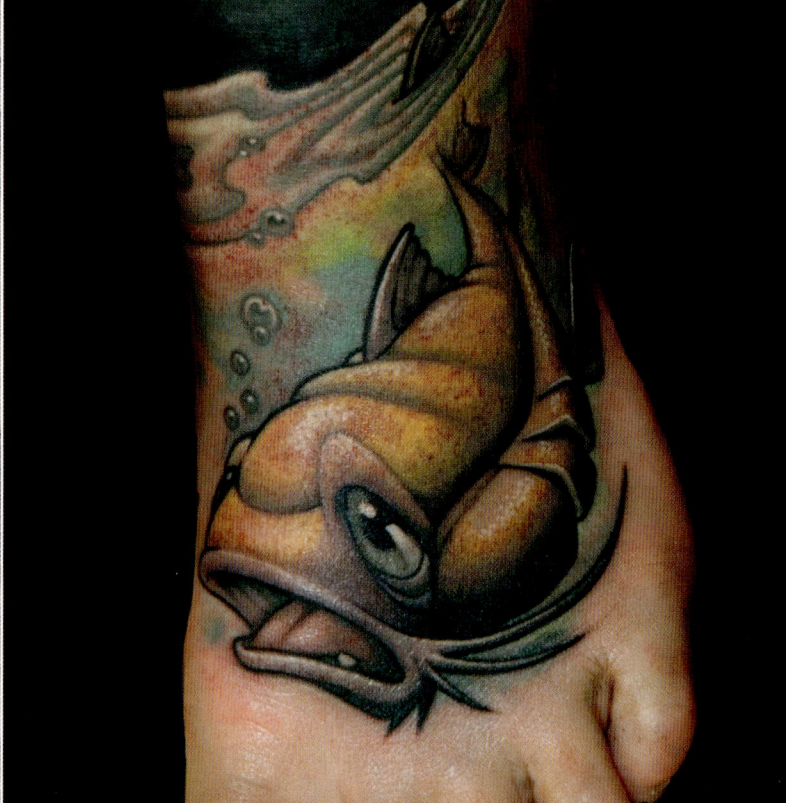

Jesse Smith

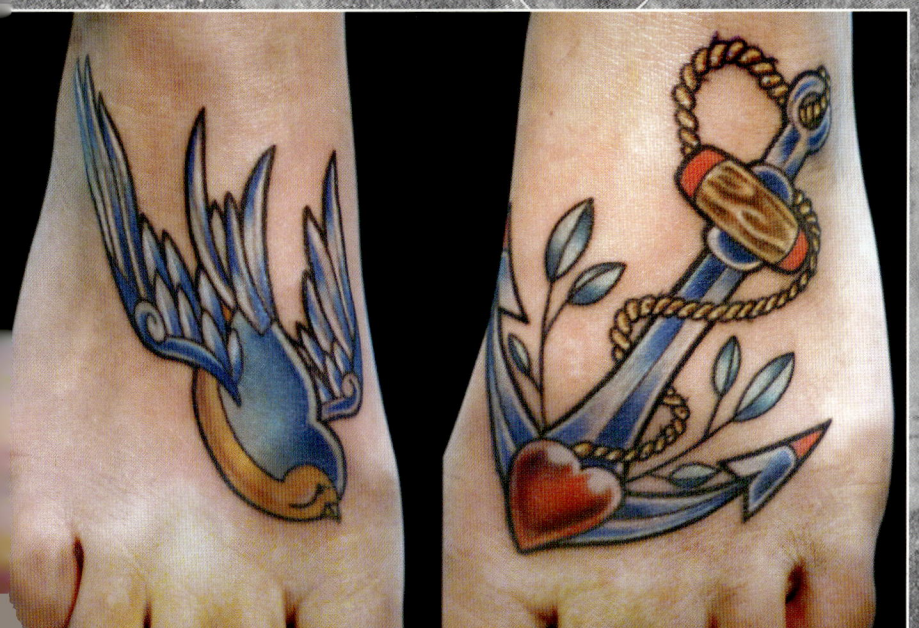

Becka MacDonald

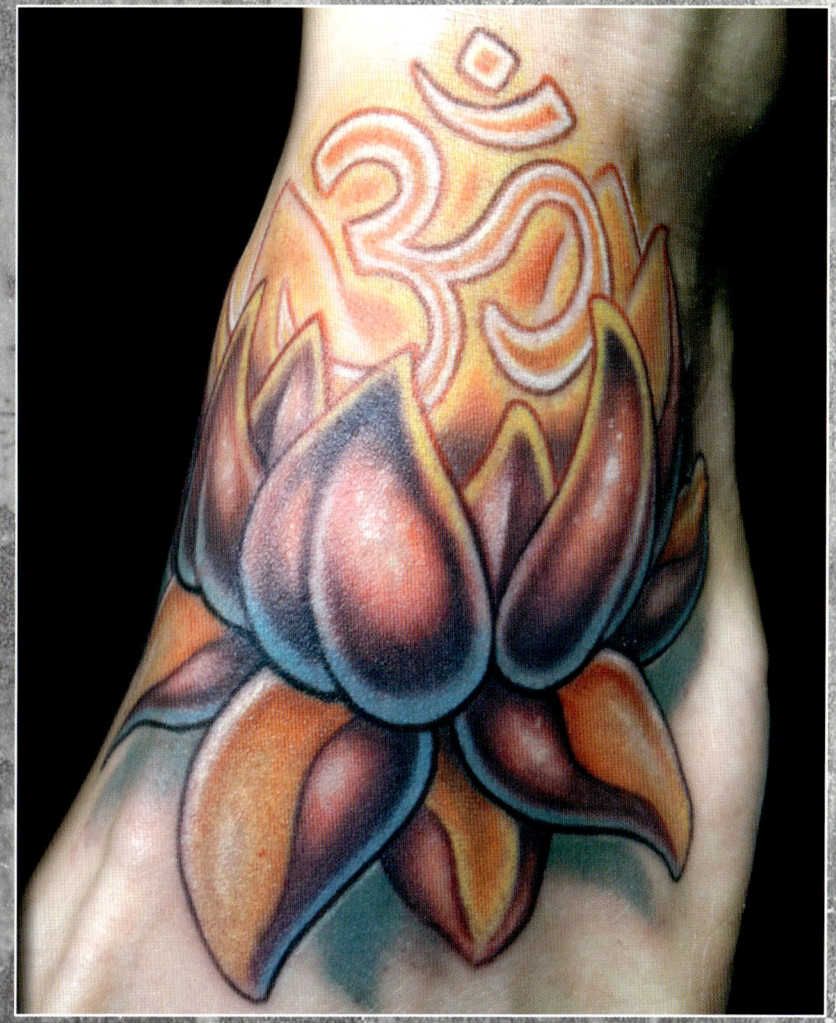

Brandon Schultheis

Jason Reeder

Durb

232

Tutti Serra

Bugs

Jeff Zuck

Canman

Brady Willmott

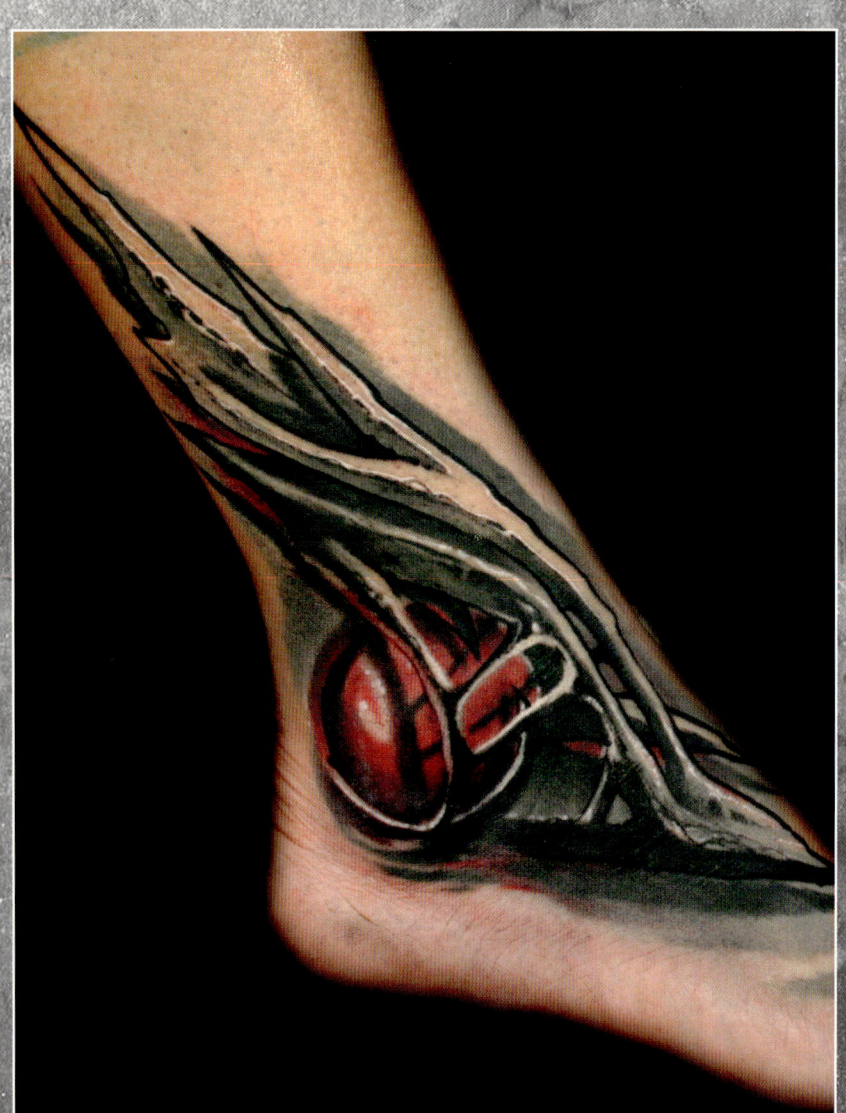

Kazimierz "KOSA" Rychlikowski

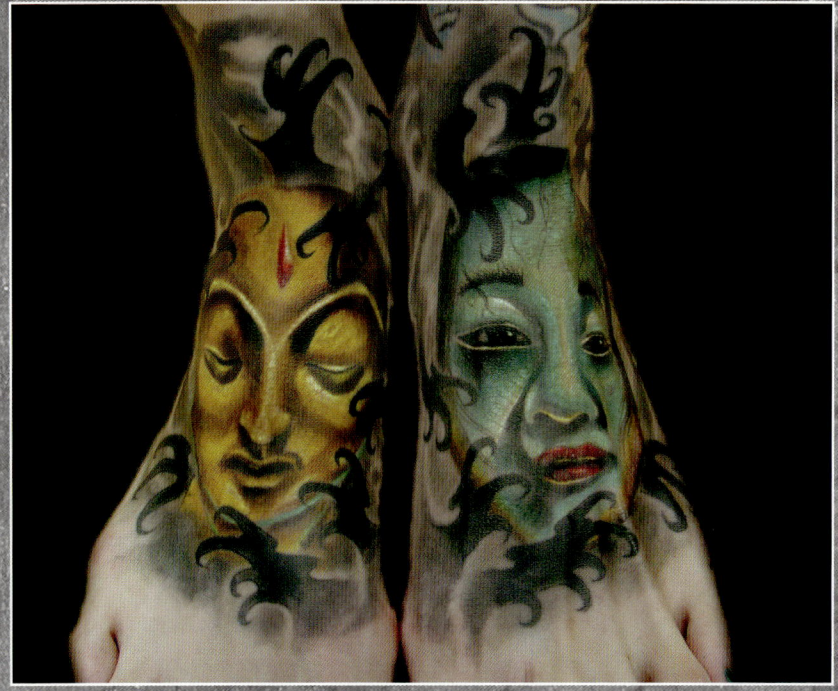

Holly Azzara

Brandon Bond

Holly Azzara

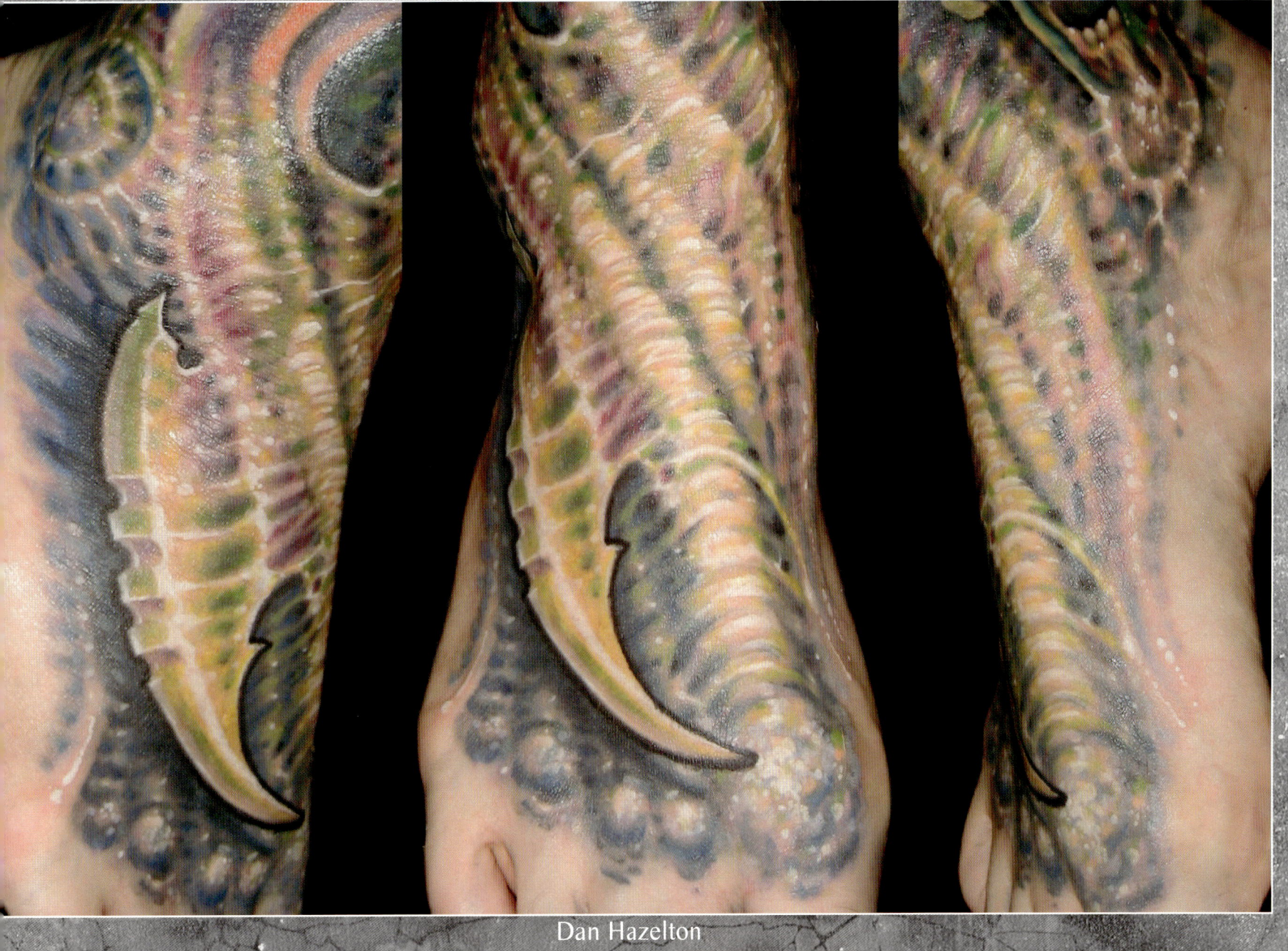

Dan Hazelton

Josh Woods

Josh Woods

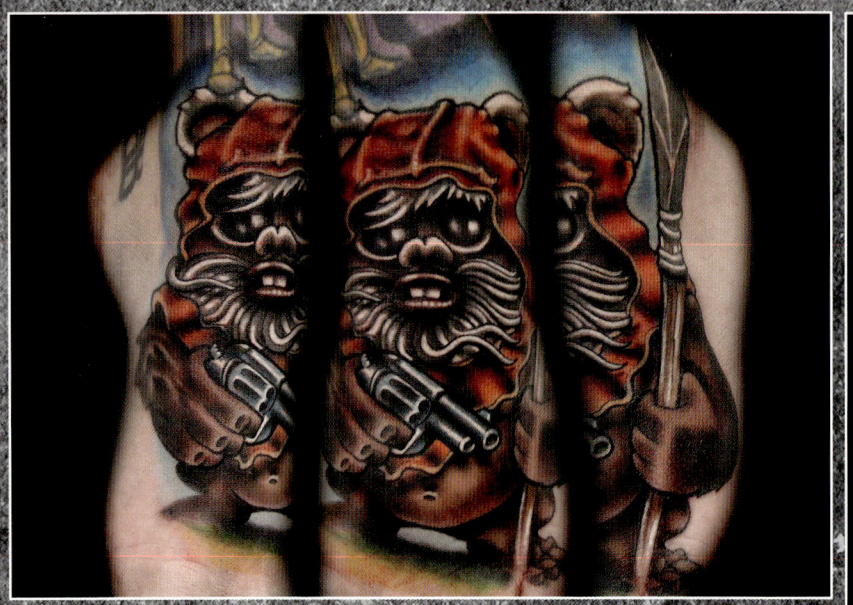

Josh Woods

Ben Ryan

Ben Ryan

Josh Woods

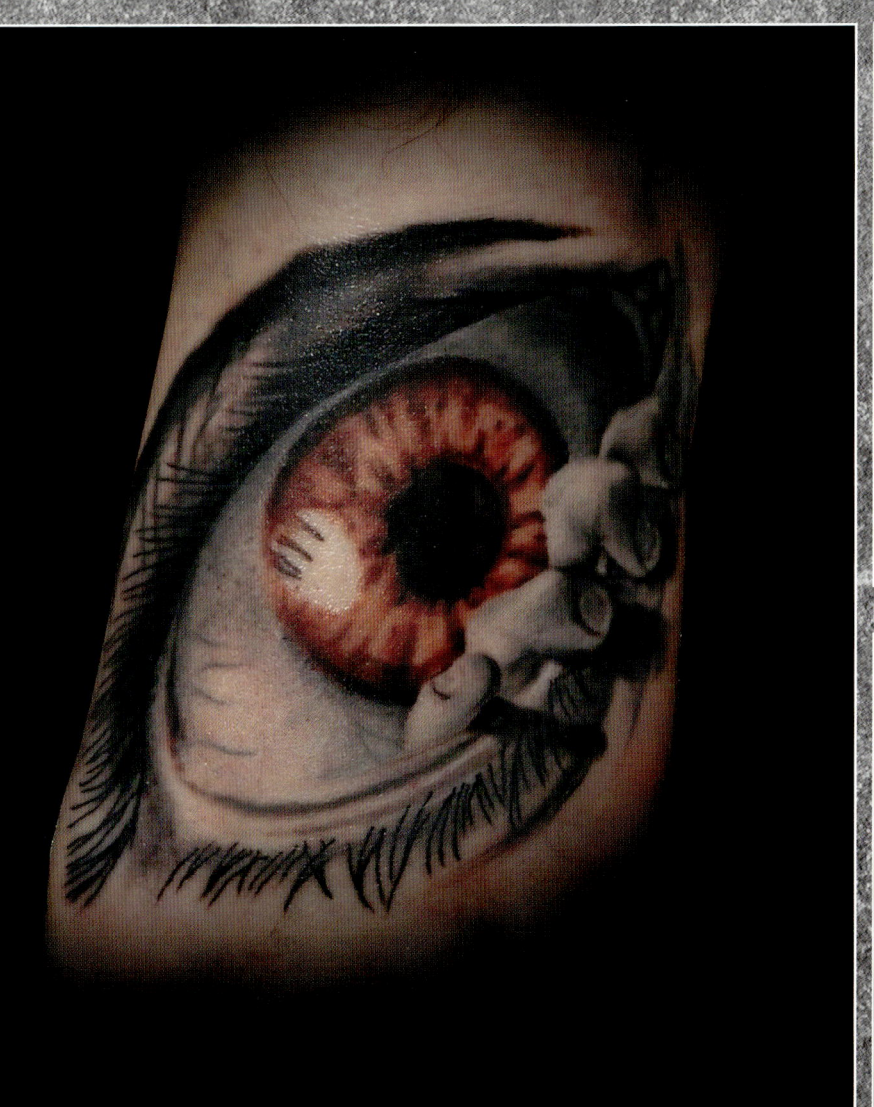

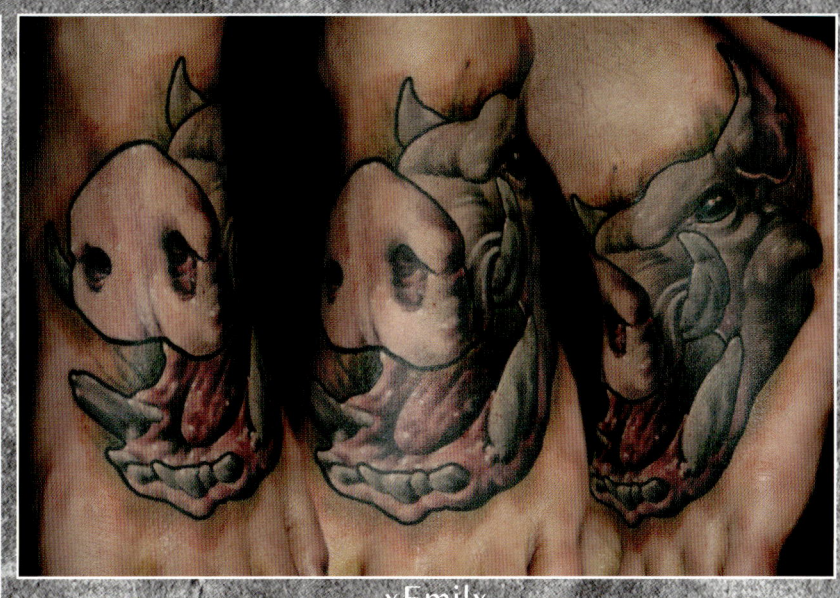

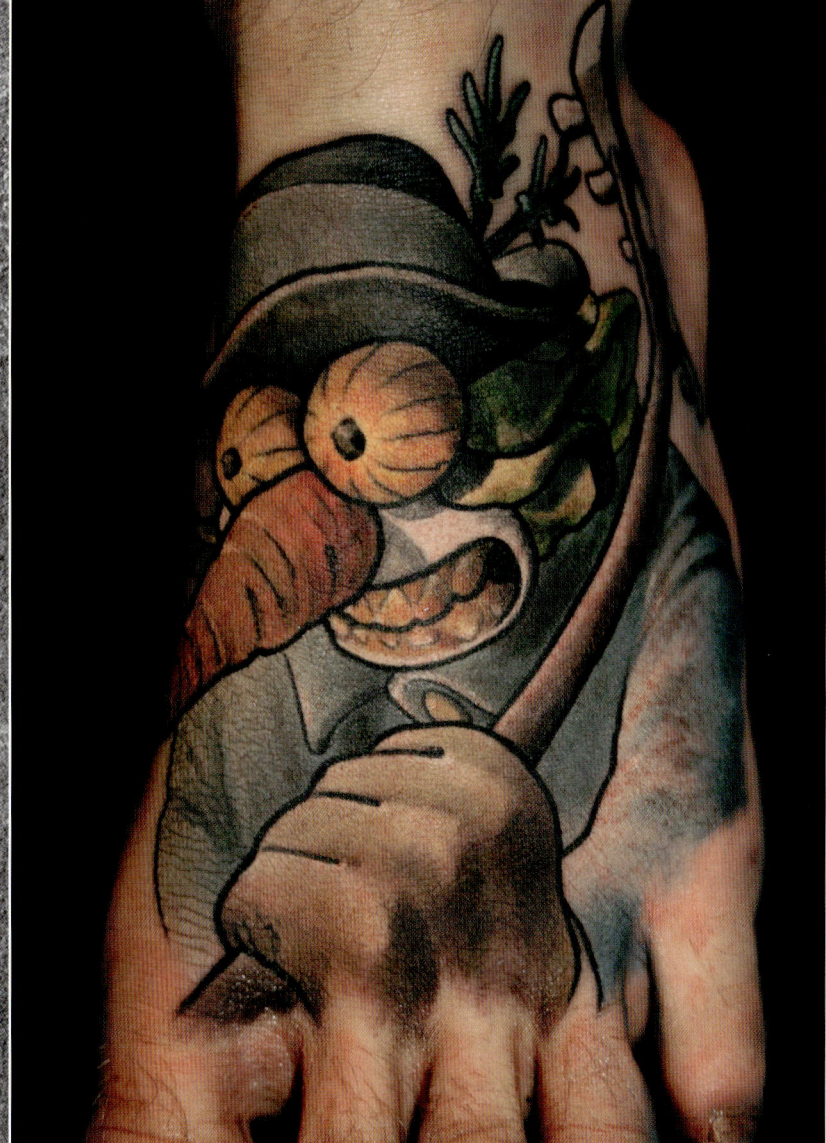

Danny Fugate

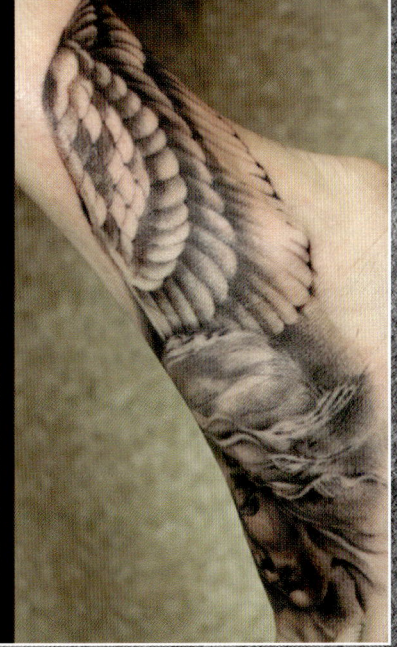

xEmilx

David Bennett

xEmilx

238

Vova Mult

Collecting, editing, and designing this book has been a labor of love. The incredible artists from around the world who contributed their work were a pleasure to correspond with and brought endless amounts of inspiration to our workspaces.

While not all tattoo collectors choose to tattoo their heads, hands, or feet, those who do so are joining a long tradition that has been used by cultures throughout history for a variety of reasons. Utilizing these areas of the body can be extremely beneficial in accomplishing magnificent artistic feats, as is evidenced by the photos contributed to the project and contained within these pages.

We would like to extend a hearty round of thanks to the multitude of artists who were a part of making the idea for this project come to fruition. It was our pleasure to work with some of the industry's best and honor their work within these pages. As with all efforts that are expressed in the artistic form, we ask that you use the images within this book as inspiration and not duplication. Learn from the contents, but please do not copy the artwork; out of respect for both artists and collectors.

As we explore the history of this great art form, learn from our forefathers, and acquire new tattooed treasures in the future, we thank you for joining us on this portion of the journey and sincerely hope that you have enjoyed the Tattoo Extremities ride. Viva la creativity!

MEMENTO
PUBLISHING

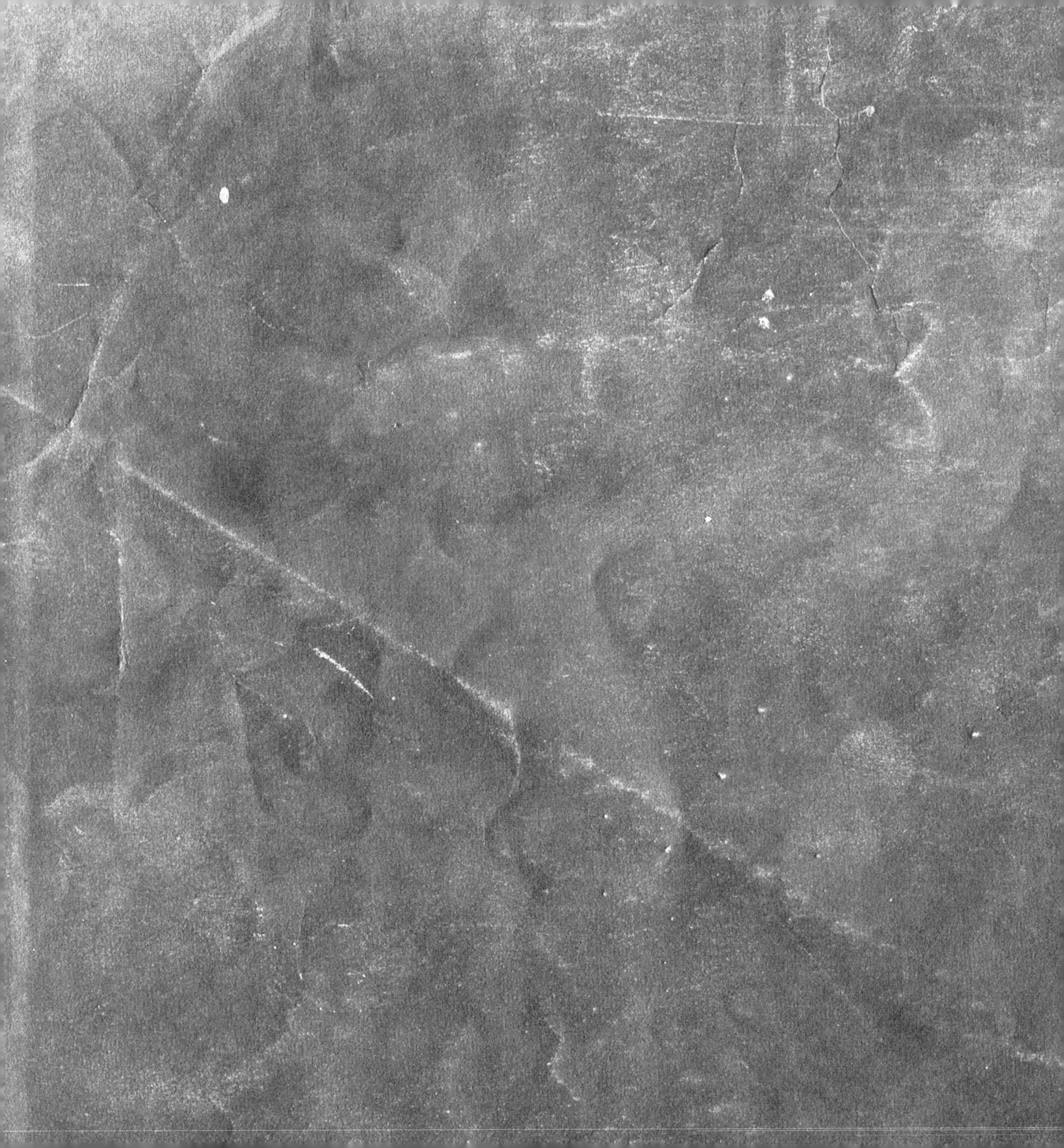